Advanced AC Circuits and Electronics:
Principles & Applications

J. Michael Jacob
Purdue University
West Lafayette, Indiana

THOMSON

DELMAR LEARNING

Australia • Canada • Mexico • Singapore • Spain • United Kingdom • United States

THOMSON

DELMAR LEARNING

Advanced AC Circuits and Electronics: Principles & Applications

By J. Michael Jacob

Vice President, Technology and Trades SBU:

Alar Elken

Editorial Director:

Sandy Clark

Acquisitions Editor:

David Garza

Senior Development Editor:

Michelle Ruelos Cannistraci

Channel Manager:

Fair Huntoon

Marketing Coordinator:

Brian McGrath

Production Director:

Mary Ellen Black

Production Manager:

Larry Main

Senior Project Editor:

Christopher Chien

Art/Design Coordinator:

Francis Hogan

Technology Project Manager:

David Porush

Technology Project Specialist:

Kevin Smith

Senior Editorial Assistant:

Dawn Daugherty

Library of Congress Cataloging-in-Publication Data:

Jacob, J. Michael
  Advanced AC circuits and electronics : principles & applications / J. Michael Jacob
    p. cm.
  ISBN 0-7668-2330-X (hardcover)
  1. Electric circuits—Alternating current—Problems, exercises, etc.
2. Electronics—Problems, exercises, etc. I. Title.
  TK454.15.A48J33 2003
  621.319'13—dc21 2003007979

NOTICE TO THE READER

Contents

iv

13
Circuit Analysis with
Nonsinusoidal Waveforms *329*

Part IV Advanced Analysis Techniques   *361*

14
Series-Parallel Analysis by
Impedance Combination *363*

Preface

Introduction to the Herrick & Jacob series

The traditional approach to teaching fundamentals in Electronics Engineering Technology begins with a course for freshmen in DC Circuit Theory and Analysis. The underlying laws of the discipline are introduced and a host of tools are presented, and applied to very simple resistive circuits. This is usually followed by an AC Circuit Theory and Analysis course. All of the topics, rules, and tools of the DC course are revisited, but this time using trigonometry and complex (i.e., real and imaginary) math. Again, applications are limited to passive (simple) resistor, capacitor, and inductor circuits. It is during this second semester, but often not until the third semester, that students are finally introduced to the world of *electronics* in a separate course or two. At this point they find what they have been looking for: amplifiers, power supplies, waveform generation, feedback to make everything behave, power amps, and radio frequency with communications examples. This approach has been in place for *decades*, and is the national model.

So, what is wrong with this approach? Obviously, it has been made to work by many people. Currently, during the DC and AC courses students are told not to worry about why they are learning the material, only that they will have to remember and apply it later (provided they survive). These two courses (DC and AC) have become tools courses. A whole host of techniques are taught one after the other, with the expectation (often in vain) that eventually when (or if) they are ever needed, the student will simply *remember*. Even for the most gifted teachers and the most dedicated students these two courses have become "weed out" classes, where the message seems to be one of "if you show enough perseverance, talent, and faith we will eventually (later) show you the *good* stuff (i.e., the electronics)." Conversely, the electronics courses are taught separately from the circuit analysis classes. It is expected that the student will quickly recall the needed circuit analysis tool (learned in the DC and AC courses) taken several semesters before when it is needed to understand how an amplifier works or a regulator is designed. This leads to several results, all of which have a negative effect. First, the student sees no connection between DC and AC circuit theory and electronics. Each is treated as a separate body of knowledge, to be memorized. If the students hang on long enough, they will eventually get to the electronics courses where the circuits do something useful; second, the teachers are frustrated because students are bored and uninterested in the first courses, where they are supposed to learn all of the fundamentals they will need. But when the students need the information in the later electronics courses, they do not remember. The result is a situation under which we have struggled for decades. This is the premise upon which the concept for this new series has been developed.

The Herrick & Jacob series offers a different approach. The approach integrates circuit theory tools with electronics, interleaving the topics as needed. Circuit analysis tools are taught on a *just-in-time* basis to support the development of the electronics circuits. Electronics are taught as applications of fundamental circuit analysis techniques, not as unique magical things with their own rules and incantations. Topics are visited and revisited in a spiral fashion throughout the series, first on a simple, first-approximation level. Later, as the students develop more sophistication and stronger mathematical underpinnings, the complex AC response and then the nonsinusoidal response of these same electronics circuits are investigated. Next, at the end of their two years of study, students probe these central electronic blocks more deeply, looking into their nonideal behavior, nonlinearity, responses to temperature, high power, and performance at radio frequencies, with many of the parasitic effects now understood. Finally, in the Advanced Analog Signal Processing book Laplace transforms are applied to amplifiers, multistage filters, phase locked loops, and other closed-loop processes. Their steady-state and transient responses as well as their stability are investigated.

The pervading attitude is "Let's do interesting and useful things right from the start. We will develop and use the circuit theory, as we need it. Electronics is *not* magic; it only requires the rigorous use of a few fundamental laws. As you (the student) learn more, we will enlarge the envelope of performance for these electronic circuits." Learn and learn again. Teach and teach again. Around and around we go, ever upward.

Introduction to this text

This is the second book in an integrated series of texts (Herrick & Jacob Series). It applies circuit theory and fundamental electronics to a *wide* variety of circuits that process sinusoidal and repetitive nonsinusoidal signals.

Organization of the text

Part I–Sinusoidal Fundamentals: This section introduces the sinusoidal waveform and its parameters. The phasor, a vector treatment of electrical quantities, is explained and impedance is derived. These quantities are applied to a host of series and parallel circuits, including audio cross-over networks and reactive networks around op amps.

Part II–Filter Applications: Filters form one of the largest classes of sinusoidal applications. This section begins with a detailed description of terminology, parameter definition, and frequency response plotting. Low-pass and high-pass filters each have their own chapter. In each, the general approach to determining filter type, pass-band gain, order, roll-off rate, gain equation derivation, and cut-off frequencies is applied to a variety of both passive and active filters.

Series and parallel resonance circuits, both passive and active, also are treated in a separate chapter. Full parameter derivations and audio, power, and communication examples are given.

A separate chapter deals with an op amp's frequency response, gain bandwidth, and slew rate, as well as the op amp integrator and differentiator.

Part III–Fourier Series Analysis: This treatment of nonsinusoidal analysis is unique. It offers three chapters to explain the concepts, allowing the student to extend sinusoidal analysis techniques to *all* repetitive wave shapes. Other texts offer one chapter or less.

The superposition theorem is fundamental to constructing nonsinusoidal signals. It is given an entire chapter, and waveform reconstruction is visualized with *Excel, MATLAB, PSpice,* and *MultiSIM.*

The Fourier series and its coefficients for a wide variety of standard waveforms are developed, and then summarized in an interactive table.

An entire chapter is dedicated to Circuit Analysis with Nonsinusoidal Signals. A Fourier Analysis Worksheet is used to analyze the performance of low- and high-pass filters (both passive and active), differentiators, and integrators to steps, ramps, and rectified sinusoids.

Part IV–Advanced Analysis Techniques: There are a host of advanced analysis techniques. This section has been restricted to three of the most widely applied. The chapter on Impedance Combination focuses on visualization and solution organization. Mesh and nodal analysis are introduced as extensions of Kirchhoff's laws. The reader is then taught to write the simultaneous equations *by inspection* for a variety of filters, amplifiers, and electrical power systems.

Part V–Power Systems: The foundation for a more detailed study of power (in the next text of the series) is laid down by introducing average power as the average of instantaneous voltage times current. This is then extended to resistors (W), inductors and capacitors (VARs), and complex impedances (VA). The power factor is explained, and its impact on industrial loads is studied. Full and partial power factor corrections are illustrated.

A separate chapter presents three-phase generation. Wye and delta loads are studied using traditional analysis rules, rather than developing a host of specialized formulas.

Features

Pedagogy

Each chapter begins with *performance-based learning objectives*. These are rigorously implemented throughout that chapter.

There are over 100 *examples*, each providing fully worked out and explained numerical illustration of the immediately preceding information. Every example is followed by a practice exercise that requires the reader to extend the techniques just illustrated with a different set of numbers, perhaps in a slightly different direction. Numerical results are given to allow students to check the accuracy of their own work.

The *wide margins* have been reserved to serve as visual keys for important information contained in the adjacent text. Significant effort has been made to keep these margins uncluttered. The occasional figure or table supports the text beside it. Icons, notes, equations and key specifications are placed so that the reader can quickly glance along the margins to find the appropriate details explained in the adjacent text. This is very much the way a good student annotates a book as she/he studies. And, plenty of room has been left for precisely that purpose.

Computer Software

Electronics Workbench's *MultiSIM* ® and Cadence's *PSpice* are used to illustrate the applications *and weaknesses* of electronics simulations. There are considerable differences between the two packages. Each has its own strength and weaknesses. To rely only on one may mislead the reader.

how big can you dream?™

MATLAB and Excel are also threaded throughout the text, presented at that point where it is important to the reader's understanding to see the equations implemented and the results plotted as a function of time or frequency. A facility with these math programs strengthens the student's grasp of the fundamental relationships without the tedium and errors of manual calculations and plotting. These two math programs unveil much of the computational "magic" buried within simulators.

MATLAB®

All four programs are *integrated* throughout the text, at the points in the instruction where they are needed to illustrate a point. They are *not* just included at the end of each chapter as add-ons, as if an afterthought. The appropriate icon is placed in the margin beside each example, as a key. About one fourth of the examples use MultiSIM, one fourth use Cadence, about one fourth use MATLAB, and one fourth are completed using Excel. Files for these examples are available on the accompanying CD.

Most of the end-of-chapter problems can be solved, or the manual calculations verified by using one of the software packages. For faculty adopting this text, these files are also available.

The education version of MultiSIM 2001, the evaluation version of Cadence's PSpice 9.2, the student version of MATLAB 5.3, and Microsoft Office 2000 are used. Although later versions of each were released during the writing of this text, these do not alter the instructional value of the examples.

MultiSIM is a product of Electronics Workbench.
http://www.electronicsworkbench.com

PSpice is a product of Cadence. Information is available from
http://www.orcad.com

MATLAB is a product of The MathWorks.
http://www.mathworks.com/

Problems
End-of-chapter problems are organized in sections that correspond to the sections within the chapter. Problems are presented in pairs. Answers to the odd-numbered problems are given at the end of the text. Readers must solve the even-numbered problems without a target.

Laboratory Exercises
These are included at the end of most chapters. They are modifications of those used by over a thousand students and a variety of teachers through several revisions.

CD
The textbook version of MultiSIM is provided on the accompanying CD. Also, the MultiSIM, PSpice, MATLAB, and Excel files used to create the examples and figures throughout the text are included there.

Supplements
Online Companion Visit the textbook's companion Web site at

http://www.electronictech.com

There you will find:
- MultiSIM, PSpice, MATLAB, and Excel files for the examples and problem solutions

- A link to automated homework: a unique set with random values is generated each time you access the problems. Your answers are checked as soon as you enter the result of each step.
- Over thirty downloadable PowerPoint presentations.
- Links to streamed video of the author teaching AC Electronics to his students at Purdue University's School of Technology.
- Link to the author's Purdue University course Web site.
- Sample lab results.
- Text updates.

Instructor's Guide
The guide contains solutions to all of the end-of-chapter problems.
ISBN number: 07668-2331-8

About the author

Mike Jacob is a test and control engineer with experience in a variety of industries. He has designed an IR-based rotational torque gage and control system for tightening military tank transmission bolts, a PC-based instrumentation and control system for the testing of hydraulic steering gear on over-the-road trucks, and electronic drive and interface circuitry for laser testing of the space shuttle tiles and for measuring the blood flow rate in a portable artificial kidney. He has designed, programmed, and installed automated manufacturing equipment for the artificial kidney, automotive controls, and residential watt-hour meter calibration.

Mike Jacob is the current McNelly Professor and an award-winning teacher. He has received the CTS Microelectronics Outstanding Undergraduate Teaching award as the best teacher in the Electrical and Computer Engineering Technology Department seven times. He has won the Dwyer Undergraduate Teaching Award as the top teacher in the School of Technology three times. He also received the Purdue University's undergraduate teaching award (the Amoco award), the Paradigm Award from the Minority Technology Association, and the Joint Services Commendation Medal (for excellence in instruction) from the Secretary of Defense. In 1999 he was listed in Purdue University's **Book of Great Teachers,** which holds the top 225 faculty ever to teach at Purdue University. He has taught at Purdue for 25 years and at a community college in South Carolina for seven years.

He is also a passionate bicycle tourist, logging 8,000 miles each year on his recumbent bike. He has ridden that bike from Seattle to Indiana, from Indiana across New England to Nova Scotia and back, and across the Great Plains, through the Rocky Mountains and down the west coast to Southern California. So keep your eyes open; you may just meet him some summer on a back road.

Acknowledgements

The author and Delmar Publishers wish to express their gratitude to the reviewers and production team that made this textbook possible. These include:

Rabah Aoufi, DeVry University
John Blankenship, DeVry Universtiy, Atlanta, GA
Harold Broberg, Indiana Universtiy–Purdue University Indianapolis
Parker Sproul, DeVry University, Phoenix, AZ
Andrzej M. Trzynadloswki, University of Nevada

The author would also like to thank Delmar Publishers for their belief and commitment to this project, particularly Greg Clayton, Acquisitions Editor, Michelle Ruelos Cannistraci, Development Editor, and the rest of their team. Their willingness to venture down untraditional paths in pedagogy and text production has provided the essential elements necessary to convert this unique idea into a viable contribution in the spread of learning.

Avenue for feedback

No system ever performs exactly as intended. In fact, the quality of the results is more a function of the quality of the negative feedback. So please contact either the author or the publisher with your suggestions.

J. Michael Jacob
175 Knoy Hall
Purdue University
W. Lafayette, IN 47907
jacobm@purdue.edu

Michelle Ruelos Cannistraci
Delmar Learning
5 Maxwell Drive
Clifton Park, NY 12065-2919
michelle.cannistraci@delmar.com

Dedication

This text is dedicated to college teachers, who are spending their lives building the intellect and character of our youth. The influence for good that we have as teenagers mature into professional associates is awesome. Use it wisely.

J. Michael Jacob (Mike)   April, 2003

Sinusoidal Fundamentals

The sinusoid is a fundamental waveform. Commercial electrical power is produced and distributed worldwide as a sine wave. All audio signals are just combinations of a variety of different sinusoidal tones. Communications by radio, television, even digital pagers and cellular phones are carried by high-frequency sine waves. In fact, any waveform, even digital pulses, can be broken into and synthesized from sine waves.

Therefore, a detailed study of the elements of a sine wave is important. In Chapter 1, Sinusoidal Waveforms, you will review the sine function's basic mathematical equation. The definition and impact of amplitude, frequency, and phase shift are related to the waveform. You will learn to create sine waves using a spreadsheet and two different simulators. Amplitude will be expressed in peak and rms. Angles must be converted between degrees and radians. Frequency is expressed as either hertz or radians per second.

Linear circuits only affect the amplitude and the starting point (the phase) of a sine wave. These effects can be calculated by keeping track of the magnitude (length) and phase (angle). This is *just* like the vectors you used in physics. In electrical circuits these vectors are called **phasors**. In Chapter 2 you will learn to use trigonometry and vectors to represent these sinusoidal signals. As with vectors, there will be lots of diagrams.

Resistors, capacitors, and inductors respond to sine waves differently than they do to dc. You combine them differently. Using the concepts of phasors, however, you can still apply *all* of the laws and techniques you already know to the opposition that these components present to ac. That opposition is called **impedence**, the subject of Chapter 3

Chapter 4, Series Circuits, applies these phasors to circuits made with resistors, capacitors, inductors, and even op amps, connected end-to-end. You will be able to predict the magnitude and phase of all of the voltages and currents. Again, expect to draw many diagrams and to use manual calculations as well as two simulators.

Every time you plug something into an outlet, you are connecting it in parallel with all of the other loads connected to that line. In Chapter 5, you will learn that the analysis of an ac parallel circuit is just like that of a dc parallel circuit. But, with ac, you must keep track of the magnitude and phase of all of the signals and components by using phasors.

1

Sinusoidal Waveforms

Introduction

Electrical power is commonly produced and distributed in the shape of a sine wave. Sound consists of a combination of a variety of sine waves. Sine waves also carry every communications signal, from pagers to cellular phones to digital television. If fact, any repetitive waveform can be created by adding a series of different sinusoids. So, if you know how a circuit treats a sine wave, you can figure out how it will process *any* wave. Finally, sine waves are the only signal whose *shape* is not altered by a linear circuit. This makes the sinusoid an ideal test signal.

In this chapter you will review the sine function's basic mathematical equation. Angles may be expressed in either degrees or radians, and you must be able to convert between these units. Amplitude is expressed in both peak and root-mean-squared. You will learn the definitions of each and their relationship. Both frequency and angular velocity describe how rapidly a sine wave changes. When the signal starts is important as well, and is defined by its phase. How much the sine lags or leads is a critical element in describing what has happened to a signal.

You will create sine waves using a math software program, a spreadsheet, and two different simulators. These tools allow you to relate the key parameters of amplitude, frequency, and phase to the signal's shape.

Objectives

Upon completion of this chapter, you will be able to do the following:

- Write the general equation for a sinusoidal signal based on its amplitude, frequency, and phase shift.

- Manipulate the general equation for a sinusoidal signal to determine its amplitude, frequency, phase shift, and value at any instant.

- Define and calculate angles in radians, root-mean-squared amplitude, angular velocity, and phase angle.

- Use math software, a spreadsheet, and simulation to create and evaluate a sinusoidal signal.

1.1 The Basic Sine Function

The sine function is used extensively in mathematics, and describes the behavior of many phenomena. Its equation is

$$y = \sin(\alpha)$$

Figure 1-1 is a plot of this y as the angle α varies from 0° to 360°.

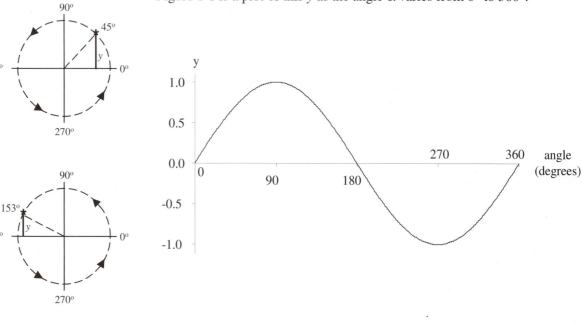

Figure 1-1 Basic sine function

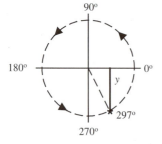

Figure 1-2 A point revolving in a circle

The sine function describes the height above the x axis, as a point travels counterclockwise in a circle. This is shown in Figure 1-2. The function (Figure 1-1) and the point (Figure 1-2) start at a value of 0 at 0°, and rise smoothly to a maximum of 1 at 90°. They then fall, just as they rose, back to 0 at 180°, half-way around the circle. On the second half of the revolution, the point moves below the x axis, so the value of the sine function is negative. The negative peak (−1) is reached three-quarters of the way around the circle, at 270°. The point and the function then return symmetrically to 0, when the circle is completed at 360° = 0°.

Many technical operations measure angles in **radians** instead of degrees. A radian is defined as the angle whose arc length is equal to its radius. An **arc length** is the distance along the arc of a circle from the origin to the end of the angle. These terms are shown in Figure 1-3.

How does the radian relate to the degree? First figure out how many radians there are in a full circle. The circumference of a circle is the distance around it.

$$C = \pi d$$

where d is the diameter. In terms of the radius, r,

$$C = 2\pi r$$

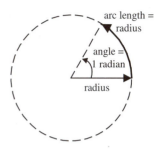

Figure 1-3 Definition of radian angle of measure

There are 2π radii around the edge of the circle, so there are 2π radians around the circle. There are also 360° around the circle.

$$360° = 2\pi \, \text{rad}$$

This produces the unit multipliers

$$\frac{2\pi \, \text{rad}}{360°} \quad \text{or} \quad \frac{360°}{2\pi \, \text{rad}} \quad \text{or} \quad \frac{57.3°}{\text{rad}}$$

Figure 1-4 shows the angles of a circle and the sine function measured both in degrees and in radians.

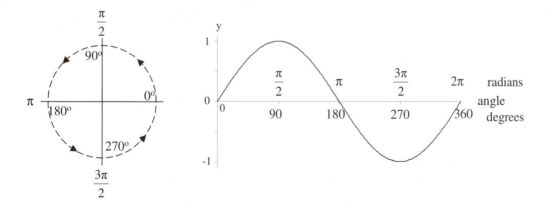

Figure 1-4 Degrees and radians of a circle and the sine function

So far the sine function has been generated traveling to the right along the x axis, with increasingly more positive angle. The circles have all assumed counterclockwise rotation. Movement in the opposite direction, for both the sine function and the circle, is defined as *negative*. This is illustrated in Figure 1-5.

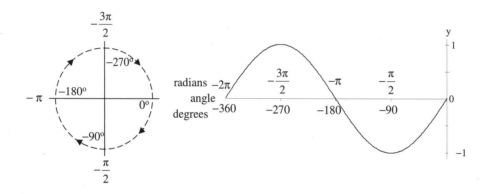

Figure 1-5 Negative angles

Since angles can be expressed as either positive or negative, it is standard practice to express them between −180° and +180°. Any point on the sine function or on the circle can be represented in this range. Look at Figure 1-6.

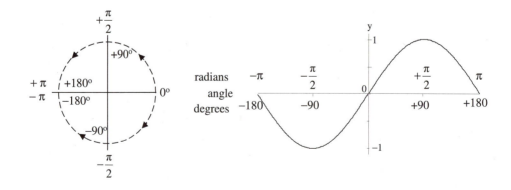

Figure 1-6 Angles between ±180°

Traveling around the circle or along the sine wave for 360° or 2π radians in one direction gets you to precisely the same value as where you started. Therefore you can add or subtract 360° or 2π radians to any angle to translate it into the ±180°, (±π) range that is preferred.

$$\alpha = \alpha \pm 360° \quad \text{or} \quad \alpha \pm 2\pi \, \text{rad}$$

Example 1-1

Convert each of the following into ±180° or ±π rad. Also convert the radians to degrees.

    a. 245°   **b.** −290°   **c.** 3π rad   **d.** −1.2π rad

Solution

    a. $\alpha = 245° - 360° = -115°$ **b.** $\alpha = -290° + 360° = 70°$

    c. $\alpha = 3\pi \, \text{rad} - 2\pi \, \text{rad} = \pi \, \text{rad} = 180°$

    d. $\alpha = -1.2\pi \, \text{rad} + 2\pi \, \text{rad} = 0.8\pi \, \text{rad} = 2.51 \, \text{rad}$

$$\alpha = 2.51 \, \text{rad} \times \frac{57.3°}{\text{rad}} = 143.8°$$

Practice: Convert each of the following into ±180° or ±π rad. Also convert the radians to degrees.

    a. 300°   **b.** −200°   **c.** 5.5 rad   **d.** −1.75π rad

Answers: a. −60° **b.** 160° **c.** −0.78 rad = −44.7° **d.** 0.78 rad = 44.7°

The program MATLAB allows you to easily plot a sine wave. In the m file below, the first line creates an array of values from 0, in steps of 1, to 360. The second line converts these angles to radians. The sine is then calculated for each of the angles, in radians. The last line plots the function, with the x axis in degrees, and the function v on the y axis. The plot is shown in Figure 1-7.

```
» degrees=0:1:360;
» radians=degrees/57.3;
» v=sin(radians);
» plot(degrees,v)
```

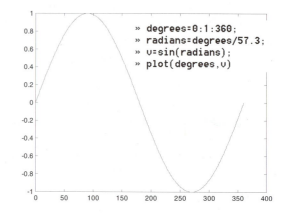

```
» degrees=0:1:360;
» radians=degrees/57.3;
» v=sin(radians);
» plot(degrees,v)
```

Figure 1-7 MATLAB plot of a sine wave

1.2 Amplitude

Peak

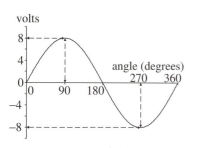

Figure 1-8 Effect of amplitude on $\sin(\alpha)$

The sine function varies from −1 to +1. To convert this into the voltages and currents that you measure in the laboratory, you just multiply it by the largest amplitude the signal reaches, its peak. For a voltage,

$$v(\alpha) = V_p \sin\alpha$$

An 8 V_p sine wave is shown in Figure 1-8.

Example 1-2

   a. What is the voltage at 45°, for a 170 V_p sine wave?

   b. At what angle does a sine wave with a 350 mV_p amplitude equal 250 mV? Answer in radians.

Solution

   a. $v(45°) = 170\,V_p \sin(45°) = 120\,V$

   b. To complete this calculation, first, divide both sides by V_p.

$$\frac{v(\alpha)}{V_p} = \sin(\alpha)$$

Taking the inverse sine of each side removes the sine function. Then move α to the left.

$$\alpha = \sin^{-1}\left(\frac{v(\alpha)}{V_p}\right)$$

$$\alpha = \sin^{-1}\left(\frac{250\,\text{mV}}{350\,\text{mV}}\right) = 0.796\,\text{rad}$$

Practice: For a sine wave with a peak amplitude of 12 A$_p$, find the

    a. current at 0.33 rad   **b.** angle in degrees where $i(\alpha) = 5$ A

Answers: **a.** $i(0.33\,\text{rad}) = 3.89$ A   **b.** $\alpha = 24.6°$

Root-Mean-Squared Values

The peak amplitude clearly defines the sine wave's amplitude. But, how can you compare the *effect* that a sinusoidal signal has to the effect of a dc signal from a steady supply? Do they both deliver the same power?

Example 1-2

Compare the power delivered to a light bulb by a 100 V$_{dc}$ steady supply and a 100 V$_p$ sine wave.

Solution

The results of a MultiSIM simulation are shown in Figure 1-10. Each supply drives a lamp rated at 120 V, 100 W. The oscilloscope shows that the dc supply is indeed a steady 100 V$_{dc}$, while the sine wave has a 100 V$_p$ peak amplitude.

Wattmeters measure the power delivered to each load. The 100 V$_{dc}$ signal delivers *twice* the power to the load that the 100 V$_p$ sine wave delivers. This should seem reasonable since the dc signal is 100 V all of the time, but the sine wave is only 100 V at its peaks. Most of the time, the sine wave is below 100 V.

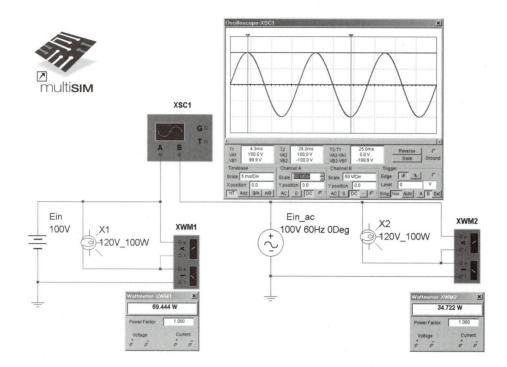

Figure 1-9 Power comparison of dc and ac for Example 1-2

Practice: Compare these values to the power delivered by a 100 V_p 25% pulse.

Answer: $P_{dc} = 69.4$ W, $P_{sine} = 34.7$ W, $P_{pulse} = 17.4$ W

The average value is indicated by dc measuring instruments. It gives a good measure of the performance you can expect from dc rated loads, such as a dc motor. It is a very poor way, though, to compare the effectiveness of differently shaped signals, as you saw in Example 1-2.

The **root-mean-squared** value (**rms**) of a signal is a much better indicator of the power a signal can deliver, regardless of wave shape.

Signals of the same rms value deliver the same power to a resistive load.

$$P = \frac{V_{rms}^{\ 2}}{R}$$

To begin the calculation of rms, the voltage function must be squared.

$$V_{rms} \propto v^2(t)$$

When a sinusoid is applied to a permanent magnet dc motor, for one half-cycle the shaft tries to turn one direction, and for the other half-cycle, it tries to turn the opposite direction. On the average, no work is done. So, it is the *average* value of the power that is of interest.

To calculate the average value of a function, you must first integrate it across one cycle of the function (0 to T). This gives the area under the function. The average height, then, is this area divided by its length, T.

$$V_{rms} \propto \frac{1}{T} \int_0^T v^2(t)\,dt$$

Now, however, the units in the relationship are wrong. There are volts on the left, and v^2 on the right. To correct this and produce an equality, you must take the square root of the right side.

$$V_{rms} = \sqrt{\frac{1}{T} \int_0^T v^2(t)\,dt}$$

Definition of root-mean-squared

The general equation for a simple sine wave is

$$v(\alpha) = V_p \sin\alpha$$

The rms value is defined as

$$V_{rms} = \sqrt{\frac{1}{T} \int_0^T v^2(t)\,dt}$$

Substitute, and adjust the limits.

$$V_{rms\,sine} = \sqrt{\frac{1}{2\pi} \int_0^{2\pi} (V_p \sin\alpha)^2\,d\alpha}$$

Square both terms.

$$V_{rms\,sine} = \sqrt{\frac{1}{2\pi} \int_0^{2\pi} V_p^2 \sin^2\alpha\,d\alpha}$$

Move the constant out of the integral.

$$V_{rms\,sine} = \sqrt{\frac{V_p^2}{2\pi} \int_0^{2\pi} \sin^2\alpha\,d\alpha}$$

There are several forms for the integral of $\sin^2\alpha$. A convenient one is

$$\int \sin^2 \alpha \, d\alpha = \frac{1}{2}\alpha - \frac{1}{4}\sin 2\alpha$$

Apply this integral solution to the rms calculation.

$$V_{\text{rms sine}} = \sqrt{\frac{V_p^2}{2\pi}\left(\frac{1}{2}\alpha - \frac{1}{4}\sin 2\alpha\right)\Big|_0^{2\pi}}$$

Evaluate this result at the two limits.

$$V_{\text{rms sine}} = \sqrt{\frac{V_p^2}{2\pi}\left[\left(\frac{1}{2}2\pi - \frac{1}{4}\sin 4\pi\right) - \left(\frac{1}{2}0 - \frac{1}{4}\sin 0\right)\right]}$$

$$V_{\text{rms sine}} = \sqrt{\frac{V_p^2}{2\pi}\left[\left(\pi - \frac{1}{4}0\right) - (0 - 0)\right]}$$

This is getting simpler. Now, just finish the algebra.

$$V_{\text{rms sine}} = \sqrt{\frac{V_p^2}{2\pi}\pi}$$

The rms of a sine wave

$$V_{\text{rms sine}} = \frac{V_p}{\sqrt{2}}$$

Example 1-3

Calculate the peak amplitude of a 100 V_{rms} sine wave.

Solution

$$V_{\text{rms sine}} = \frac{V_p}{\sqrt{2}}$$

Rearranging to solve for V_p,

$$V_p = 100\,V_{\text{rms}}\,\sqrt{2} = 141.12\,V_p$$

Practice: What peak amplitude is required of a 25% pulse to provide 100 V_{rms}? (Hint: use simulation software and trial-and-error.)

Answer: $V_{p\,25\%\,\text{pulse}} = 200\,V_p$

1.3 Frequency and Angular Velocity

So far, the horizontal axis of the sine function graphs has been scaled in terms of angle, in either degrees or radians. But in the lab, you measure the signals as voltage versus *time*. The length of time it takes to complete one cycle (360°, 2π rad) is called the **period** (**T**). Normally, it is measured from the positive-going zero crossing on one wave to the same point on the next. But it may be measured from any point on one wave to the same point on the next wave. Measuring between the peaks is convenient. Several ways to measure period are shown in Figure 1-10.

The period indicates the time for one cycle of the signal. That is

$$T = \frac{seconds}{cycle}$$

Period

Symbol: *T*
Units: seconds

Definition: time for one complete cycle

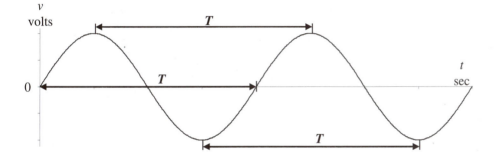

Figure 1-10 Period of a sine wave

The number of cycles that occur in a single second is the **frequency.**

$$f = \frac{cycles}{second}$$

This is just the reciprocal of period.

$$f = \frac{1}{T} \quad \text{or} \quad T = \frac{1}{f}$$

Frequency

Symbol: *f*
Units: hertz (Hz)

Definition: number of cycles in a second

The units of measure for frequency are cycles/second. This is called **hertz**, abbreviated **Hz**. Table 1-1 lists the frequency and period of several sinusoids with which you may be familiar.

Table 1-1 Common frequencies and periods

	frequency (Hz)	period (s)
one minute	0.01667	60
audio bass	< 300	> 3.33 m
commercial power	50 to 60	20 m to 16.7 m
music middle C	262	3.82 m
middle of speech band	1 k	1.00 m
bottom of AM radio band	530 k	1.89 μ
top of FM radio band	108 M	9.26 n
television channel 18	510 M	1.96 n
cellular phone	850 M	1.18 n
portable phone	900 M	1.11 n
pager	950 M	1.05 n
police radar	10 G	100 p

The sine wave may be calculated and plotted with angle on the horizontal axis (the *independent* variable). However, to repeat, you usually measure a sinusoidal signal with *time* as the horizontal axis. One full cycle takes T sec and covers 2π rad. The rate at which the sine wave moves through the angles is the **angular velocity (ω)**.

Angular velocity

Symbol: ω
Units: rad/s

Definition: rate a sine function moves through the angles along the horizontal axis

$$\omega = \frac{2\pi\,\text{rad}}{T} = 2\pi\,\text{rad} \times f$$

The angle at any instant in time is α.

$$\omega = \frac{\alpha}{t}$$

But we have just defined ω in terms of the signal's frequency. Making that substitution gives

$$2\pi f = \frac{\alpha}{t} \quad \text{or} \quad \alpha = 2\pi ft$$

In terms of *time* and *frequency*, the general sine equation becomes

Sine wave equation including   amplitude and frequency

$$v(t) = V_\text{p} \sin(2\pi ft)$$

Remember that the angle is in radians, not degrees.

Example 1-4

    a. Write the equation for a sine wave with an 8 V_p amplitude and a frequency of 20 kHz.

    b. Plot one cycle of this signal.

Solution

    a. $v(t) = 8\,V_p \sin(2\pi \times 20\,\text{kHz} \times t)$

    b. To plot one cycle, first calculate the period.

$$T = \frac{1}{20\,\text{kHz}} = 50\,\mu\text{s}$$

The partial spreadsheet and the plot are shown in Figure 1-11.

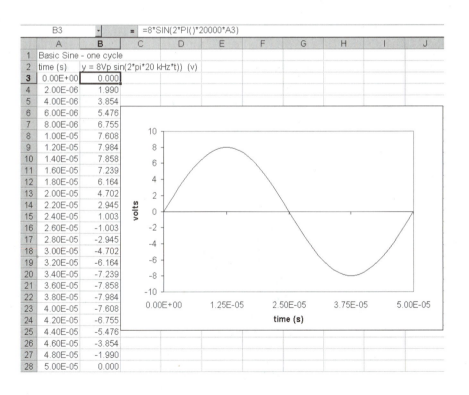

Figure 1-11 Plot of 8 V_p, 20 kHz sine wave

how big can you dream?™

Example 1-5

   a. Write the equation for a 120 V_{rms}, 60 Hz sine wave.

   b. Determine the voltage at 3.3 ms with your calculator.

   c. Determine the first time that the sine wave passes 100 V.

   d. Verify your answers with a plot.

Solution

   a. $V_p = 120\,V_{rms}\,\sqrt{2} = 170\,V_p$

$$v(t) = 170\,V_p \sin(2\pi \times 60\,Hz \times t)$$

   b. At t = 3.3 ms,

$$v(3.3\,ms) = 170\,V_p \sin(2\pi \times 60\,Hz \times 3.3\,ms)$$

Remember to put your calculator in radians mode.

$$v(3.3\,ms) = 161\,V$$

   c. This is very similar to Example 1-2 with more information.

$$\frac{100\,V}{170\,V} = \sin(2\pi \times 60\,Hz \times t)$$

$$2\pi \times 60\,Hz \times t = \sin^{-1}\left(\frac{100\,V}{170\,V}\right) = 0.629\,rad$$

$$t = \frac{0.629\,rad}{2\pi\,\dfrac{rad}{cycle} \times 60\,\dfrac{cycles}{s}} = 1.67\,ms$$

   d. The simulation set-up is shown in Figure 1-12. The signal generator is **VSIN**. You must enter the *peak* amplitude, 170 V_p. The same load used in Examples 1-2 and 1-3 is used.

      The common symbol must be *renamed*. Double click on the symbol, and change the name to 0 (the number zero). This establishes the 0 node for the simulation. The dialog box is shown beside the schematic.

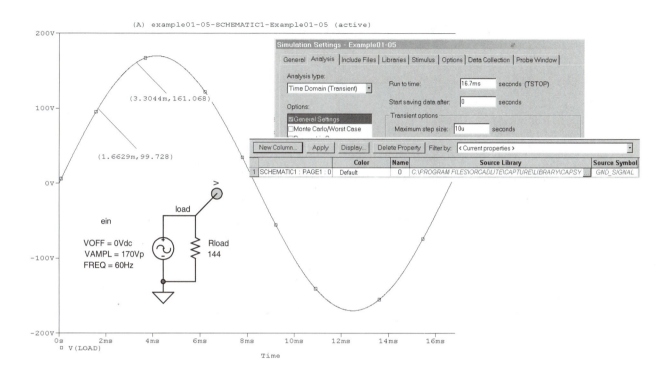

Figure 1-12 Simulation results for Example 1-5

The transient analysis parameters are shown in the dialog box to the right of the schematic. The **Run to time** is set at 16.7 ms to produce a full cycle. The **Maximum step size** is set to 10 μs to force the simulator to take steps no larger than 10 μs.

Figure 1-12 shows the resulting **Probe** display. The first cursor shows that at 1.66 ms the voltage is 99.7 V. This matches the calculations. The second cursor indicates that at 3.3 ms the voltage is 161 V. This also confirms the manual calculations.

Practice: For a 0.25 V_{rms}, 3 kHz signal, determine:

  a. the voltage at 200 μs.

  b. the first time the signal passes 100 mV.

Answer: a. $v(200 \ \mu s) = -208$ mV   **b.** $t = 15.2 \ \mu s$

1.4 Phase Shift

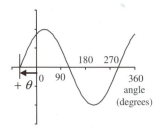

(a) Positive, leading phase

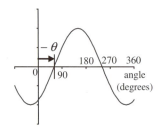

(b) Negative, lagging phase

Figure 1-13 Lag and lead phase shift

The starting point of a sine wave is when it crosses 0 V, going positive. All of the signals, so far, have started at $t = 0$. They have been just-in-time, neither too early, nor too late.

However, many of the circuits you will deal with either advance or delay the signal's starting point. *When* the signal starts is determined by its **phase shift**. That phase shift is measured, in degrees, from $\alpha = 0°$ to the starting point of the wave. If the signal starts early, before $\alpha = 0°$, it is said to **lead**, and is given a *positive* angle. If the signal starts after $\alpha = 0°$, the signal **lags**. Lagging signals have a negative phase shift. Both phase shifts are shown in Figure 1-13.

Be very careful. It is easy to get the polarity of the phase angle backwards. Signals with a positive phase shift start early, to the *left* of $\alpha = 0°$. These signals have a positive value at $\alpha = 0°$.

Negative phase shift indicates that the signal's starting point has been delayed, moving it to the *right* of $\alpha = 0°$. These signals have a negative value at $\alpha = 0°$.

Phase shift is included in the general equation of a sinusoid by adding it to the angle created by $2\pi ft$.

$$v(t) = V_p \sin(2\pi ft \pm \theta)$$

It is traditional to express the phase shift, θ, in degrees, even though $2\pi ft$ results in an angle in radians. So, when you are evaluating this equation, before you can take the sine, you must either convert $2\pi ft$ from radians to degrees, or you must convert θ from degrees to radians.

Example 1-6

    a. Write the equation for a sine wave with an 8 V_p amplitude, a frequency of 20 kHz, and a phase shift of $-75°$.

    b. Plot one cycle of this signal.

Solution

    a. $v(t) = 8\,V_p \sin(2\pi \times 20\,\text{kHz} \times t - 75°)$

    b. This is an extension of Example 1-4. Look back at that problem to determine the step size. Since the signal has been delayed $-75°$, extend the cycle another

$$\frac{75°}{360°} \times 50\,\mu s \approx 10\,\mu s$$

Figure 1-14 is the plots of this general sine wave.

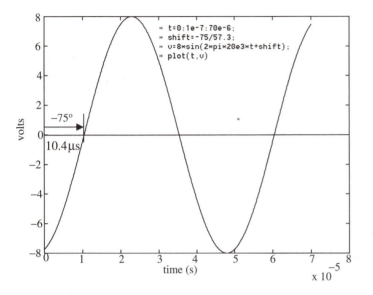

Figure 1-14 MATLAB plot of a general sine wave for Example 1-6

Practice: Determine the voltage at 5 μs for a 12 V_p, 30 kHz sine wave with a +50° phase shift. Use a spreadsheet plot to verify the calculation.

Answer: $v(t) = 12\,V_p \sin(2\pi \times 30\,\text{kHz} \times 5\,\mu s + 50°) = 11.6\,V$

The oscilloscope is often used to measure phase shift. But, the horizontal axis of an oscilloscope is scaled in *time* not in *phase*. The conversion is a simple ratio. Look at Figure 1-15.

First, the oscilloscope must be able to define when *t* = 0 occurs. It needs some reference on which to trigger. Usually this is the input signal, the signal from the generator. Place this on channel A (or channel

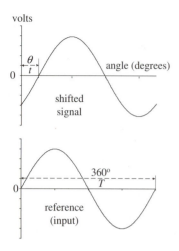

Figure 1-15 Lagging
phase shift measurement

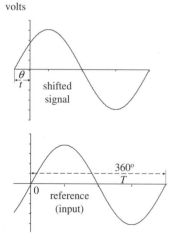

Figure 1-16 Leading
phase shift measurement

1) and move it to the lower half of the screen. In Figure 1-15, this is the lower, reference waveform.

Adjust the oscilloscope's *trigger source* to this signal, channel A. Set the *trigger level* to 0 V. As you move the trigger level, watch the starting point of the input signal's display. Move the trigger level up and down until the trace begins at 0 V, the center line for that signal.

Place the signal whose phase you want to measure on channel B. Move its zero line until it is positioned in the center of the upper half of the display. This is shown as the upper trace in Figure 1-15.

The oscilloscope allows you to measure the period of the signals, T and the time from the beginning of the trace until the shifted signal crosses 0 V going positive, t. Cursors on the display make this simple. Otherwise, count the divisions and multiply by the time/div setting.

These times can then be converted to angles by recognizing that the relationships are proportional.

$$\frac{\theta}{t} = \frac{360°}{T} \quad \text{or} \quad \theta = \frac{360°}{T} \times t$$

If you already know the frequency, this simplifies to

$$\theta = 360° \times f \times t$$

If the signal you are measuring leads, with a positive phase shift, the oscilloscope display looks like Figure 1-16. The shifted signal starts, crossing 0 V going positive, while the input is still negative. For these waves, lower the *trigger level* to display the starting point.

Example 1-7

Calculate the lag *time* for a 120 V_{rms}, 60 Hz sine wave with a phase shift of −30°.

Solution

$$\theta = 360° \times f \times t \quad \text{or} \quad t = \frac{\theta}{360° \times f}$$

$$t = \frac{30°}{360° \times 60\,\text{Hz}} = 1.39\,\text{ms lag}$$

Practice: Calculate the lead *time* for a 0.25 V_{rms}, 3 kHz, +60 signal.

Answer: $t = 55.6\ \mu s$ Set the lag to 60° − 360° = 300° lag.

Multiplying the magnitude by −1 negates every point, making those that are normally positive, negative. Those that are usually negative become positive. The signal has just been flipped over. Or, it has been shifted ±180°. Look at Figure 1-17.

$$-V_\text{p} \sin(2\pi ft \pm \theta) = V_\text{p} \sin(2\pi ft \pm \theta \pm 180°)$$

Although either form is mathematically correct, it is preferred to remove the − sign, by adding or subtracting 180° to the phase angle.

In Figure 1-18, the sine wave is shown as a vector. The length of the vector indicates the magnitude of the sine wave, and the angle (as measured from the +x axis) is the phase shift. The entire next chapter is dedicated to investigating the characteristics of this electrical vector, called a **phasor**. Multiplying the magnitude by −1 sends the phasor in the exactly opposite direction. In Figure 1-18, $V_\text{p}\sin(2\pi ft+\theta)$ is shown in the first quadrant. So, $-V_\text{p}\sin(2\pi ft+\theta)$ is in the third quadrant. It extends in exactly the opposite direction. It has been shifted ±180°.

The cosine is also a sinusoidal function.

$$\cos(\alpha) = \sin(\alpha + 90°)$$

In later chapters, it is assumed that the signal is a sine function. So if you are given a signal in terms of cosines, add 90° to the phase in order to convert the cosine signal to sine.

Finally, there are occasions when the magnitude of a cosine wave may be inverted. The negative magnitude can be removed by subtracting 180° from the phase angle.

$$-\cos(\alpha) = \cos(\alpha - 180°)$$

Then you must add 90° to convert the cosine into a sine.

$$-\cos(\alpha) = \sin(\alpha - 180° + 90°)$$

$$-\cos(\alpha) = \sin(\alpha - 90°)$$

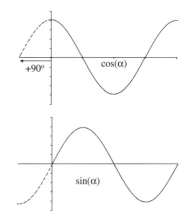

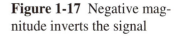

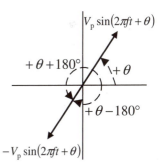

Figure 1-17 Negative magnitude inverts the signal

Figure 1-18 Negative magnitude shifts the phase angle ±180

Example 1-8

a. Write the equation for a sine wave and a cosine wave with a −8 V_p amplitude, and a frequency of 20 kHz.

b. Plot one cycle of each of these signals.

Solution

a. For the sine wave,

$$v(t) = -8\,V_p\sin(2\pi \times 20\,\text{kHz} \times t) = 8\,V_p\sin(2\pi \times 20\,\text{kHz} \times t - 180°)$$

For the cosine wave

$$v(t) = -8\,V_p\cos(2\pi \times 20\,\text{kHz} \times t) = 8\,V_p\sin(2\pi \times 20\,\text{kHz} \times t - 90°)$$

b. The plots are shown in Figure 1-19.

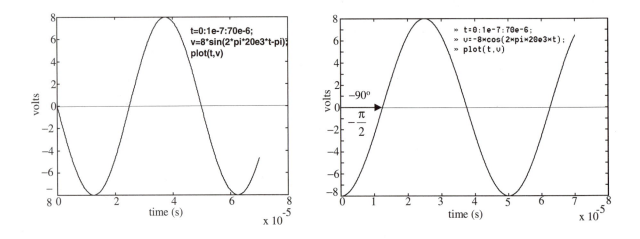

Figure 1-19 MATLAB plots for a sine and cosine wave of $-8\,V_p$ and 20 kHz

---

Practice: Determine the voltage at 5 μs for a 12 V_p, 30 kHz cosine wave. Plot the equation to verify the calculation.

Answer: $v(t) = 12\,V_p\cos(2\pi \times 30\,\text{kHz} \times 5\,\mu s) = 7.05\,V$

Summary

The simplest way to express a sine wave is $y = \sin(\alpha)$. The variable, α, is an angle and varies from $0°$ through a full circle of $360°$. Angle may be expressed in degrees or radians. A radian is defined as that angle whose arc length is equal to the radius of the circle. Sine waves extend infinitely in both the positive and negative x direction, repeating every $360°$, or 2π rad. Express angles in the range of $-180° \leq \alpha \leq 180°$.

Since the sine function varies between ± 1, multiply it by the peak amplitude to express y in terms of voltage or current. A 10 V_p sine wave delivers more power to a resistive load than does a 10 V_p triangle wave, but less than a 10 V steady dc voltage. Signals with equal root-mean-squared values *do* provide the same power to a resistive load, regardless of wave shape. The relationship between peak and rms values is different for different waves. Most meters indicate the root-mean-squared value of the applied signal.

How long (in seconds) it takes for a voltage or current sine wave to complete a single cycle is the period (T). The number of cycles completed in a second is the frequency (f). The unit of measure for the number of cycles per second is hertz. Period and frequency are reciprocals.

Angular velocity (ω) is the number of radians crossed each cycle. With a little algebra, you can replace the angle α with time, t. Circuits may delay (lag) or advance (lead) the start of a sinusoidal signal. This is accounted for by adding (lead) or subtracting (lag) a phase angle in the function. Be careful to remember that $2\pi ft$ indicates an angle in radians, while θ is generally expressed in degrees.

There are precise steps you must follow in measuring the phase angle either with an oscilloscope or with simulation software.

Be sure to get the polarity of the phase shift correct. If the signal crosses 0 V going positive *before* $t = 0$, the phase shift is positive and is said to lead. If the signal crosses 0 V going positive *after* $t = 0$, the phase shift is negative and is said to lag.

If the magnitude is negative, this is the same as adding or subtracting $180°$ ($\pm\pi$ rad) to the phase.

$$-V_p \sin(2\pi ft \pm \theta) = V_p \sin(2\pi ft \pm \theta \pm 180°)$$

A cosine is just a sine shifted ahead $90°$

$$\cos(\alpha) = \sin(\alpha + 90°)$$

$$-\cos(\alpha) = \sin(\alpha - 90°)$$

Problems

The Basic Sine Function

1-1   Calculate the value of the sine function at the following angles:
    a. 60°   **b.** −45°   **c.** 135°   **d.** −150°

1-2   Calculate the value of the sine function at the following angles:
    a. −75°   **b.** 140°   **c.** 30°   **d.** −125°

1-3   Determine the angle whose sine produces the following value:
    a. 0.50 (0° to 90°)   **b.** −0.71 (−90° to −180°)

1-4   Determine the angle whose sine produces the following value:
    a. 0.10 (90° to 180°)   **b.** −0.33 (0° to −90°)

1-5   Convert the following from degrees to radians, or radians to degrees. Express your answer between ±180° or ±π rad.
    a. 135°   **b.** −200°   **c.** 320°   **d.** 2 rad   **e.** −4 rad   **f.** 5 rad

1-6   Convert the following from degrees to radians, or radians to degrees. Express your answer between ±180° or ±π rad.
    a. 70°   **b.** −160°   **c.** 220°   **d.** 3 rad   **e.** −4.5 rad   **f.** −5.3 rad

Amplitude

1-7   Calculate the current at 45° of a sine wave with $I_p = 12$ A$_p$.

1-8   Calculate the voltage at −120° of a sine wave with $I_p = 294$ V$_p$.

1-9   Calculate the root-mean-squared value of a 294 V$_p$ sine wave.

1-10   Calculate the peak voltage of a 120 V$_{rms}$ sine wave.

Frequency and Angular Velocity

1-11   Calculate the period and angular velocity of a 60 Hz sine wave.

1-12   Calculate the f and ω of a sine wave with a period of 50 μs.

1-13   For the signal described by $40\,\text{V}_p \sin(2\pi \times 2\,\text{kHz} \times t)$, calculate
    a. V_{rms} **b.** f **c.** T **d.** ω **e.** $v(200\,\mu s)$

1-14   For the signal described by $622\,\text{V}_p \sin(377\ t)$, calculate
    a. V_{rms} **b.** f **c.** T **d.** ω **e.** $v(6.7\,ms)$

1-15   For the signal described by $40\,\text{V}_p \sin(2\pi \times 2\,\text{kHz} \times t)$, calculate the first *time after* $t = 0$ s when the signal equals 10 V.

1-16   For the signal described by 622 V_p sin(377*t*), calculate the first *time after t* = 0 s when the signal equals 200 V.

Phase Shift

1-17   For the signal described by $40\,V_p \sin(2\pi \times 2\,\text{kHz} \times t - 60°)$

    a.   Determine when the signal crosses *v* = 0 V going positive.

    b.   Is this lead or lag?

1-18   For the signal described by 622 V_p sin(377*t* + 120°):

    a.   Determine when the signal crosses *v* = 0 V going positive.

    b.   Is this lead or lag?

1-19   Accurately plot the equation in Problem 1-17. Highlight the value of *v*(200 μs).

1-20   Accurately plot the equation in Problem 1-18. Highlight the value of *v*(6.7 ms).

1-21   For the signal described by $-40\,V_p \sin(2\pi \times 2\,\text{kHz} \times t - 60°)$:

    a.   Rewrite the equation to remove the negative magnitude.

    b.   Determine when the signal crosses *v* = 0 V going positive.

1-22   For the signal described by –622 V_p cos(377 *t* + 120°):

    a.   Rewrite the equation to remove the negative magnitude and convert the function to sine by adjusting the phase shift

    b.   Determine when the signal crosses *v* = 0 V going positive.

    c.   Is this lead or lag?

    d.   Calculate the voltage at *t* = 6.7 ms.

1-23   Accurately plot the equation in Problem 1-21.

1-24   Accurately plot the equation in Problem 1-22.

1-25   Write the equation for the signal shown in Figure 1-20.

1-25   Write the equation for the signal shown in Figure 1-21.

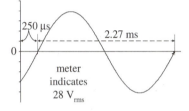

Figure 1-20 Signal for Problem 1-25

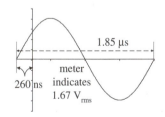

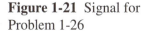

Figure 1-21 Signal for Problem 1-26

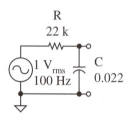

Figure 1-22 Phase shift schematic (Note: capacitance in μF)

Phase Angle Measurement Lab Exercise

A. Calculations

1. The goal of this exercise is to practice making phase angle measurements with an oscilloscope and with a simulator. There is *no* expectation that you understand the circuit or why it shifts the phase as it does. The circuit theory will be explained in detail in the following chapters.

2. The circuit in Figure 1-22 delays the phase of the output voltage, with respect to the source. Calculation of this phase shift requires two steps. First calculate the capacitor's reactance, X_C.

$$X_C = \frac{1}{2\pi f C}$$

Knowing this reactance, you can then calculate the phase shift of the output voltage.

$$\theta = -90° - \arctan\left(-\frac{X_C}{R}\right)$$

3. Complete these calculations for frequencies of 100 Hz, 150 Hz, 300 Hz, 750 Hz, 1200 Hz, and 3000 Hz. A spreadsheet table is handy. Be sure that the angles are converted to degrees.

B. Measurements

1. Build the circuit carefully. Keep the leads from the signal generator and the leads to each component as short as practical.

2. Set the signal generator's amplitude to 1 V_{rms}, and its frequency to 100 Hz.

3. Place channel A (or channel 1) across the signal generator and move that trace to the lower half of the screen. This is the reference waveform.

4. Adjust the oscilloscope's *trigger source* to this signal, channel A. Set the *trigger level* to 0 V. As you move the trigger level,

watch the starting point of the input signal's display. Move the trigger level up and down until the trace begins at 0 V, the center line for that signal.

5. Place channel B of the oscilloscope across the capacitor. Be *sure* that both oscilloscope probe commons are tied *together.*

6. Move channel B's zero line until it is positioned in the center of the upper half of the display.

7. The oscilloscope allows you to measure the time from the beginning of the trace until the shifted signal crosses 0 V going positive, *t*. Cursors on the display make this simple. Otherwise, count the divisions and multiply by the time/div setting.

8. This lag time, *t*, can then be converted to the phase shift angle.

$$\theta_{measured} = 360° \times f \times t$$

Record this *measured* phase in the table from section A.

9. If your measurement of the phase in step B8 is within ±4° of the calculated angle from step A3, continue to section C. If not, double check your calculations and your measurement until you find the error.

C. Simulation
   1. Repeat each of the steps in Section B to build and measure the phase shift of the voltage across the capacitor in Figure 1-22 with your simulator. Refer to the examples earlier in the chapter for suggestions in using MultiSIM and Cadence PSpice.

   2. Calculate and record this simulated phase in a cell in your data table beside the calculated phase and the measured phase.

   3. Continue only when the calculated, measured, and simulated phase angles agree to within ±4°.

2

Phasors

Introduction

From the previous chapter you learned to define, calculate, and express in a single equation the magnitude, frequency, and phase shift of a variety of sinusoidal signals. Linear circuits, containing resistors, capacitors, inductors, transformers, transistor amplifiers, and op amps do *not* alter the shape of these signals, nor their frequency. Applying a 10 MHz sine wave to a linear circuit produces a 10 MHz sine wave at its output. So, once these two pieces of information are stated, they remain constant throughout the circuit and throughout your calculations.

However, linear circuits *do* change a signal's amplitude, (amplifying or attenuating), and shift its phase (causing the output to lead or lag the input). These are the two quantities that you must keep track of as you determine how a circuit affects your signal, or which components are needed to provide the amplitude and phase shift required at the output.

Phasors represent a sinusoidal signal's magnitude and phase shift in a simple, compact form. They are similar to the vectors you may have seen in your basic physics course, and allow the same mathematical operations. By representing sinusoidal signals as phasors, you can then apply all of the circuit analysis laws and procedures that you learned when studying dc electronics. With this technique, the effects on both the magnitude and the phase shift of the voltage, current, and opposition can be accounted for.

Objectives

Upon completion of this chapter, you will be able to do the following:

- Convert between time domain and phasor notation.
- Represent a phasor in polar or rectangular notation.
- Convert between polar and rectangular notation.
- Add, subtract, multiply, and divide phasors.

2.1 Polar Notation

A sinusoidal voltage can be represented by

$$v(t) = V_\text{p} \sin(2\pi f t + \theta)$$

The **magnitude** indicates how "tall" it is. This is set by V_p. The signal's **frequency** (f) indicates how rapidly it changes. *When* the sine wave starts is determined by its **phase shift**, θ. The independent variable is time, t. These relationships were developed in Chapter 1.

Linear circuits comprise the majority of analog electronics. Considerable effort is made to assure that a change at the input produces a proportional change at the output. Circuits made of resistors, capacitors, inductors, and transformers are usually linear. Amplifiers that do not distort the signal are also linear. So, many bipolar junction transistor, MOSFET, and op amp circuits are considered to be linear, faithfully producing a larger version of the input at the load.

In designing or analyzing *linear* circuits, you do not have to keep track of the fact that the signal is sinusoidal in shape, or the frequency of that signal. Linear circuits do not alter the shape or frequency of the signal. They do not introduce distortion. A sine wave at the input produces a sine wave at the output. These linear circuits only change the amplitude and the phase of that sinusoid. Of the full, time-domain equation, therefore, only the magnitude (V_p) and the phase (θ) change. Only these two quantities enter into your calculations.

This is just like the vectors from physics. A vector quantity has both magnitude and direction, and is represented by an uppercase variable with a line above it. The mechanics vector

$$\overline{V} = 29\frac{\text{m}}{\text{s}} \angle 90°$$

indicates that your velocity is 29 m/s straight up (or 65 mi/hr due north). These velocities are shown in Figure 2-1.

This same convention, when applied to electrical signals, is called a **phasor**. A sinusoidal voltage can be represented by

$$\overline{V} = V_\text{rms} \angle \theta$$

The uppercase V, with a bar over it, indicates that the quantity is a phasor, having both magnitude and phase. The magnitude is usually recorded as root-mean-squared, *not* peak. The phase angle is in degrees.

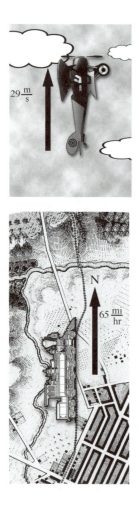

Figure 2-1 Velocity vectors

Do not forget the polarity of the phase: + means that the signal leads the reference, beginning early; − means that the signal lags the reference, beginning late.

Phasors can also be plotted, just as vectors are. This makes it easier to visualize the signal, and how several signals of different magnitudes and phase shifts combine. Just as with the sine waves in Chapter 1, the phase angle is measured from the +x axis. Counterclockwise is considered positive. The phasor of 120 V_{rms} ∠+60° is shown in Figure 2-2. Be sure to include a scale, 1 inch = 100 V_{rms} in this case.

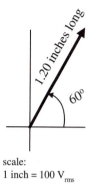

Figure 2-2 120 V_{rms}∠+60°

Example 2-1

Write the phasor notation for each of the following signals and draw their phasor diagram. Use a scale of 1 in = 100 V_{rms}.

a. $v_a = 294V_p \sin(377t - 30°)$

b. $v_b = 170V_p \sin(2\pi \times 60\,\text{Hz} \times t + 120°)$

Solution

a.
$$V_{rms} = \frac{V_p}{\sqrt{2}}$$

$$V_{rms} = \frac{294\,V_p}{\sqrt{2}} = 208\,V_{rms}$$

$$\overline{V_a} = (208\,V_{rms} \angle -30°)$$

The phasor diagram is shown in Figure 2-3(a).

b.
$$V_{rms} = \frac{170\,V_p}{\sqrt{2}} = 120\,V_{rms}$$

$$\overline{V_b} = (120\,V_{rms}\angle +120°)$$

The phasor diagram is shown in Figure 2-3(b).

Practice: Write the following as phasors:

$$i_a = 230\mu A_p \sin(2\pi \times 900\,\text{MHz} \times t + 245°)$$

$$v_b = 1.0V_p \sin(18,850 \times t - 45°)$$

Answer: $\overline{I_a} = (163\mu A_{rms}\angle -115°)$ $\overline{V_b} = (707\,mV_{rms}\angle -45°)$

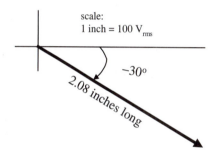

Figure 2-3(a) Example 2-1a
208 $V_{rms}\angle -30°$

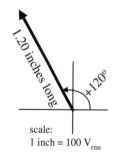

Figure 2-3(b) Example 2-1b
120 $V_{rms}\angle +120°$

2.2 Rectangular Notation

The phasor is represented by a two-part quantity. One part indicates the magnitude and the other the phase. Together, they specify a position in space. 120 $V_{rms}\angle+60°$ tells you to travel away from the origin at 60° above the horizon, for a distance of 120 V_{rms}.

This same position can be specified by giving its x and y coordinates. The horizontal, x direction is defined as **real**. The vertical, y direction is **imaginary**. This is denoted with the prefix, j. You may also see i used to identify the imaginary component. In mathematical terms,

$$j = i = \sqrt{-1}$$

Since you cannot find a real number that, when multiplied by itself, gives −1, this number must be *imagined*.

When a real number (x axis) and an imaginary number (y axis) are combined to locate a point, the result is called a **complex** number.

$$\overline{V} = (60\,V_{rms} + j104\,V_{rms})$$

is a complex phasor voltage. You reach it from the origin by moving to the right 60 V_{rms}, then vertically 104 V_{rms}. Look at Figure 2-4.

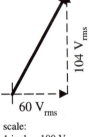

scale:
1 inch = 100 V

Figure 2-4
60 V_{rms} + j104 V_{rms}

Example 2-2
Using a scale of 1 in = 100 V_{rms}, draw the phasor diagrams of

a. $\overline{V_a} = (180\,V_{rms} - j104\,V_{rms})$ **b.** $\overline{V_b} = (-60\,V_{rms} + j104\,V_{rms})$

Solution:

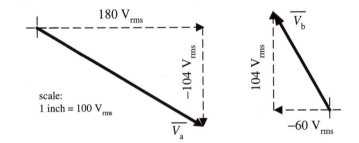

Figure 2-5 Rectangular phasor plots for Example 2-2

2.3 Conversions

Phasor voltages and currents are easily measured in their polar form. The digital multimeter provides the amplitude, and the oscilloscope indicates the phase shift. However, combining components and certain phasor calculations are easier to complete in the rectangular form.

Many scientific calculators allow you to enter phasors in either form. These calculators also complete computation with phasors and present the answer in whichever form you request. They even provide the conversion from one form to the other.

However, often the problem is to determine an *equation*. Equations give you a good general description of how the quantities interact, far better than a single number, or even a simulation. This is required when you investigate the performance of filters in the second section of this book, or when you consider the response of a circuit to a variety of sine waves, in the Fourier Analysis section.

It is *not* adequate to just let your calculator handle the phasor conversions and calculations. You must be able to complete these with *symbols* as well.

Figure 2-6 shows a right triangle. The length of the phasor, which is the hypotenuse of the triangle, is labeled C. The angle is θ. The horizontal distance is A. This is also the *real* value of the rectangular form. The *imaginary* distance, the rise, is B. This phasor may be represented in either the polar or the rectangular form.

$$\overline{V} = C\angle\theta = A + jB$$

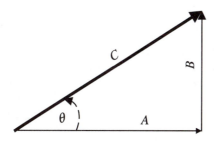

Figure 2-6 Right triangle relates polar and rectangular phasor forms

To convert from the polar form to the rectangular form of a phasor, you have $C \angle \theta$, and must find A and B. From trigonometry, you should remember that the cosine of an included angle relates the length of the adjacent side and the length of the hypotenuse.

$$\cos\theta = \frac{adjacent}{hypotenuse} = \frac{A}{C}$$

Since you know both C and θ, solve this for A, the real part of the rectangular form.

Real part of polar to rectangular conversion

$$A = C\cos\theta$$

Similarly, the sine of an included angle relates the length of the opposite side to the length of the hypotenuse.

Imaginary part of polar to rectangular conversion

$$\sin\theta = \frac{opposite}{hypotenuse} = \frac{B}{C}$$

Solve this for B, the imaginary part of the rectangular form.

$$B = C\sin\theta$$

MATLAB

```
>>Va=120*cos(60/57.3)
   +j*120*sin(60/57.3)

Va=
      60.00 + 103.92i
```

```
>>Vb=208*cos(-120/57.3)
   +j*208*sin(-120/57.3)

Vb=
   -103.97 – 180.15i
```

```
>>Vc=40e-6*cos(135/57.3)
   +j*40e-6*sin(135/57.3)

Vc=
   -2.8279e-005
   + 2.8289e-005i
```

Example 2-3

Convert each of the following polar phasors into their rectangular form.

a. $\overline{V_a} = (120\,\text{V}_{\text{rms}}\angle 60°)$ b. $\overline{V_b} = (208\,\text{V}_{\text{rms}}\angle -120°)$

c. $\overline{I_c} = (40\,\mu\text{A}_{\text{rms}}\angle 135°)$

Solution

a. $\overline{V_a} = (60\,\text{V}_{\text{rms}} + j104\,\text{V}_{\text{rms}})$ b. $\overline{V_b} = (-104\,\text{V}_{\text{rms}} - j180\,\text{V}_{\text{rms}})$

c. $\overline{I_c} = (-28.3\,\mu\text{A}_{\text{rms}} + j28.3\,\mu\text{A}_{\text{rms}})$

Practice: Convert each of the following polar phasors into their rectangular form.

a. $\overline{V_a} = (0.25\,\text{V}_{\text{rms}}\angle 45°)$ b. $\overline{I_b} = (1.4\,\text{A}_{\text{rms}}\angle 150°)$

c. $\overline{V_c} = (208\,\text{V}_{\text{rms}}\angle -150°)$

Answers: a. $\overline{V_a} = (177\,\text{mV}_{\text{rms}} + j177\,\text{mV}_{\text{rms}})$

b. $\overline{V_b} = (-1.21\,\text{A}_{\text{rms}} + j700\,\text{mV}_{\text{rms}})$ **c.** $\overline{V_c} = (-180\,\text{V}_{\text{rms}} - j104\,\text{V}_{\text{rms}})$

Conversion *from* rectangular form *to* polar requires a different set of trigonometric relationships. Given the real, *A*, and imaginary, *B*, sides, you can calculate the hypotenuse with the Pythagorean theorem.

$$C^2 = A^2 + B^2$$

Or,
$$C = \sqrt{A^2 + B^2}$$

The tangent of the included angle, θ, is the ratio of the length of the opposite side, *B*, over the adjacent side, *A*.

$$\tan \theta = \frac{B}{A}$$

Taking the inverse tangent of each side leaves θ alone on the left of the equation.

$$\theta = \tan^{-1}\left(\frac{B}{A}\right)$$

You need to be particularly careful when using this relationship. Positive *B* and positive *A* come from a phasor in the first quadrant (0° to 90°). The ratio is positive. The inverse tangent of a positive quantity produces an angle between 0° and 90°. This is correct.

However, if both *B* and *A* are negative, then the phasor is in the third quadrant (−90° to −180°). The ratio of these two negative numbers produces a positive quotient. The inverse tangent of a positive number is between 0° and 90°, even though the phasor lies in the third quadrant. You must sketch the phasor to be sure that the result makes sense. In the case of a negative *B* and a negative *A*, the positive angle that you get from the inverse tangent must be measured from the −*x* axis.

A similar problem exists when either but not both *B* and *A* is negative. When *B* is negative and *A* is positive, the phasor is in the fourth quadrant (0° to −90°). The resulting inverse tangent returns the correct angle.

But when *B* is positive and *A* is negative, the phasor is in the second quadrant (90° to 180°). The ratio is still negative, so the resulting inverse tangent is between 0° and −90°. This negative angle must be measured from the −*x* axis to provide a correct conversion to polar form. Again, a sketch is critical. Look at the four phasors in Figure 2-7.

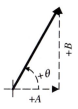
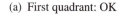
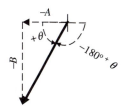

(a) First quadrant: OK

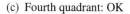

(b) Third quadrant: adjust it

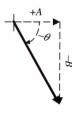

(c) Fourth quadrant: OK

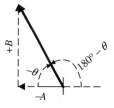

(d) Second quadrant: adjust it

Figure 2-7 Adjust the angle for rectangular to polar conversion

MATLAB

```
» Va=60+j×104;
» abs(Va)

ans =

        120.07

» angle(Va)×57.3

ans =

        60.023
```

```
» Vb=-104-j×180;
» abs(Vb)

ans =

        207.88

» angle(Vb)×57.3

ans =

        -120.03
```

```
» Ic=-28.3e-6+j×28.3e-6;
» abs(Ic)

ans =

    4.0022e-005

» angle(Ic)×57.3

ans =

        135.01
```

Example 2-4

Convert each of the following rectangular phasors into their polar form.

a. $\overline{V}_a = (60\,V_{rms} + j104\,V_{rms})$ **b.** $\overline{V}_b = (-104\,V_{rms} - j180\,V_{rms})$

c. $\overline{I}_c = (-28.3\,\mu A_{rms} + j28.3\,\mu A_{rms})$

Solution

a. This is in the first quadrant. No correction is needed.
$$\overline{V}_a = (120\,V_{rms}\angle 60°)$$

a. This is in the third quadrant. The angle should be $-90°$
$-180°$. $= +60°$. The corrected phasor is
$$\overline{V}_b = (208\,V_{rms}\angle -120°)$$

b. This is in the second quadrant. The angle should be $90°$ to $180°$. However, the inverse tangent returns $\theta = -45°$. The corrected phasor is
$$\overline{I}_c = (40\,\mu A_{rms}\angle 135°)$$

Practice: Convert each of the following rectangular phasors into their polar form.

a. $\overline{V}_a = (177\,mV_{rms} + j177\,mV_{rms})$ **b.** $\overline{I}_b = (-1.21\,A_{rms} + j700\,mV_{rms})$

c. $\overline{V}_c = (-180\,V_{rms} - j104\,V_{rms})$

Answers: a. $\overline{V}_a = (0.25\,V_{rms}\angle 45°)$ **b.** $\overline{I}_b = (1.4\,A_{rms}\angle 150°)$

c. $\overline{V}_c = (208\,V_{rms}\angle -150°)$ (Third quadrant: must be adjusted.)

Example 2-5

Write the polar form of this rectangular phasor.

$$\overline{Z} = R + jX_L$$

Solution

$$\overline{Z} = \sqrt{R^2 + X_L{}^2}\angle \tan^{-1}\left(\frac{X_L}{R}\right)$$

2.4 Mathematical Operations

Phasors can be added, subtracted, multiplied, and divided. In fact, many scientific calculators allow you to enter the phasor directly, in either polar or rectangular form, then complete standard calculator functions, delivering a phasor result.

The most intuitive phasor operation is addition.

$$\overline{V_{total}} = \overline{V_1} + \overline{V_2}$$

When adding phasors manually, or when you have symbols for the values instead of numbers, begin by converting the phasors into rectangular form.

Addition Form:
   rectangular

$$\overline{V_1} = A_1 + jB_1$$

$$\overline{V_2} = A_2 + jB_2$$

To complete the addition, add the real parts together. Then add the imaginary parts together.

Rule:
   $A_1 + A_2$
   $j(B_1 + B_2)$

$$\overline{V_{total}} = (A_1 + A_2) + j(B_1 + B_2)$$

This is very much like finding your way when traveling. For

$$(7 + j\,3) + (1 + j\,2)$$

Go east 7 miles, then north 3 miles. Next turn east again for 1 more miles then back north 2 miles. The result is that you are 8 miles east and 5 miles north of where you started. This **nose-to-tail** form of phasor addition is shown in Figure 2-8.

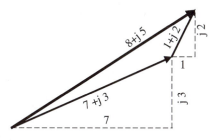

Figure 2-8 Nose-to-tail phasor addition

Example 2-6

Add $(120\ V_{rms} \angle 165°) + (120\ V_{rms} \angle 45°)$, and convert to polar.

Solution

$$(120\,V_{rms}\angle165°) = (-116\,V_{rms} + j31\,V_{rms})$$

$$(120\,V_{rms}\angle45°) = (85\,V_{rms} + j85\,V_{rms})$$

$$(-116\,V_{rms} + j31\,V_{rms}) + (85\,V_{rms} + j85\,V_{rms}) = (-31\,V_{rms} + j116\,V_{rms})$$

$$(-31\,V_{rms} + j\,116\,V_{rms}) = (120\,V_{rms}\angle105°)$$

Practice: Add the following phasors and express the sum in polar.

$$\overline{Z1} = (1.2\,k\Omega - j360\,\Omega) \qquad \overline{Z2} = (800\,\Omega\angle-90°) \qquad \overline{Z3} = (100\,\Omega\angle0°)$$

Answer: $\overline{Z_{total}} = (1.3\,k\Omega - j1.16\,k\Omega) = (1.74\,k\Omega\angle-41.7°)$

Subtraction is also done with phasors in the rectangular form.

$$\overline{V}_{total} = \overline{V}_1 - \overline{V}_2$$

Subtraction Form: rectangular

$$\overline{V}_1 = A_1 + jB_1 \qquad \overline{V}_2 = A_2 + jB_2$$

To complete the subtraction, subtract the real parts. Then subtract the imaginary parts.

Rule:
$A_1 - A_2$
$j(B_1 - B_2)$

$$\overline{V}_{total} = (A_1 - A_2) + j(B_1 - B_2)$$

Example 2-7

Subtract the phasors $(275\ \mu A_{rms}\angle-45°) - (190\ \mu A_{rms}\angle60°)$ and express the result in polar notation.

Solution

$$(275\,\mu A_{rms}\angle-45°) = (194.5\,\mu A_{rms} - j194.5\,\mu A_{rms})$$

$$(190\,\mu A_{rms}\angle60°) = (95.00\,\mu A_{rms} + j164.5\,\mu A_{rms})$$

$$(194.5\,\mu A_{rms} - j\,194.5\,\mu A_{rms}) - (95.00\,\mu A_{rms} + j164.5\,\mu A_{rms}) =$$

$$(99.50\ \mu A_{rms} - j\,359.0\ \mu A_{rms}) = (373\ \mu A_{rms}\angle-74°)$$

Practice: $\overline{Z}_{total} = \overline{Z1} - \overline{Z2} - \overline{Z3}$

$\overline{Z1} = (1.2\,k\Omega - j360\,\Omega)$ $\overline{Z2} = (800\,\Omega \angle -90°)$ $\overline{Z3} = (100\,\Omega \angle 0°)$

Answer: $\overline{Z}_{total} = (1.10\,k\Omega - j440\,\Omega) = (1.18\,k\Omega \angle 22°)$

You will multiply phasors when completing the Ohm's law calculation

$$\overline{V} = \overline{I} \times \overline{Z}$$

The impedance, $\overline{Z}$, will be explained in detail in the following chapter.

Multiplication of phasors is more easily completed when the quantities are expressed in polar notation.

$$\overline{I} = (C_1 \angle \theta_1)$$

$$\overline{Z} = (C_2 \angle \theta_2)$$

Multiply the magnitudes $(C_1 \times C_2)$ and *add* the angles $(\theta_1 + \theta_2)$.

$$(C_1 \angle \theta_1) \times (C_2 \angle \theta_2) = [(C_1 \times C_2) \angle (\theta_1 + \theta_2)]$$

Multiplication Form:
    polar

Rule:
    $C_1 \times C_2$
    $\angle (\theta_1 + \theta_2)$

Example 2-8

Calculate the phasor voltage developed by a current of $\overline{I} = (4.2\,A_{rms} \angle -32°)$ flowing through $\overline{Z} = (8\,\Omega + j4\,\Omega)$.

Solution

First the impedance must be converted into polar notation.

$$(8\,\Omega + j4\,\Omega) = (8.94\,\Omega \angle 26.6°)$$

$$\overline{V} = \overline{I} \times \overline{Z}$$

$$\overline{V} = (4.2\,A_{rms} \angle -32°) \times (8.94\,\Omega \angle 26.6°)$$
$$= (37.6\,V_{rms} \angle -5°)$$

MATLAB

```
» I=4.2×cos(-32/57.3)+j×4.2×sin(-32/57.3);
» Z=8+j×4;
» U=I×Z;
» abs(U)

ans =

        37.566

» angle(U)×57.3

ans =

        -5.433
```

Practice: Calculate the phasor voltage developed by a current of $\overline{I} = (2.34\,mA_{rms} \angle 150°)$ flowing through $\overline{Z} = (2\,k\Omega - j1\,k\Omega)$

Answer: $(5.23\,V_{rms} \angle 123°)$

You must divide phasors when completing the Ohm's Law calculation

$$\bar{I} = \frac{\bar{V}}{\bar{Z}}$$

Division of phasors is more easily completed when the quantities are expressed in polar notation.

Division Form:
  polar

$$\bar{V} = (C_1 \angle \theta_1)$$

$$\bar{Z} = (C_2 \angle \theta_2)$$

Divide the magnitudes (C_1/C_2) and *subtract* the angles $(\theta_1 - \theta_2)$.

Rule:
  C_1/C_2
  $\angle (\theta_1 - \theta_2)$

$$\frac{(C_1 \angle \theta_1)}{(C_2 \angle \theta_2)} = \frac{C_1}{C_2} \angle (\theta_1 - \theta_2)$$

Example 2-9

Calculate the phasor current produced by a voltage of $\bar{V} = (5.23\,\text{V}_{\text{rms}} \angle 123°)$ developed across $\bar{Z} = (2\,\text{k}\Omega - \text{j}1\,\text{k}\Omega)$.

Solution

First the impedance must be converted into polar notation.

$$(2\,\text{k}\Omega - \text{j}1\,\text{k}\Omega) = (2.24\,\text{k}\Omega \angle -26.6°)$$

$$\bar{I} = \frac{\bar{V}}{\bar{Z}}$$

$$\bar{I} = \frac{(5.23\,\text{V}_{\text{rms}} \angle 123°)}{(2.24\,\text{k}\Omega \angle -26.6°)} = (2.34\,\text{mA}_{\text{rms}} \angle 150°)$$

Practice: Calculate the phasor current produced by a voltage of $\bar{V} = (37.6\,\text{V}_{\text{rms}} \angle -5°)$ developed across $\bar{Z} = (8\,\Omega + \text{j}4\,\Omega)$.

Answer: $(4.2\,\text{A}_{\text{rms}} \angle -32°)$

Summary

Linear circuits affect only the magnitude and the phase shift of a sine wave. This allows the full, time-domain equation of a sinusoid to be shortened to its phasor form. V_{rms} indicates the magnitude (length) and θ gives the phase shift (direction for the x axis).

These phasors can be plotted, just as you have done with vectors in basic physics. This often helps visualize signals and how they combine.

Phasors can also be represented in rectangular form. A is the distance along the x axis, and B is the vertical (y axis) distance above or below the origin. It is also common to consider the horizontal axis as *real* and the vertical axis as *imaginary*. That is the purpose of j. The phasor, with both real and imaginary parts, is a *complex* number.

It is easier to measure a phasor voltage or current in its polar form. A digital multimeter indicates its magnitude and an oscilloscope can display the phase shift. But many calculations and drawings are better done with the rectangular form. By using a few trigonometric relationships, phasors in one form may be converted to the other. C is the magnitude of the polar form.

$$A = C\cos\theta \qquad\qquad B = C\sin\theta$$

$$C = \sqrt{A^2 + B^2} \qquad\qquad \theta = \tan^{-1}\frac{B}{A}$$

Phasors can be added, subtracted, multiplied, and divided. This allows you to use them in Kirchhoff's laws and Ohm's law. Although many calculators complete these computations automatically, when you derive the relationship of phasors in terms of symbolic circuit elements, you must be able to complete each operation manually.

Addition and subtraction are done with the rectangular form.

$$\overline{V_1} = A_1 + jB_1 \qquad\qquad \overline{V_1} + \overline{V_2} = (A_1 + A_2) + j(B_1 + B_2)$$

$$\overline{V_2} = A_2 + jB_2 \qquad\qquad \overline{V_1} - \overline{V_2} = (A_1 - A_2) + j(B_1 - B_2)$$

Multiplication and division are done using the polar form.

$$\overline{C_1} = (C_1\angle\theta_1) \qquad (C_1\angle\theta_1)\times(C_2\angle\theta_2) = [(C_1\times C_2)\angle(\theta_1 + \theta_2)]$$

$$\overline{C_2} = (C_2\angle\theta_2) \qquad \frac{(C_1\angle\theta_1)}{(C_2\angle\theta_2)} = \frac{C_1}{C_2}\angle(\theta_1 - \theta_2)$$

Problems

Polar Notation

2-1   Convert each of the following time-domain expressions into a polar notation phasor.

    a. 311 V_p sin (377t)

    b. 55 V_p sin$(2\pi \times 440\,\mathrm{Hz} \times t - 36°)$

    c. 350 mV_p cos (6.3 k)

    d. 7.2 A_p cos$(2\pi \times 60\,\mathrm{Hz} \times t + 120°)$

    e. $-17\,\mu A_p$ sin$(587\mathrm{E}6t + 60°)$

2-2   Convert each of the following time-domain expressions into a polar notation phasor.

    a. 6.8 kV_p sin (377t)

    b. 32 V_p sin$(2\pi \times 1000\,\mathrm{Hz} \times t + 76°)$

    c. 120 mV_p cos (12000t)

    d. 15 A_p cos$(2\pi \times 60\,\mathrm{Hz} \times t - 160°)$

    e. $-50\,\mu A_p$ sin$(820\mathrm{E}6t - 45°)$

2-3   Write the time-domain expression for each of the following phasors.

    a. (8.3 V_{rms} ∠ 32°), f = 700 Hz

    b. (120 V_{rms} ∠ −45°), f = 60 Hz

    c. (1.5 A_{rms} ∠ 135°), f = 540 kHz

    d. (32 μA_{rms} ∠ −100°), f = 105.6 MHz

2-4   Write the time-domain expression for each of the following phasors.

    a. (7.2 V_{rms} ∠ 53°), f = 900 Hz

    b. (240 V_{rms} ∠ −60°), f = 50 Hz

    c. (5.2 A_{rms} ∠ 150°), f = 820 kHz

    d. (76 μA_{rms} ∠ −130°), f = 99.7 MHz

2-5   Accurately draw the phasor diagram for each of the quantities in Problem 2-3. Choose and indicate an appropriate scale.

2-6   Accurately draw the phasor diagram for each of the quantities in Problem 2-4. Choose and indicate an appropriate scale.

Rectangular Notation

2-7   Accurately draw each of the following phasors. Indicate an appropriate scale. Plot these phasors manually.

   a. $(3.2 \text{ V}_{rms} + j2.1 \text{ V}_{rms})$

   b. $(63 \text{ V}_{rms} - j100 \text{ V}_{rms})$

   c. $(-1.5 \text{ A}_{rms} + j130 \text{ A}_{rms})$

   d. $(-570 \text{ V}_{rms} - j1.2 \text{ kV}_{rms})$

2-8   Accurately draw each of the following phasors. Indicate an appropriate scale. Plot these phasors manually.

   a. $(6.6 \text{ V}_{rms} + j2.9 \text{ V}_{rms})$

   b. $(-230 \text{ V}_{rms} + j70 \text{ V}_{rms})$

   c. $(3.6 \text{ A}_{rms} - j3.7 \text{ A}_{rms})$

   d. $(-28 \text{ } \mu\text{A}_{rms} - j71 \text{ } \mu\text{A}_{rms})$

Conversions

2-9   Convert each of the following polar phasors into rectangular form.

   a. $(9.4 \text{ V}_{rms} \angle 32°)$

   b. $(110 \text{ V}_{rms} \angle -45°)$

   c. $(2.2 \text{ A}_{rms} \angle 135°)$

   d. $(45 \text{ } \mu\text{A}_{rms} \angle -100°)$

2-10   Convert each of the following polar phasors into rectangular form.

   a. $(2.7 \text{ V}_{rms} \angle 35°)$

   b. $(208 \text{ V}_{rms} \angle -45°)$

   c. $(2.5 \text{ A}_{rms} \angle 130°)$

   d. $(67 \text{ } \mu\text{A}_{rms} \angle -145°)$

Mathematical Operations

2-11   Perform each of the following operations. Be sure to show all of your intermediate steps.

   a. $(8.3 \text{ V}_{rms} \angle 32°) + (9.4 \text{ V}_{rms} \angle -48°)$

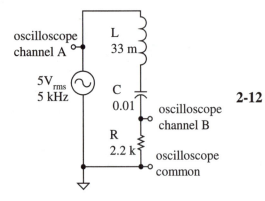

Figure 2-9

Measurement of $\overline{V}_R$

b. $(1.5\ \text{A}_{rms}\ \angle135°) - (-2\ \text{A}_{rms} - j1.2\ \text{A}_{rms})$

c. $(76\ \mu\text{A}_{rms}\ \angle-130°) \times (56\ \text{k}\Omega - j32\ \text{k}\Omega)$

d. $(120\ \text{V}_{rms}\ \angle+30°) \div (45\ \Omega + j20\ \Omega)$

2-12 Perform each of the following operations. Be sure to show all of your intermediate steps.

    a. $(3.8\ \text{V}_{rms}\ \angle23°) + (4.9\ \text{V}_{rms}\ \angle-84°)$

    b. $(5.1\ \text{A}_{rms}\ \angle150°) - (-3\ \text{A}_{rms} - j2.1\ \text{A}_{rms})$

    c. $(67\ \mu\text{A}_{rms}\ \angle-110°) \times (47\ \text{k}\Omega + j22\ \text{k}\Omega)$

    d. $(120\ \text{V}_{rms}\ \angle+90°) \div (23\ \Omega + j18\ \Omega)$

Phasor Measurements Lab Exercise

A. Low-Frequency Measurements

  1. Build the circuit in Figure 2-9.

  2. Assure that the oscilloscope is connected as indicated.

  3. With a digital multimeter, verify that the signal generator is providing the correct amplitude and frequency.

  4. Move the digital multimeter across the resistor. Record the magnitude of the voltage across the resistor.

  5. With the oscilloscope, measure the phase of the voltage across the resistor. Combine this with the magnitude from step A4 as the phasor representation of the resistor's voltage at 5 kHz. Also record the rectangular form of the voltage.

  6. Exchange the resistor and the capacitor, as shown in Figure 2-10. This allows you to measure the phase of the voltage across the capacitor. The oscilloscope must be tied to circuit common at the signal generator. To place channel B across the capacitor, the capacitor must also be tied to circuit common. This does *not* affect the circuit, since the components are in series.

  7. Place the digital multimeter across the capacitor. Record the magnitude of the voltage across the capacitor.

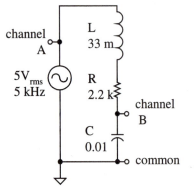

Figure 2-10

Measurement of $\overline{V}_C$

8. With the oscilloscope, measure the phase of the voltage across the capacitor. Combine this with the magnitude from step A7 as the phasor representation of the capacitor's voltage at 5 kHz. Also record the rectangular form of the voltage.

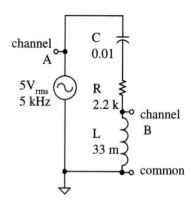

9. Exchange the inductor and the capacitor, as shown in Figure 2-11. This allows you to measure the phase of the voltage across the inductor.

10. Place the digital multimeter across the inductor. Record the magnitude of the voltage across the inductor.

11. With the oscilloscope, measure the phase of the voltage across the inductor. Combine this with the magnitude from step A10 as the phasor representation of the inductor's voltage at 5 kHz. Also record the rectangular form of the voltage.

Figure 2-11

Measurement of $\overline{V_L}$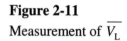

B. Low-Frequency Calculations

1. Using a protractor and a scale, carefully draw a separate phasor diagram for each of the voltages measured in Section A. Be sure to record the scale factor.

2. Using your calculator, add $\overline{V_R} + \overline{V_C} + \overline{V_L}$.

3. Using a protractor and a scale, carefully add the three voltage phasors nose-to-tail, as shown in Figure 2-8. Your answer from step B2 should match the answer from step B3. Repeat your work until the magnitudes of the two sums agree to within 5% and the phase angles of the two sums agree within 4°.

C. Midrange-Frequency Measurements and Calculations

1. Set the signal generator's frequency to 10 kHz.

2. Repeat the measurements of Section A and then the calculations in Section B.

D. Upper-Frequency Measurements and Calculations

1. Set the signal generator's frequency to 30 kHz.

2. Repeat the measurements of Section A and then the calculations in Section B.

3

Impedance

Introduction

From the previous chapter you learned that sinusoidal voltage and current can be expressed both in the time domain, and as phasors.

Voltage divided by current is opposition. For dc circuits, this opposition is called resistance. Ohm's law applies equally for sine waves. Placing a sinusoidal voltage across a resistor results in a sinusoidal current, in phase with the applied voltage.

However, inductors oppose a change in current. The voltage across an inductor depends on the *rate of change* of the current through it. Similarly, the current through a capacitor depends on how rapidly the voltage across the capacitor *changes*. Derivatives are involved.

So, to begin, you must learn to take the derivative of the time-domain sinusoidal voltage and current. This can be done by using rules of calculus, or it can be done graphically by looking at the signal's *slope*.

The result of these derivatives is that an inductor and capacitor not only affect the amplitude of the signal passing through it, but also alter its phase. Effects on amplitude and phase are included in a phasor. The impedance of a resistor, inductor, and capacitor are phasors, containing both magnitude (in ohms that tells of the opposition) and phase shift. Impedance diagrams help visualize how these elements combine.

Objectives

Upon completion of this chapter, you will be able to do the following:

- Find the derivative of a time-domain sinusoidal voltage or current.

- Determine the relationship between the sinusoidal voltage and current in a resistor, an inductor, and a capacitor.

- Define and calculate the impedance of a resistance and the reactance and impedance of an inductor and a capacitor. Draw their impedance diagrams.

- Determine the impedance of a wire, a cable, and a motor, using manufacturers' specifications.

3.1 Derivative of a Sine Wave

Ohm's law defines the relationship between the voltage across a resistor and the current through it.

$$v_R = i R$$

An inductor opposes a *change* in current. The voltage across the inductor depends on how rapidly the current through it changes.

$$v_L = L \frac{di}{dt}$$

where $i = I_p \sin(2\pi ft)$

The capacitor opposes a *change* in voltage. That is, the current through a capacitor depends on how quickly the voltage changes.

$$i_C = C \frac{dv}{dt}$$

where $v = V_p \sin(2\pi ft)$

You must take the derivative of a sine wave in order to figure out how an inductor and a capacitor respond to it in an ac circuit.

It's Just Slope: a Graphical Approach

The value of the derivative of a function at a point is just the *slope* of the curve at that point.

$$\frac{d f(x)}{dx} = \frac{\Delta y}{\Delta x} = \frac{\Delta v}{\Delta t} = \text{slope}$$

A basic sine wave is shown in Figure 3-1(a). The slope is indicated every 90°. At 0°, point a, the slope is positive and steep. At 90° (point b), and 270° (point d) the slopes are horizontal. And at 180° the slope is decreasing (negative), just as steeply as it was at the origin.

These slopes are plotted in Figure 3-1(b). The intervening points have been added in Figure 3-1(c). The result is a cosine wave.

$$\frac{d}{dt} V_p \sin(2\pi ft) \propto \cos(2\pi ft)$$

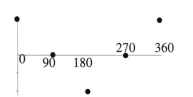

(a) Basic sine wave with slopes

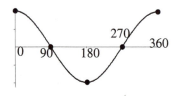

(b) Slopes of a sine

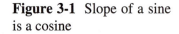

(c) Derivative of sine is cosine

Figure 3-1 Slope of a sine is a cosine

The amplitude of the sine wave, V_p, also affects the slope. Look at Figure 3-2. There are two sine waves. Each has the same frequency and zero phase shift. The only difference between the two is that wave **a** is twice as tall as wave **b**.

$$V_{pa} = 2\,V_{pb}$$

To go twice as far in the same time, wave **a** must go twice as fast as wave **b**. That is, wave **a** must have twice the slope as wave **b**. So, the slope, or derivative, of a sine wave is proportional to its amplitude.

$$\frac{d}{dt}V_p \sin(2\pi ft) \propto V_p$$

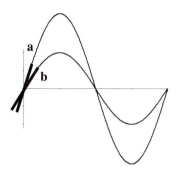

Figure 3-2 Slope of a sine depends on amplitude

The other variable is frequency. In Figure 3-3, wave **c** has twice the frequency of wave **d**.

$$2\pi f_c = 2 \times (2\pi f_d)$$

Both waves have the same amplitude. However, sine wave **c** reaches its peak in half the time that it takes sine wave **d**. Sine wave **c** must travel more quickly. It must have a steeper slope. So, the derivative of a sine wave is proportional to the frequency of the signal.

$$\frac{d}{dt}V_p \sin(2\pi ft) \propto 2\pi f$$

From a graphical interpretation of the derivative of a sine wave being dictated by its slope, you have determined that the derivative is shaped like a cosine, and is proportional to both its amplitude and its frequency.

$$\frac{d}{dt}V_p \sin(2\pi ft) \propto 2\pi f V_p \cos(2\pi ft)$$

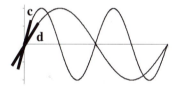

Figure 3-3 Slope of a sine depends on frequency

In reality, those are the only factors that determine the derivative. So this proportionality can be changed to an equality. Finally, change the cosine function to a sine.

$$\frac{d}{dt}V_p \sin(2\pi ft) = 2\pi f\, V_p \cos(2\pi ft)$$

Derivative of sinusoidal voltage

$$= 2\pi f\, V_p \sin(2\pi ft + 90°)$$

The Calculus Approach

Determining the derivative of a sine wave requires four calculus rules. The derivative of a constant times a function is that constant times the derivative of the function.

Derivative of a constant multiplier

$$\frac{d}{dt}[k \times f(t)] = k\frac{df(t)}{dt}$$

The chain rule applies to the derivative of the function of a function.

Chain rule

$$\frac{d}{dt}f(t) = \frac{df(\alpha)}{d\alpha}\frac{d\alpha}{dt}$$

Finally, the derivative of the sine is the cosine.

Derivative of sine

$$\frac{d}{d\alpha}\sin\alpha = \cos\alpha$$

The equation of a sinusoidal current is

$$i(t) = I_p \sin(2\pi ft)$$

The derivative of this current is

$$\frac{d\,i(t)}{dt} = \frac{d}{dt}\left[I_p \sin(2\pi ft)\right]$$

Move the constant, I_p, out in front of the derivative.

$$\frac{d\,i(t)}{dt} = I_p \frac{d}{dt}\left[\sin(2\pi ft)\right]$$

You now must take the derivative of the function (sin) of a function ($2\pi ft$). This uses the chain rule. Substitute $\alpha = 2\pi ft$.

$$\frac{d\,i(t)}{dt} = I_p \frac{d}{d\alpha}\sin\alpha\frac{d\alpha}{dt}$$

The derivative of the sine is the cosine.

$$\frac{d\,i(t)}{dt} = I_p \cos\alpha\frac{d\alpha}{dt}$$

Now replace α with $2\pi ft$.

$$\frac{d\,i(t)}{dt} = I_p \cos(2\pi ft)\frac{d(2\pi ft)}{dt}$$

Move the constant outside of the derivative.

$$\frac{d\,i(t)}{dt} = 2\pi f\, I_p \cos(2\pi f t)\frac{dt}{dt}$$

Finally, the derivative of a variable with respect to itself is unity.

$$\frac{d\,i(t)}{dt} = 2\pi f\, I_p \cos(2\pi f t)$$

Put the cosine function in terms of the sine.

$$\frac{d\,i(t)}{dt} = 2\pi f\, I_p \sin(2\pi f t + 90°)$$

Derivative of sinusoidal current

3.2 Impedance of a Resistance

Impedance is opposition to a sinusoidal current and voltage. It is a phasor. Its basic equation is

$$\overline{Z} = \frac{\overline{V}}{\overline{I}}$$

Ohm's law gives the value of the voltage across a resistor when a sinusoidal current flows through it.

$$v(t) = i(t) \times R$$

where

$$i(t) = I_p \sin(2\pi f t)$$

Substitute for i.

$$v(t) = RI_p \sin(2\pi f t)$$

Convert the time-domain form of v and i into phasors.

$$\overline{V} = R \times (I_{rms}\angle 0°)$$

Substitute this into the definition of impedance.

$$\overline{Z}_R = \frac{(R \times I_{rms}\angle 0°)}{(I_{rms}\angle 0°)}$$

To divide phasors, divide the magnitudes and subtract the angles.

$$\overline{Z}_R = (R\angle 0°)$$

Impedance of a resistor:
magnitude = R
phase = 0°

voltage and current are *in phase*

The impedance of a resistor has the magnitude of the resistor. So, its units are ohms. The 0° phase shift indicates that the voltage across the resistor is in phase with the current through it.

The resistor's impedance has no imaginary part.

$$\overline{Z}_R = (R + j0)$$

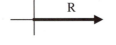

Figure 3-4 Resistor's impedance diagram

The **impedance diagram** of a resistor is a phasor whose length is R that lies entirely along the real ($+x$) axis. This is shown in Figure 3-4.

Example 3-1

Calculate the phasor current through a 145 Ω resistor that has a voltage of (115 $V_{rms} \angle 0°$) applied across it.

Solution

$$\overline{I} = \frac{(115\ V_{rms}\angle 0°)}{(145\Omega\angle 0°)} = (793\ mA_{rms}\angle 0°)$$

Practice: Calculate the phasor current through an 8 Ω resistance with a 35 V_{rms} applied voltage.

Answer: $\overline{I}_R = (4.38\ A_{rms}\angle 0°)$

3.3 Impedance of an Inductance

1. List 10 electrical/electronic items you used today.

Most electrical "muscle" is produced by electromagnets, converting current into a magnetic field that pushes or pulls iron and steel. Motors, relays, solenoids, transformers, and even loudspeakers are all electromagnets. In one form or another, electromagnets are in most electrical and electronic systems. An electromagnet is a coil of wire wrapped around a core. That also forms an inductor. Inductors are common in the processing and application of electrical signals.

2. Which of these (if any) do *not* have some form of electromagnet (inductor)?

An inductor opposes a *change* in current. The voltage across the inductor depends on how rapidly the current through it changes.

$$v_L = L\frac{di}{dt}$$

where

$$i = I_p \sin(2\pi ft) = \left(I_{rms}\angle 0°\right)$$

In Section 3.1 the derivative of a sinusoidal current was calculated.

$$\frac{di}{dt} = 2\pi f\, I_p \sin(2\pi ft + 90°)$$

Substitute this into the basic equation for the inductor.

$$v_L = L\big[2\pi f\, I_p \sin(2\pi ft + 90°)\big]$$

Regroup the constants.

$$v_L = (2\pi fL)\, I_p \sin(2\pi ft + 90°)$$

This equation points out that there is a phase shift of 90°. The current is at 0°, but the resulting voltage drop across the inductor is at +90°.

> **Voltage (E)** across an **inductor (L)** *leads* **current (I)** by 90°.   E L I

The voltage equation can be expressed as a phasor.

$$\overline{V_L} = (2\pi fL) \times (I_{rms} \angle 90°)$$

The quantity, $2\pi fL$, is the **inductive reactance**. Its units are ohms.

$$X_L = 2\pi f\, L$$

Impedance is

$$\overline{Z} = \frac{\overline{V}}{\overline{I}}$$

The voltage across the inductor depends on inductive reactance and the rms value of the current through it, but is shifted ahead 90°. The current was defined at the beginning of the problem as having a magnitude of I_{rms} and 0° phase shift.

$$\overline{Z_L} = \frac{(X_L I_{rms} \angle 90°)}{(I_{rms} \angle 0°)}$$

Divide the magnitudes, and subtract the angles.

$$\overline{Z_L} = (X_L \angle 90°)$$

where $X_L = 2\pi fL$

These equations indicate that the opposition an inductor presents to a sinusoidal current is directly proportional to the size of the inductor (L) and how *rapidly* that current is changing (f). The voltage is shifted 90° ahead of the current.

Figure 3-5 Inductor's impedance diagram

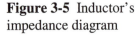

In rectangular notation, the inductor's impedance is all + imaginary. It has no real part.

$$\overline{Z}_L = (0 + jX_L)$$

The **impedance diagram** of an inductor is a phasor whose length is X_L that lies entirely along the imaginary (+y) axis, shown in Figure 3-5.

Example 3-2

a. Calculate the phasor current through a 334 mH inductance that has a 60 Hz voltage of $(115\ V_{rms}\angle0°)$ applied across it.

b. Verify your calculations with a simulation.

Solution

a. $X_L = 2\pi fL$

$X_L = 2\pi \times 60\,Hz \times 334\,mH = 126\,\Omega$

$$\overline{I} = \frac{\overline{V}}{\overline{Z}}$$

$$\overline{I} = \frac{(115\ V_{rms}\angle0°)}{(126\,\Omega\angle90°)} = (913\,mA_{rms}\angle-90°)$$

The current *lags* the voltage by 90° (shown as –90°).

b. The MultiSIM results are shown in Figure 3-6

- The source voltage is set to 163 V_p, that is 115 V_{rms}.

- The ammeter is set to AC. The current is in A_{rms}.

- The part in the dashed box that is labeled V_1V_per_A is a **current-controlled-voltage-source**. Each ampere of current that flows through the rectangle in the direction of the arrow causes the diamond to output one volt. To change this scale factor, change the 1Ohm. This produces a voltage for the oscilloscope to measure. It has no effect on the circuit.

- The oscilloscope has been set as directed in Example 1-7. The current through the inductor lags its voltage by ¼ cycle, –90°. That current peaks at 1.3 $V_p \rightarrow$ 1.3 $A_p =$ 919 mA_{rms}.

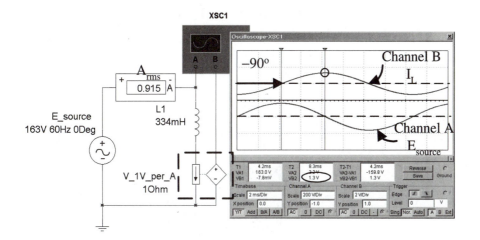

Figure 3-6 MultiSIM results for Example 3-2

Practice: Determine the reactance, impedance, and the phasor current through a 20 nH inductance across a 35 mV$_{rms}$, 900 MHz source.

Answer: $X_L = 113\,\Omega$, $\overline{Z_L} = (113\Omega\angle 90°)$, $\overline{I_L} = (310\mu A_{rms}\angle -90°)$

3.4 Impedance of a Capacitance

An inductor opposes a change in current. A capacitor does the opposite. It opposes a change in voltage. The correct value capacitor, then, can compensate for the nonideal characteristic of electromagnets (motors, relays, solenoids, transformers, and loudspeakers). Since capacitors oppose a change in voltage, they are slow to charge. This makes them work well for timing circuits and circuits that pass certain frequency tones while blocking others (such as loudspeaker cross-over networks). Capacitors are used to block the dc between amplifier stages, but pass the signal. In fact, any time two conductors run side-by-side, capacitance is formed. Coaxial cables and printed circuit board runs all exhibit

capacitance that may have an effect on sinusoidal signals trying to pass
along them.

The current through a capacitance depends on how rapidly the voltage across it changes.

$$i_C = C\frac{dv}{dt}$$

where
$$v = V_p \sin(2\pi ft) = V_{rms}\angle 0°$$

In Section 3.1 the derivative of a sinusoidal voltage was calculated.

$$\frac{dv}{dt} = 2\pi f\, V_p \sin(2\pi ft + 90°)$$

Substitute this into the basic equation for the capacitor.

$$i_C = C[2\pi f\, V_p \sin(2\pi ft + 90°)]$$

Regroup the constants.

$$i_C = (2\pi fC)V_p \sin(2\pi ft + 90°)$$

This equation points out that there is a phase shift of 90°. The voltage is at 0°, but the resulting current through the capacitance is at +90°.

I C E   **Current (I)** through a **capacitor (C)** *leads* **voltage (E)** by 90°.

The current equation can be expressed as a phasor.

$$\overline{I_C} = (2\pi fC)\times(V_{rms}\angle 90°)$$

Capacitive reactance is defined as

$$X_C = \frac{1}{2\pi fC}$$

The phasor current equation can be rewritten in terms of phasor voltage and reactance.

$$\overline{I_C} = \frac{(V_{rms}\angle 90°)}{X_C} = \left(\frac{V_{rms}}{X_C}\angle 90°\right)$$

The current through the capacitor depends on capacitive reactance and the rms value of the voltage across it, but is shifted ahead 90°. The voltage was defined at the beginning of the problem as having a magnitude of V_{rms} and 0° phase shift.

Impedance is

$$\overline{Z} = \frac{\overline{V}}{\overline{I}}$$

$$\overline{Z}_C = \frac{(V_{rms} \angle 0°)}{\left(\dfrac{V_{rms}}{X_C} \angle 90° \right)}$$

Divide the magnitudes, and subtract the angles.

$$\overline{Z}_C = (X_C \angle -90°)$$

where

$$X_C = \frac{1}{2\pi f C}$$

Capacitive:
Impedance

Reactance

ICE

These equations indicate that the opposition a capacitance presents to a sinusoidal voltage is inversely proportional to the size of the capacitance (C) and how *rapidly* that current is changing (f). The current is shifted 90° ahead of the voltage. Look back carefully at the derivation of inductive impedance. Capacitive and inductive impedances are opposites in all respects.

In rectangular notation, the capacitor's impedance is all − imaginary.

$$\overline{Z}_C = (0 - jX_C)$$

The **impedance diagram** of a capacitor is a phasor whose length is X_C that lies entirely along the imaginary (−y) axis, shown in Figure 3-7.

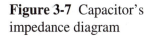

Figure 3-7 Capacitor's impedance diagram

Example 3-3

    a. Calculate the phasor current through a 2.3 µF capacitance that has a 5 kHz voltage of (40 $V_{rms} \angle$ 0° applied across it.

    b. Verify your calculations with a simulation.

Solution

    a. $X_C = \dfrac{1}{2\pi \times 5\,\text{kHz} \times 2.3\,\mu\text{F}} = 13.8\,\Omega$

        $\overline{I} = \dfrac{40\ V_{rms} \angle 0°}{13.8\,\Omega \angle -90°} = (2.9\,A_{rms} \angle 90°)$

        The current *leads* the voltage by +90°.

b. The results of the simulation are shown in Figure   3-8. The current through the capacitor is

$$I_C = 4.087 \text{ A}_p = 2.89 \text{ A}_{rms}$$

The current waveform leads the voltage.

$$\theta = (200\,\mu s - 149.7\,\mu s) \times 360° \times 5\,kHz = 90.5°$$

Together, $\overline{I_C} = (2.89 \text{ A}_{rms} \angle 90.5°)$, a good match to part a.

Practice: Determine the reactance, impedance, and the phasor current through a 3.3 pF capacitance across a 35 mV$_{rms}$, 900 MHz source.

Answer: $X_C = 53.6 \; \Omega$, $\overline{Z_C} = (53.6\,\Omega \angle -90°)$, $\overline{I_L} = (653\,\mu A_{rms} \angle 90°)$

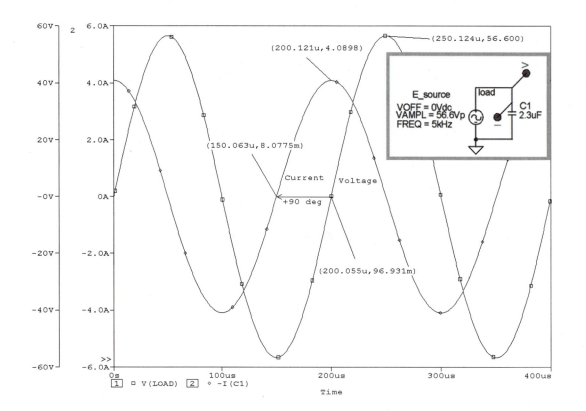

Figure 3-8   Simulation for Example 3-3

3.5 Practical Impedances

All electrical devices have impedance. It is tempting to view everything as a simple resistance. However, many common loads alter the amplitude of a signal based on the frequency of that signal. *And*, they shift the phase of the current through them with respect to the voltage across them. To determine the effect these elements have, you must consider their *impedance*.

Inductance of a Wire

A single straight copper wire has a small resistance. This resistance depends on its diameter, as specified by its wire gage (see Table 3-1) and its length. Except in very long runs, or in very precise instrumentation measurements, this resistance can be ignored.

When current flows through that wire an electromagnetic field is created. These field lines expand and contract with the current. As they move, they cut the wire. A magnetic field cutting a wire generates an opposing voltage. This is the principle upon which generators and inductors are built. The following equation is valid across the frequencies commonly used for radio communications, 540 kHz to 2 GHz.

$$L \approx 2 \times 10^{-7} \frac{\mathrm{H}}{\mathrm{m}} \, l \left[2.303 \log_{10}\left(\frac{4l}{d}\right) - 1 \right] \quad ^*$$

where l = length of the wire
d = diameter of the wire

Table 3-1
Wire gage diameters

Wire Size AWG	Diameter millimeter (10^{-3} m)
2	6.543
4	5.189
6	4.115
8	3.264
10	2.588
12	2.052
14	1.628
16	1.290
18	1.024
20	0.813
22	0.643
24	0.511
26	0.404
28	0.320
30	0.254
32	0.203
34	0.160
36	0.127

Example 3-4

   a. Calculate the impedance of a 10-inch length of 22 AWG wire.

   b. If the source has a 50 Ω output resistance and the load, at the other end of the 10-inch wire, has a 50 Ω input impedance, describe the effect of the impedance of the wire.

   c. Calculate the impedance of a 0.01 µF capacitor at 100 MHz, ignoring its leads.

   d. Calculate the impedance of a 0.01 µF capacitor at 100 MHz including its two 26 AWG 2-inch leads.

   * *Handbook of Chemistry and Physics,* C.D. Hodgman, 1961, Cleveland, Chemical Rubber Publishing Co., 44th edition, page 3394

Solution

a. The wire's length is

$$l = 10\,\text{inches} \times \frac{2.54\,\text{cm}}{\text{inch}} \times \frac{1\,\text{m}}{100\,\text{cm}} = 0.254\,\text{m}$$

From Table 3-1

$$d = 0.643 \times 10^{-3}\,\text{m}$$

Calculate the inductance.

$$L \approx 2 \times 10^{-7}\,\frac{\text{H}}{\text{m}}\,0.254\,\text{m}\left[2.303\log_{10}\left(\frac{4 \times 0.254\,\text{m}}{0.643 \times 10^{-3}\,\text{m}}\right) - 1\right]$$

$$L = 323\ \text{nH}$$

Now, calculate the reactance at 100 MHz.

$$X_{\text{L}} = 2\pi \times 100\,\text{MHz} \times 323\,\text{nH} = 203\,\Omega$$

$$\overline{Z_{\text{wire}}} = (203\,\Omega \angle 90°)$$

b. Compared to the source and the load impedance, most of the voltage will be dropped across the wire's impedance, and a serious phase shift will be introduced. That is, the signal will enter one end of the 10-inch wire, but very little of it will come out the other end!

c. At 100 MHZ, a 0.01 µF capacitor has a reactance of

$$X_C = \frac{1}{2\pi \times 100\,\text{MHz} \times 0.01\,\mu\text{F}} = 0.159\,\Omega$$

$$\overline{Z_C} = (0 - j0.159\,\Omega)$$

This would act as a good short to the 100 MHz signal, passing it, while blocking any dc bias voltages.

d. Now add the effect of the leads. The inductance of two 26 AWG 2-inch leads is

$$l = 4\,\text{inches} \times \frac{2.54\ \text{cm}}{\text{inch}} \times \frac{1\,\text{m}}{100\ \text{cm}} = 0.102\ \text{m}$$

From Table 3-1

$$d = 0.404 \times 10^{-3}\,\text{m}$$

Calculate the inductance.

$$L \approx 2 \times 10^{-7}\,\frac{\text{H}}{\text{m}}\,0.102\,\text{m}\left[2.303\log_{10}\left(\frac{4 \times 0.102\,\text{m}}{0.404 \times 10^{-3}\,\text{m}}\right) - 1\right]$$

$$L = 121\,\text{nH}$$

Now, calculate the reactance at 100 MHz.

$$X_L = 2\pi \times 100\,\text{MHz} \times 121\,\text{nH} = 76\,\Omega$$

The impedance of that capacitor's leads is

$$\overline{Z_{\text{leads}}} = (0 + j76\,\Omega)$$

Combine this with the impedance of the capacitor itself.

$$\overline{Z_C} = (0 + j76\,\Omega) + (0 - j0.159\,\Omega)$$

$$\overline{Z_C} = (0 + j76\,\Omega)$$

So, the effect of the capacitor disappears. Its leads are dominant. Instead of passing through the capacitor, most of the signal may be dropped across its leads, and a significant phase shift will be introduced. The circuit may break into oscillations, because of the 2-inch *leads* to the capacitor. This is one of the reasons that surface mount technology is used for high-frequency circuits.

Practice: Calculate the impedance of a 15-inch length of 32 AWG wire-wrap wire carrying a 10 MHz TTL clock signal.

Answer: $L_{\text{wire}} = 604\,\text{nH}$, $\overline{Z_{\text{wire}}} = (38\,\Omega\angle 90°)$, expect some distortion

---

Capacitance of a Cable

A coaxial cable, like the microphone cable shown in Figure 3-9, has a center conductor (or two) surrounded by a shield. Each conductor has resistance and inductance. These are in *series*. At frequencies below rf they may be combined as a single resistance and a single inductance.

The conductors are separated by insulation. Two conductors separated by insulation form a capacitor. The longer the cable, the more conductor area there is, and the higher the capacitance becomes. This

Figure 3-9 Coaxial cable (*courtesy of Belden Wire*)

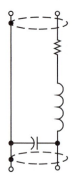

Figure 3-10
Low-frequency model
of coaxial cable

capacitance is *between* the center conductor and the shield. It is in *parallel* with the cable. Look at Figure 3-10.

Example 3-5

An inexpensive microphone may have an output resistance of as much as 10 kΩ. To prevent loading, you built an amplifier with an input resistance of 100 kΩ. The system sounds fine with the microphone connected *directly* to the amplifier. However, when the microphone is moved on stage, and connected to the amplifier through a 100-foot-long coaxial microphone cable, the sound is garbled and hard to understand.

Use simulation software to determine the voltage at the input of the amplifier under the following conditions:

a.   with the microphone connected directly to the amplifier.

b.   with the cable *capacitance* included at 400 Hz and 8 kHz.

Solution

a.   The schematic for the microphone and amplifier input *without* the cable is given in Figure 3-11. All of the meters have been set to ac. They display the rms value of the voltage. The signal that arrives at the amplifier is smaller than that from the microphone. However, it is the same at 400 Hz and 8 kHz. The amplifier can make up for this reduction.

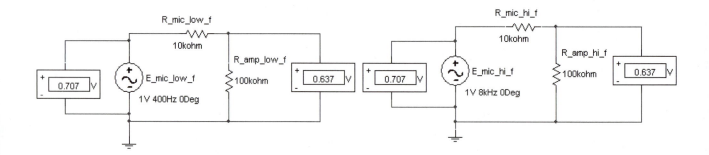

Figure 13-11 Simulation results ignoring cable capacitance

b. Typical cable specifications are given in Figure 3-12. They indicate that there is 28 pF/ft of capacitance. For a 100-foot cable this means that there is a total capacitance of 2.8 nF. This capacitance is added to the simulation in Figure 3-13.

MAX. OPERATING VOLTAGE	5000 V DC
CONTINUOUS CURRENT PER CONDUCTOR	
@ 25 DEG. C (10DEG C RISE)	1.9 A
NOM. CONDUCTOR OF DC RESISTANCE	
@ 20 DEG. C	45.1 OHMS/1000 FT.
NOM. SHIELD DC RESISTANCE	
@ 20 DEG. C	9.12 OHMS/1000 FT.
NOM. IMPEDANCE	57 OHMS
NOM. CAPACITANCE BETWEEN	
CONDUCTOR & SHIELD 1 KHZ	28 PF/FT.

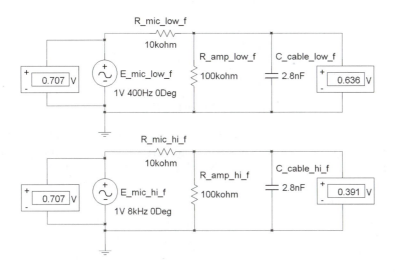

Figure 3-12
Microphone cable specifications (*courtesy of Belden Wire*)

Figure 3-13 Simulation of cable capacitance

At 400 Hz the amplitude delivered to the amplifier through the cable is the same as it was without the cable. But the high-frequency signal is attenuated much more. This should make sense. At 400 Hz

$$X_C = \frac{1}{2\pi \times 400\,\text{Hz} \times 2.8\,\text{nF}} = 142\,\text{k}\Omega$$

Even in parallel with the 100 kΩ of the amplifier, most of the voltage appears across this higher opposition. Little of the signal is dropped by the 10 kΩ output resistance of the microphone. However, when the frequency rises to 8 kHz, the capacitor's reactance becomes

$$X_C = \frac{1}{2\pi \times 8\,\text{kHz} \times 2.8\,\text{nF}} = 7.1\,\text{k}\Omega$$

Figure 3-14 AC motor (*courtesy of Reliance Electric*)

This lowers the impedance into the amplifier below that of the microphone. You should expect that less than half of the signal makes it to the amplifier. Most of it is dropped across the microphone's internal resistance.

Practice: Suggest two possible corrections to this cable capacitance loading problem. Use the simulator to verify your answers.

Answer: 1. Shorten the cable to 10 feet.   2. Use a mic, $R_{out} = 100\ \Omega$.

AC Motor's Complex Impedance

Motors "make the world go around." AC motors are found throughout your world, from the tiny, sub-fractional horsepower motor in the fan on your desk to the monsters that turn the rolls used to crush steel ingots into sheets. A typical industrial grade, fractional horsepower motor is shown in Figure 3-14.

Since these machines are rotating electromagnets, they have a significant inductance. However, current is converted to work by resistance, so, there is also resistance. When running steadily, under full mechanical load, the motor can be represented as a resistance in series with an inductance. This is shown in Figure 3-15.

Motor specifications include the full load voltage (V_{rms}), current (A_{rms}), and the power factor. A sample is shown in Figure 3-16.

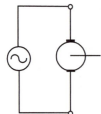

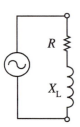

Figure 3-15
Simple model of steady-state AC motor

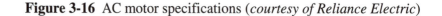

HP	Full Load RPM	Encl	Type	Amperes @230V			Efficiency Load			Power Factor Load			S.F.	Torque Lb. Ft.			Perf. Data Design
				FL	LR	Code	4/4	3/4	2/4	4/4	3/4	2/4		FL	BD	LR	
							Single-Phase - TENV, TEFC										
1/6	1140	TENV	CS	2.20	9.20	N	48.0	42.0	36.0	48.0	43.0	36.0	1.15	0.74	1.83	2.70	M2800E
1/4	1725	TENV	CS	2.60	6.20	M	58.1	53.9	45.2	59.2	51.6	43.3	1.15	0.75	2.50	2.83	M2270J
	1140	TENV	CS	2.90	13.00	L	58.0	53.0	43.0	50.0	44.0	36.0	1.15	1.12	3.56	3.81	M3129

Figure 3-16 AC motor specifications (*courtesy of Reliance Electric*)

The Reliance M2270J delivers ¼ hp at 1725 rev/min. It requires a supply of 230 V$_{rms}$. When running at full load (**FL**), it requires a current of 2.60 A$_{rms}$. The power factor under full load (**4/4**) is 59.2%.

Power factor may be interpreted to indicate the angle between the phasor voltage applied and the phasor current flowing through the motor.

$$power\ factor = \cos\theta$$

The voltage is assumed to be the reference ($\angle 0°$). Since the motor is inductive, **E L I**, the current lags the voltage. Its phase angle is *negative*.

$$\theta = -\cos^{-1}(power\ factor)$$

For the M2270J,　　　$\overline{V} = 230\,V_{rms}\angle 0°$

$$\theta = -\cos^{-1}(0.592) = -53.7°$$

$$\overline{I} = 2.60\,A_{rms}\angle -53.7°$$

$$\overline{Z} = \frac{\overline{V}}{\overline{I}}$$

For the M2270J,　　　$\overline{Z} = \dfrac{(230\,V_{rms}\angle 0°)}{(2.60\,A_{rms}\angle -53.7°)} = 88.46\Omega\angle 53.7°$

$$\overline{Z} = 88.46\Omega\angle 53.7° = 52.4\Omega + j71.3\Omega$$

The motor may be modeled as a resistance, R = 52.4 Ω, and an inductive reactance, X_L = 71.3 Ω. Figure 3-17 is the motor's impedance diagram.

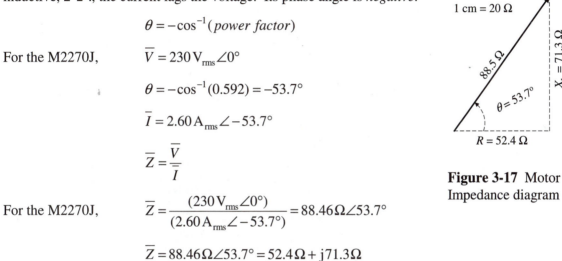

1 cm = 20 Ω

$\theta = 53.7°$

$R = 52.4\ \Omega$

$X_L = 71.3\ \Omega$

88.5 Ω

Figure 3-17 Motor Impedance diagram

Example 3-6

　a. Determine the impedance for the ⅙ hp motor in Figure 3-18. Also calculate the inductance, assuming f = 60 Hz.

　b. Verify the current magnitude and phase shift by simulation.

Solution

　a. The motor specifications needed are:

$$V = 230\ V_{rms},\ I = 2.20\ A_{rms},\ pf = 0.48$$

　　The current lags the voltage by

$$\theta = -\cos^{-1}(0.48) = -61.3°$$

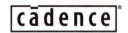

cadence

how big can you dream?™

The phasor forms of the voltage and current are

$$\overline{V} = (230\,\text{V}_{\text{rms}}\angle 0°) \qquad \overline{I} = (2.20\,\text{A}_{\text{rms}}\angle -61.3°)$$

The motor's impedance is

$$\overline{Z} = \frac{\overline{V}}{\overline{I}}$$

$$\overline{Z} = \frac{(230\,\text{V}_{\text{rms}}\angle 0°)}{(2.20\,\text{A}_{\text{rms}}\angle -61.3°)} = 104.6\,\Omega\angle 61.3°$$

$$\overline{Z} = 104.6\,\Omega\angle 61.3° = 50.2\,\Omega + j91.7\,\Omega$$

The resistor is 50.2 Ω and the inductor has a reactance of 91.7 Ω. The inductance is related to its reactance by

$$X_{\text{L}} = 2\pi f L$$

Solve for *L*.

$$L = \frac{X_{\text{L}}}{2\pi f} = \frac{91.7\,\Omega}{2\pi \times 60\,\text{Hz}} = 243.2\,\text{mH}$$

b. The simulation schematic and its probe display are shown in Figure 3-18. Notice that the current and voltage begin together at *t* = 0. By the second cycle current lags voltage properly. So, it is important to run the transient analysis long enough to allow the start-up transients to settle out. It is also important to set the minimum step size no larger than $\frac{1}{1000}$ of the total time. This assures enough data points to obtain a reasonable measurement of the time between when the voltage wave crosses ground going positive and when the current wave starts. That lag time is

$$t = 16.672 \text{ ms} - 19.456 \text{ ms} = -2.78 \text{ ms}$$

This gives a phase shift of

$$\theta = t \times 360° \times f$$

$$\theta = -2.78\,\text{ms} \times 360° \times 60\,\text{Hz} = -60.1°$$

The current's peak amplitude is 3.13 A$_{\text{p}}$. The rms value is

$$I_{rms} = \frac{3.13 \mathrm{A_p}}{\sqrt{2}} = 2.21 \mathrm{A_{rms}}$$

giving $\overline{I} = 2.21 \mathrm{A_{rms}} \angle -60.1°$

Practice: Determine the impedance for the last motor in Figure 3-18.

Answer: $\overline{Z} = 39.7\Omega + \mathrm{j}68.7\Omega$

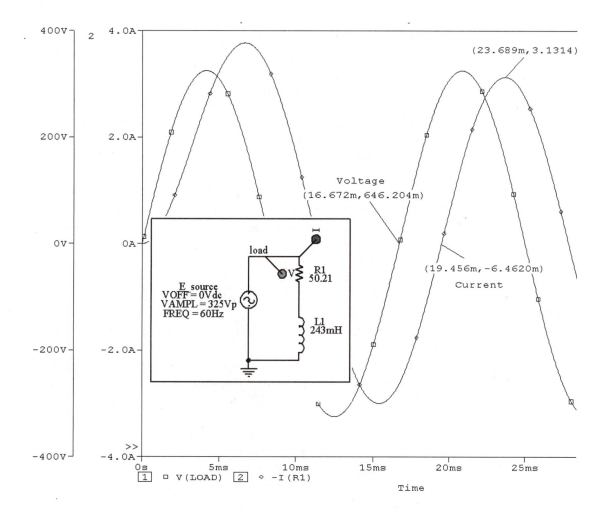

Figure 3-18 Schematic and simulation results for Example 3-6

Summary

Impedance is the opposition an element offers to a sinusoidal current. It is a phasor quantity. The current and voltage relationships for inductors and capacitors depend on the *rate of change*. There are derivatives involved. The derivative of a sinusoidal voltage can be taken graphically or with calculus. Either approach leads to

$$\frac{d}{dt}V_p \sin(2\pi ft) = 2\pi f V_p \sin(2\pi ft + 90°)$$

Ohm's law defines the relationship between current and voltage in a resistive circuit. So, current and voltage are in phase, and the resistive impedance phasor points horizontally to the right.

Every machine that converts electricity to motion uses an electromagnet, which is just an inductor. The voltage to current relationship for an inductor involves a derivative. Assuming a sinusoidal current and taking its derivative produces an impedance and an inductive reactance.

This impedance indicates that voltage leads current through an inductor (ELI), and its magnitude is proportional to frequency. The phasor points vertically straight up.

Capacitors behave in a manner opposite to the inductor. For a sinusoidal voltage, this defines capacitive impedance and reactance. This impedance indicates that current leads voltage though a capacitor (ICE), and its magnitude is inversely proportional to frequency. The phasor points vertically straight down.

Practical components exhibit impedance that is some combination of these three ideal cases. Even a simple wire has inductance. In fact, this inductive impedance becomes the dominant characteristic at radio frequencies.

At audio frequencies, the capacitive impedance of a coaxial cable may seriously attenuate a signal. This depends on the length of the cable, the cable's capacitive specifications, and the source and load impedances.

AC motors, when operating under steady voltage and shaft load, appear as a series combination of resistance and inductance. The manufacturers' specifications indicate the full-load voltage, current, and power factor. This power factor gives the angle between the voltage and current sinusoids. The voltage is generally assumed to be at 0°. So, θ is the lag angle of the current. You can then calculate the impedance of the motor. The rectangular form gives $R + jX_L$, the two components of the AC motor model.

Problems

Derivative of a Sine Wave

3-1   Using graphical techniques:
  a. Plot $V_p\cos(2\pi ft)$.

  b. Take the derivative of $V_p\cos(2\pi ft)$. Determine the effect of V_p, the cosine function, and $2\pi ft$.

3-2   Using graphical techniques:
  a. Plot $-V_p\cos(2\pi ft)$.

  b. Take the derivative of $V_p\cos(2\pi ft)$. Determine the effect of V_p, the $-$cosine function, and $2\pi ft$.

3-3   Using the rules of calculus, take the derivative of $V_p\cos(2\pi ft)$. Carefully identify each step and each rule as you apply it.

3-4   Using the rules of calculus, take the derivative of the function $V_p\cos(2\pi ft-90°)$. Carefully identify each step and each rule.

Impedance of a Resistance

3-5   Current through a circuit element is measured at 2.5 A_{rms}, in phase with the 40 V_{rms} voltage across the part.
  a. Calculate the impedance, in polar and rectangular forms.

  b. Accurately draw its impedance diagram and the phasor diagram of the voltage and the current.

3-6   Current through a circuit element is 416 μA_p $\sin(4000t - 90°)$. The applied voltage is 5.2 V_p $\sin(4000t - 90°)$.
  a. Calculate the impedance, in polar and rectangular forms.

  b. Accurately draw its impedance diagram and the phasor diagram of the voltage and the current.

Impedance of an Inductance

3-7   Current through a circuit element is measured at 2.5 A_{rms}, and it lags the 40 V_{rms} voltage across the part by 90°.
  a. Calculate the impedance, in polar and rectangular forms.

  b. Accurately draw its impedance diagram and the phasor diagram of the voltage and the current.

3-8   Current through a circuit element is 416 μA$_p$ sin(4000t – 90°). The applied voltage is 5.2 V$_p$ sin(4000t).

   a. Calculate the impedance, in polar and rectangular forms.

   b. Accurately draw its impedance diagram and the phasor diagram of the voltage and the current.

Impedance of a Capacitance

3-9   Current through a circuit element is measured at 2.5 A$_{rms}$, and it leads the 40 V$_{rms}$ voltage across the part by 90°.

   a. Calculate the impedance, in polar and rectangular forms.

   b. Accurately draw its impedance diagram and the phasor diagram of the voltage and the current.

3-10   Current through a circuit element is 416 μA$_p$ sin(4000t + 45°). The applied voltage is 5.2 V$_p$ sin(4000t – 45°).

   a. Calculate the impedance, in polar and rectangular forms.

   b. Accurately draw its impedance diagram and the phasor diagram of the voltage and the current.

Practical Impedances

3-11   a. Calculate the impedance of a 36-inch length of 20 AWG wire at 50 MHz, in polar and rectangular forms.

   b. Accurately draw its impedance diagram.

3-12   a. Calculate the impedance at 200 MHz of a 52 Ω resistor with 2-inch, 26 AWG leads on each end.

   b. Accurately draw the impedance diagram.

3-13   a. Calculate the impedance of a 500-foot length of coaxial microphone cable at 1 kHz. Include both the resistive and the capacitive impedance.

   b. Accurately draw the impedance diagram.

   c. If the source impedance is (1 kΩ +j0 Ω), and the load is resistance is 10 kΩ, do you expect this cable to degrade the signal? Explain why or why not.

3-14   a. Calculate the impedance of a 30-foot length of coaxial microphone cable. Include both the resistance and the capacitive impedance.

   b. Accurately draw the impedance diagram.

    c. If the source impedance is (150 Ω +j0 Ω), and the load resistance is 2 kΩ, do you expect this cable to degrade the signal? Explain why or why not.

3-15   **a.** An AC motor is rated at 115 V_{rms}, 600 mA_{rms}, with a 0.75 power factor. Calculate the motor's impedance.

    b. Accurately draw the impedance diagram.

3-16   **a.** An AC motor is rated at 230 V_{rms}, 4.70 A_{rms}, with a 0.58 power factor. Calculate the motor's impedance.

    b. Accurately draw the impedance diagram.

Impedance Measurements Lab Exercise

A. Low-Frequency Measurements
   1. Build the circuit in Figure 3-19.

   2. Assure that the oscilloscope is connected as indicated.

   3. With a digital multimeter, verify that the signal generator is providing the correct amplitude and frequency.

   4. Move the digital multimeter across the resistor. Record the magnitude of the voltage across the resistor.

   5. With the oscilloscope, measure the phase of the voltage across the resistor. Combine this with the magnitude from step A4 as the phasor representation of the resistor's voltage at 2 kHz. Also record the rectangular form of the voltage.

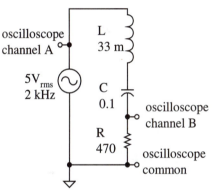

Figure 3-19
Measurement of $\overline{V_R}$

   6. Calculate the current through the circuit. Since this is a series circuit, there is only one current. The current through the resistor is the current through each of the elements, and is the total current through the circuit.

   7. Calculate and record the resistor's impedance.

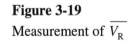

$$\overline{Z_{L\,@\,2kHz}} = \frac{\overline{V}_{L\,@\,2kHz}}{\overline{I}_{@\,2kHz}}$$

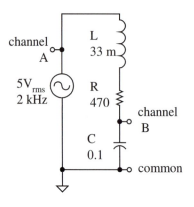

Figure 3-20

Measurement of $\overline{V_C}$

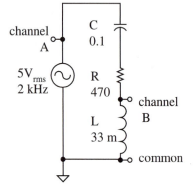

Figure 3-21

Measurement of $\overline{V_L}$

8. Calculate the circuit's *total* impedance.

$$\overline{Z_{total\,@\,2\,kHz}} = \frac{5\,V_{rms}\angle 0°}{\overline{I}_{@\,2\,kHz}}$$

9. Exchange the resistor and the capacitor, as shown in Figure 3-20. This allows you to measure the phase of the voltage across the capacitor. The oscilloscope must be tied to circuit common at the signal generator. To place channel B across the capacitor, the capacitor must also be tied to circuit common. This does *not* affect the circuit. The components are in series.

10. Place the digital multimeter across the capacitor. Record the magnitude of the voltage across the capacitor.

11. With the oscilloscope, measure the phase of the voltage across the capacitor. Combine this with the magnitude from step A10 as the phasor representation of the capacitor's voltage at 2 kHz. Also record the rectangular form of the voltage.

12. Calculate and record the capacitor's impedance.

$$\overline{Z_{C\,@\,2\,kHz}} = \frac{\overline{V_{C\,@\,2kHz}}}{\overline{I}_{@\,2kHz}}$$

13. Exchange the inductor and the capacitor, as shown in Figure 3-21. This allows you to measure the phase of the voltage across the inductor.

14. Place the digital multimeter across the inductor. Record the magnitude of the voltage across the inductor.

15. With the oscilloscope, measure the phase of the voltage across the inductor. Combine this with the magnitude from step A14 as the phasor representation of the inductor's voltage at 2 kHz. Also record the rectangular form of the voltage.

16. Calculate the inductor's impedance.

$$\overline{Z_{L\,@\,2\,kHz}} = \frac{\overline{V_{L\,@\,2kHz}}}{\overline{I}_{@\,2kHz}}$$

B. Low-Frequency Calculations

   1.   Using your calculator, add $\overline{Z_R} + \overline{Z_C} + \overline{Z_L}$. Compare this to the measured total impedance from step A8.

   2.   Using a protractor and a scale, carefully add the impedance phasors for each of the three components, nose-to-tail. If you have worked carefully, your answer from step B1 should match this drawing, as well as the measured total impedance in step A8. Compare your results and repeat your work until the magnitudes of the sum, the drawing, and the measured total impedance all agree to within 5% and the phase angles agree to within 4°.

C. Midrange-Frequency Measurements and Calculations

   1.   Set the signal generator's frequency to 3 kHz.

   2.   Repeat the measurements of Section A and then the calculations in Section B.

D. Upper-Frequency Measurements and Calculations

   1.   Set the signal generator's frequency to 5 kHz.

   2.   Repeat the measurements of Section A and then the calculations in Section B.

4

Series Circuits

Introduction

From the previous chapters you learned how to handle the complex nature of sinusoidal voltages and currents with phasors and to mathematically account for both the magnitude and the phase shift introduced into a circuit by inductors and capacitors. Often several of these components are connected end-to-end in a series circuit.

AC series circuits obey the same circuit laws as do dc circuits. You just have to use phasors to keep track of the effects on both magnitude and phase as you apply Kirchhoff's current and voltage laws, Ohm's law, and the voltage divider law. Phasor diagrams help you visualize how voltages and currents relate.

Simple ac series circuits are found in a wide variety of applications. The cross-over networks that send the proper signals to the correct drivers in a loudspeaker array may be simple series connections of inductors and/or capacitors with the electromagnet. Any motor, in its simplest version, is a series combination of an inductor and a resistor. Firing and protecting power switches require RCL series circuits. When you couple a signal and power into or out of an amplifier, often capacitors are used to block the dc bias, while passing the ac signal. However, the non-ideal effects of these components must not be ignored.

These are all ac series circuits. Applying the basic circuit laws allows you to determine their performance and to select the appropriate elements.

Objectives

Upon completion of this chapter, you will be able to do the following:

- Apply Kirchhoff's current and voltage laws, Ohm's law, and the voltage divider law to series ac circuits.

- Draw the phasor diagram for the impedances and the voltages in a series circuit.

- Apply series circuit analysis techniques to cross-over networks, power switch circuits, ac amplifiers, and power supply busses.

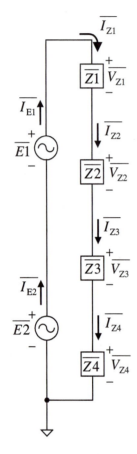

Figure 4-1
General series
circuit

4.1 Review of Fundamental Concepts

A series circuit contains a mix of elements all connected end-to-end. Figure 4-1 shows a generalized ac series circuit. Only two sources and four loads are shown, but these can be increased or decreased as needed. It helps the analysis to group the sources on the left and the loads on the right. However, since they are all connected in a single chain, each may be placed anywhere in the circuit without altering performance. This is very important when measuring voltages with an oscilloscope.

All of the component values, all of the voltages and all of the currents are *phasors*. Be sure to label each with a bar over it to remind anyone using the schematic to include both the magnitude and phase.

Although the voltages vary constantly, it helps in the analysis to assign polarities, + to −. You can think of these as the polarities at a particular instant in time. The sources provide a rise in potential from common, while the loads each produce a drop.

The currents *into* each of the loads and *out of* each of the sources are also labeled. Remember, they too are phasors, having both a direction and a magnitude.

In general, Kirchhoff's current law indicates that the current flowing into a node must equal the current flowing out of that node. That is, what goes in must come out.

$$\sum \overline{I_{in}} = \sum \overline{I_{out}}$$

In the series circuit, Figure 4-1, there is only one path into and out of each node. All of the currents are the same.

$$\overline{I} = \overline{I_{E1}} = \overline{I_{E2}} = \overline{I_{Z1}} = \overline{I_{Z2}} = \overline{I_{Z3}} = \overline{I_{Z4}} = ...$$

Remember, these are *phasors*. So you must include both the magnitude and the phase angle.

Kirchhoff's voltage law states that the sum of the voltage rises and drops around a circuit must be zero. When you start at a given point, regardless of how many stairs you climb, you must descend exactly the same amount to get back to where you started.

$$\sum \overline{E_{rises}} = \sum \overline{V_{drops}}$$

For the general circuit in Figure 4-1 this becomes

$$\overline{E_1} + \overline{E_2} + ... = \overline{V_1} + \overline{V_2} + \overline{V_3} + \overline{V_4} + ...$$

Kirchhoff's voltage law

This is just what you did with dc circuits. With ac circuits, the voltage rises and falls have both magnitude and phase. This is critical. You will see *many* ac circuits in which the magnitudes of the voltage drops are far larger than the magnitudes of the sources. They only add up correctly when you include the phase shifts.

The total impedance is the combined effect of all of the impedances acting together. Using Ohm's law,

$$\overline{Z_{total}} = \frac{\overline{E_1} + \overline{E_2}}{\overline{I}}$$

Or, rearranging

$$\overline{E_1} + \overline{E_2} = \overline{Z_{total}} \times \overline{I}$$

Similarly, each voltage drop is caused by a current flowing through an impedance.

$$\overline{Z_{total}} \times \overline{I} = \left(\overline{Z_1} \times \overline{I_1}\right) + \left(\overline{Z_2} \times \overline{I_2}\right) + \left(\overline{Z_3} \times \overline{I_3}\right) + \left(\overline{Z_4} \times \overline{I_4}\right)$$

Since the same current flows through each element

$$\overline{Z_{total}} = \overline{Z_1} + \overline{Z_2} + \overline{Z_3} + \overline{Z_4} + ...$$

Total impedance

As with series dc circuits, the total opposition is the sum of the individual oppositions. Just remember to include both the magnitude and the phase when working with components in an ac circuit.

The total current is the total voltage divided by the total impedance.

$$\overline{I} = \frac{\overline{E_1} + \overline{E_2} + ...}{\overline{Z_1} + \overline{Z_2} + \overline{Z_3} + \overline{Z_4} + ...}$$

The voltage across any element in the series circuit is

$$\overline{V_{Zx}} = \overline{I} \times \overline{Z_x}$$

$$\overline{V_{Zx}} = \frac{\overline{E_1} + \overline{E_2} + ...}{\overline{Z_1} + \overline{Z_2} + \overline{Z_3} + \overline{Z_4} + ...} \times \overline{Z_x}$$

Often, this is rearranged to group the impedances together.

$$\overline{V_{Zx}} = \left(\overline{E_1} + \overline{E_2} + ...\right) \times \frac{\overline{Z_x}}{\overline{Z_1} + \overline{Z_2} + \overline{Z_3} + \overline{Z_4} + ...}$$

Voltage divider law

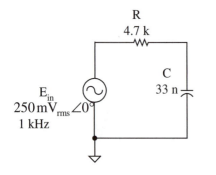

Figure 4-2
RC series circuit

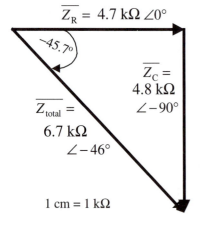

Figure 4-3 Impedance
diagram for Figure 4-2

4.2 *RC* and *RL* Circuits

Any time resistance, capacitance, and inductance are connected end-to-end, a series ac circuit is formed. All of the laws from the previous section apply, and each circuit may be analyzed using the same techniques.

The circuit in Figure 4-2 is a series *RC* circuit. The first step in calculating all of the impedances, currents and voltages is to determine the capacitor's reactance.

$$X_C = \frac{1}{2\pi \times 1\,\text{kHz} \times 33\,\text{nF}} = 4.82\,\text{k}\Omega$$

The impedance of the capacitor is

$$\overline{Z_C} = \left(4.82\,\text{k}\Omega\angle -90°\right) = \left(0 - \text{j}4.82\,\text{k}\Omega\right)$$

The impedance of the resistor is

$$\overline{Z_R} = \left(4.7\,\text{k}\Omega\angle 0°\right) = \left(4.7\,\text{k}\Omega + \text{j}0\right)$$

The circuit's total impedance is

$$\overline{Z_{\text{total}}} = \overline{Z_R} + \overline{Z_C}$$

$$\overline{Z_{\text{total}}} = \left(4.7\,\text{k}\Omega + \text{j}0\right) + \left(0 - \text{j}4.82\,\text{k}\Omega\right)$$

$$\overline{Z_{\text{total}}} = \left(4.7\,\text{k}\Omega - \text{j}4.82\,\text{k}\Omega\right) = \left(6.73\,\text{k}\Omega\angle -45.7°\right)$$

So, the total impedance is, indeed, the sum of the resistor's and the capacitor's impedances. However, even though each has a magnitude of about 4.7 kΩ, the total impedance's magnitude is just 6.7 kΩ. Remember, each impedance has an *angle* that you must consider. The impedance diagram in Figure 4-3 illustrates how these impedances combine.

Ohm's law allows you to calculate the total current in the circuit.

$$\overline{I_{\text{total}}} = \frac{\overline{E_{\text{total}}}}{\overline{Z_{\text{total}}}}$$

$$\overline{I_{\text{total}}} = \frac{\left(250\,\text{mV}_{\text{rms}}\angle 0°\right)}{\left(6.73\,\text{k}\Omega\angle -45.7°\right)} = \left(37.1\,\mu\text{A}_{\text{rms}}\angle 45.7°\right)$$

Since this is a series circuit, the current is the same everywhere.

$$\overline{I_{total}} = \overline{I_R} = \overline{I_C} = 37.1\mu A_{rms}\angle45.7°$$

Ohm's law also allows you to determine the voltage across the resistor and the voltage across the capacitor.

For the resistor $\overline{V_R} = \overline{I_R}\times\overline{Z_R}$

$$\overline{V_R} = (37.1\mu A_{rms}\angle45.7°)\times(4.7k\Omega\angle0°) = (174\,mV_{rms}\angle45.7°)$$

The capacitor $\overline{V_C} = \overline{I_C}\times\overline{Z_C}$

$$\overline{V_C} = (37.1\mu A_{rms}\angle45.7°)\times(4.82k\Omega\angle-90°) = (179\,mV_{rms}\angle-44.3°)$$

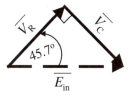

0.5 inch = 100 mV$_{rms}$

Figure 4-4 is the voltage phasor diagram. It helps visualize how the voltages combine. The voltage across the resistor leads the input signal by 45.7°. The voltage across the capacitor *always* lags the voltage across the resistor by 90°. Look for that right angle. The resistance of the resistor is almost equal to the reactance of the capacitor. So the *magnitudes* of the voltages across these two elements are nearly equal. The vector sum of the two voltage drops equals the applied signal.

Figure 4-4 Phasor diagram for Figure 4-2

If you only want the voltage across the capacitor, and do not want to solve Ohm's law twice, the voltage divider law can be applied.

$$\overline{V_C} = \overline{E_{in}}\times\frac{\overline{Z_C}}{\overline{Z_R} + \overline{Z_C}}$$

$$\overline{V_C} = (250\,mV\angle0°)\times\frac{(4.82k\Omega\angle-90°)}{(4.7k\Omega - j4.82k\Omega)} = (179\,mV_{rms}\angle-44.3°)$$

Simple, Passive Cross-Over Networks

Loudspeaker arrays are built from several speakers, called **drivers**. Each driver is able to operate over only a limited range of frequencies. Those with small cones can move rapidly, responding to high pitches, the treble notes, but cannot generate the low pitches that require much larger motion. These are called **tweeters**. On the other end of the scale, **woofers** have large cones to produce the low frequency notes. However, they cannot move quickly enough to create the high pitches.

Drivers cool themselves by moving their cones back and forth. If you apply high frequency notes to a woofer, the cone does not vibrate, even though the coil dissipates power. The result is that the coil overheats and may be damaged. So, it is important to pass to each driver only those frequencies to which it can respond. That is the job of the **cross-over network**.

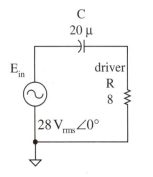

Figure 4-5 Schematic for Example 4-1

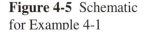

```
E=28;
f=[100;1e3;10e3];
C=20e-6;
R=8;

X_C=1./(2×pi×f×C);
Z_C=0-j×X_C;
Z_R=R+j×0;

U_R=E×Z_R./(Z_R+Z_C);
out=[f abs(U_R) angle(U_R)×57.3]

out =

         100      2.8007     84.265
        1000     19.851      44.852
       10000     27.862       5.681
```

Example 4-1

   a. Calculate the voltage across the driver for the circuit in Figure 4-5, at 100 Hz, 1 kHz, and 10 kHz.

   b. Confirm your calculations with a simulation.

   c. Is this a low-pass cross-over to be used with a woofer, or a high-pass cross-over to be used with a tweeter? Explain.

Solution

   a. First you must calculate the reactance of the capacitor at each frequency.

$$X_C = \frac{1}{2\pi f C}$$

$$X_C = 79.6\ \Omega \quad \text{at } f = 100\text{ Hz}$$

$$= 7.96\ \Omega \quad \text{at } f = 1\text{ kHz}$$

$$= 0.796\ \Omega \quad \text{at } f = 10\text{ kHz}$$

Use the voltage divider law.

$$\overline{V_R} = \overline{E_{in}} \frac{\overline{Z_R}}{\overline{Z_R} + \overline{Z_C}}$$

$$\overline{V_R} = (28\text{ V}_{rms}\angle 0°)\frac{(8\,\Omega\angle 0°)}{(8\,\Omega - jX_C)}$$

At 100 Hz, this becomes

$$\overline{V_R} = (28\text{ V}_{rms}\angle 0°)\frac{(8\,\Omega\angle 0°)}{(8\,\Omega - j79.6\,\Omega)} = (2.80\text{ V}_{rms}\angle 84.3°)$$

At 1 kHz, $\overline{V_R} = (19.8\text{ V}_{rms}\angle 44.9°)$

At 10 kHz, $\overline{V_R} = (27.9\text{ V}_{rms}\angle 5.7°)$

b. The simulation schematic is shown in Figure 4-6. There are two new parts. The signal generator is **VAC**, to allow an **AC Analysis**.

To print the results in the output file, the printer symbol, **VPRINT** has been connected above the resistor. Double clicking on the printer symbol brings up its dialog box. Enable the printer by entering the word "**true**" in the cell under **AC**. **MAG**, and **PHASE**.

To run an **AC Analysis**, select the **PSpice/New Simulation Profile** tabs on the tool bar. Once you name the profile, complete the **Simulation Settings** dialog box,

Run the simulation by selecting that tab on the tool bar. Once the simulation is complete, select **View/Output File**. Scroll to the bottom of that text

c. This is a high-pass circuit, a tweeter crossover network.

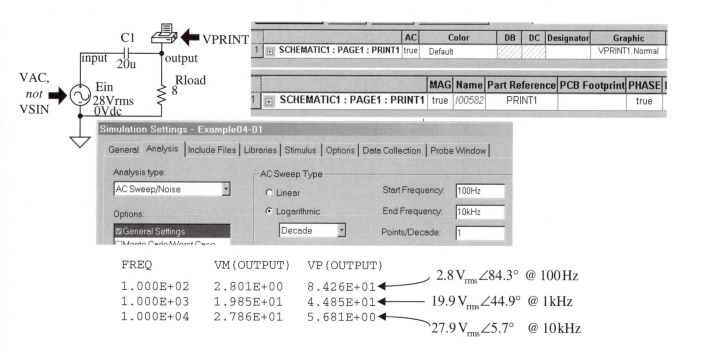

Figure 4-6 Simulation schematic for Example 4-1

Practice: Calculate the circuit's total impedance, current, and the voltage across the capacitor at 1 kHz.

Answer: $\overline{Z_{total}} = \left(11.29\,\Omega\angle - 44.9°\right),\quad \overline{I} = \left(2.48\,\text{A}_{rms}\angle 44.9°\right),$
$\overline{V_C} = \left(19.7\,\text{V}_{rms}\angle - 45.1°\right)$

The tweeter cross-over network of Example 4-1 is designed by placing a capacitor in series with the driver's resistance. It passes the high frequencies. A cross-over for a woofer must pass low frequencies. It can be built by placing an inductor in series with the driver's resistance. This is shown in Figure 4-7.

The value of the inductor can be selected by setting its inductive reactance equal to the driver's resistance at the driver's highest frequency. At lower frequencies, the inductor's reactance drops, so more of the signal is passed to the driver. At higher frequencies, the inductor's reactance increases, taking more of the input, leaving less for the driver.

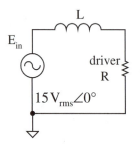

Figure 4-7 Woofer cross-over network (low pass filter)

Example 4-2

    a. Calculate the inductance needed to create a woofer cross-over, f_{max} = 300 Hz, with an 8 Ω driver.

    b. Calculate the total impedance, circuit current, and driver voltage at 150 Hz and 600 Hz.

    c. Verify the results from step b with a simulation.

Solution

    a. At 300 Hz, the inductive reactance should match the driver's resistance, 8 Ω

$$X_L = 2\pi f L$$

or $\quad L = \dfrac{X_L}{2\pi f} = \dfrac{8\,\Omega}{2\pi \times 300\,\text{Hz}} = 4.24\,\text{mH}$

    b. The reactance of the inductor is

$$X_{L\,@\,150\,Hz} = 2\pi f L = 2\pi \times 150\,\text{Hz} \times 4.24\,\text{mH} = 4.00\,\Omega$$

$$X_{L\,@\,600\,Hz} = 16.0\,\Omega$$

The total impedances at the two frequencies are

$$\overline{Z_{\text{total @ 150 Hz}}} = (8\,\Omega + \text{j}4\,\Omega)$$

$$\overline{Z_{\text{total @ 600 Hz}}} = (8\,\Omega + \text{j}16\,\Omega)$$

$$\overline{I_{\text{@ 150 Hz}}} = \frac{\overline{E_{\text{in}}}}{Z_{\text{total}}} = \frac{(15\,\text{V}_{\text{rms}}\angle 0°)}{(8\,\Omega + \text{j}4\,\Omega)} = (1.68\,\text{A}_{\text{rms}}\angle -26.6°)$$

$$\overline{I_{\text{@ 600 Hz}}} = \frac{\overline{E_{\text{in}}}}{Z_{\text{total}}} = \frac{(15\,\text{V}_{\text{rms}}\angle 0°)}{(8\,\Omega + \text{j}16\,\Omega)} = (839\ \text{mA}_{\text{rms}}\angle -63.4°)$$

The voltage across the driver's resistance is

$$\overline{V_{\text{R}}} = \overline{I} \times \overline{Z_{\text{R}}}$$

$$\overline{V_{\text{R @ 150 Hz}}} = (1.68\,\text{A}_{\text{rms}}\angle -26.6°)\times(8\,\Omega + \text{j}0)$$

$$= (13.4\,\text{V}_{\text{rms}}\angle -26.6°)$$

$$\overline{V_{\text{R @ 600 Hz}}} = (839\,\text{mA}_{\text{rms}}\angle -63.4°)\times(8\,\Omega + \text{j}0)$$

$$= (6.71\,\text{V}_{\text{rms}}\angle -63.4°)$$

c. The simulation results are shown in Figure 4-8. MultiSIM requires that the signal generator voltage be set in V_{p}.

Practice: Confirm Kirchhoff's voltage law at 150 Hz.

Answer: $\overline{V_{\text{L}}} = 6.72\,\text{V}_{\text{rms}}\angle 63.4°$ $\overline{V_{\text{L}}} + \overline{V_{\text{R}}} = 14.991\,\text{V}_{\text{rms}}\angle 0.03°$

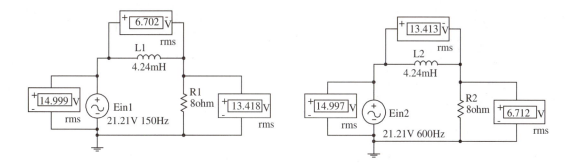

Figure 4-8 Simulation results for Example 4-2

RC Coupled Amplifiers

AC signals go both positive and negative. So the circuits used to amplify them must be able to send their outputs both up and down, in response to the ac input. It is simplest if these amplifiers are powered from a dual supply, such as ±5 V_{dc}. However, in our increasingly unhooked world, much of the electronics you use are powered from a single battery.

The simplest solution is to alter the biasing on the amplifier, so that when the input signal is at 0 V, the output of the amplifier is at half of its supply voltage. When the input signal drives the output up, it increases from this mid-point toward the supply voltage. When the output is driven down, it goes from this mid-point down toward circuit common.

Capacitors block dc ($X_C \rightarrow \infty$) but pass high-frequency sine waves ($f\uparrow \Rightarrow X_C\downarrow$). They combine with the input resistance and the load resistance to couple sine waves into and out of single-supply amplifiers.

Figure 4-9 gives the pin out of an LM324, a common op amp that works well from a single supply as small as +3 V_{dc}. An inverting amplifier is shown in Figure 4-10. Only a single supply voltage is applied, +V. The other power pin is connected to circuit common.

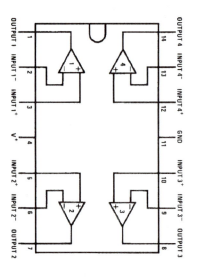

Figure 4-9

LM324 pin out *(courtesy of National Semiconductor)*

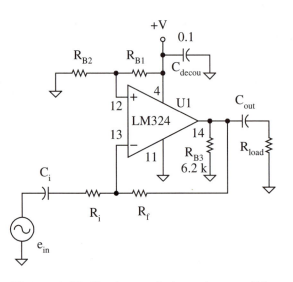

Figure 4-10 Single-supply inverting amplifier

The input resistor, R_i, is between the signal generator and the op amp's inverting input. The negative feedback resistor, R_f, is between the op amp's output pin and its inverting input. These two resistors assure that *negative* feedback is present, necessary for stable operation. As in the dual-supply version of this amplifier, R_f and R_i also control the gain.

Resistors R_{B1}, R_{B2}, and R_{B3} are for *biasing*. The two attached to the noninverting input pin establish the output dc level, half of the supply voltage. Resistor R_{B3} is required to minimize distortion The two capacitors are used to block the dc bias voltage from entering the signal generator or the load resistor. They *should*, however, pass the ac signal.

The dc bias levels are easy to see by remembering that the capacitors are open to dc. That removes C_i with its series R_i, and C_{out} with its series R_{load}. Figure 4-11 shows this simplified bias circuit. There is no significant current flowing into the inputs of an op amp. So, R_{B1} and R_{B2} are in series. The voltage at the op amp's noninverting input is set by

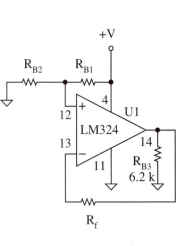

Figure 4-11 Single-supply amplifier biasing

$$V_{NI\,dc} = +V\frac{R_{B2}}{R_{B1} + R_{B2}}$$

For battery power applications, it is appropriate to set these resistors in the 100 kΩ range. Making them equal places $V_{NI\,dc} = \frac{1}{2}\,(+V)$.

Since there is negative feedback through R_f, the op amp drives its output to whatever voltage is necessary to keep the two input pins at the same voltage.

$$V_{INV\,dc} = V_{NI\,dc}$$

With no significant current into the inverting input pin, there is no current through R_f, so it drops no voltage.

$$V_{out\,dc} = V_{INV\,dc}$$

Stringing these relationships together, the output dc bias voltage is driven to the level set by R_{B1} and R_{B2}. R_i and R_{load} have no effect on bias because they are isolated by the capacitors in series with them.

Example 4-3

Calculate the dc and phasor voltages and currents for the amplifier in Figure 4-10 given:

$e_{in} = 250\ \text{mV}_{rms}$, $R_i = 10\ \text{k}\Omega$, $R_f = 33\ \text{k}\Omega$, $R_{load} = 300\ \Omega$

$f = 440\ \text{Hz}$, $C_i = 47\ \text{nF}$, $C_{out} = 1\ \mu\text{F}$

$+V = 5\ \text{V}_{dc}$, $R_{B1} = 330\ \text{k}\Omega$, $R_{B2} = 220\ \text{k}\Omega$

Solution

dc
$$V_{\text{NI dc}} = 5\,V_{\text{dc}} \frac{220\,\text{k}\Omega}{220\,\text{k}\Omega + 330\,\text{k}\Omega} = 2\,V_{\text{dc}}$$

$$V_{\text{INV dc}} = V_{\text{out dc}} = 2\,V_{\text{dc}}$$

ac

R_{B2} is connected directly to common, while R_{B1} is connected to the dc power supply bus, which has no ac voltage.

$$\overline{V_{\text{NI}}} = 0\,V_{\text{rms}}$$

The inverting input pin is held at ac virtual ground.

$$\overline{V_{\text{INV}}} = 0\,V_{\text{rms}}$$

This means that the input signal, e_{in}, is dropped across the *series* combination of C_i and R_i.

$$X_C = \frac{1}{2\pi \times 440\,\text{Hz} \times 47\,\text{nF}} = 7.70\,\text{k}\Omega$$

$$\overline{Z_i} = (10\,\text{k}\Omega - j7.70\,\text{k}\Omega)$$

$$\overline{I_i} = \frac{(250\,\text{mV}\angle 0^\circ)}{(10\,\text{k}\Omega - j7.70\,\text{k}\Omega)} = (19.8\,\mu A_{\text{rms}}\angle 37.6^\circ)$$

$$\overline{V_{\text{Ri}}} = (19.8\,\mu A_{\text{rms}}\angle 37.6^\circ) \times (10\,\text{k}\Omega\angle 0^\circ)$$

$$= (198\,\text{mV}_{\text{rms}}\angle 37.6^\circ)$$

About 20% of e_{in} never makes it into the amplifier. It is lost across the input capacitor. *And*, the phase is shifted 38°.

$$\overline{I_{\text{Rf}}} = \overline{I_{\text{Ri}}} = (19.8\,\mu A\angle 37.6^\circ)$$

$$\overline{V_{\text{op amp out}}} = -\overline{I_{\text{Rf}}} \times (R_f\angle 0^\circ)$$

The current flows from left to right through R_f, even though the op amp's inverting input is at ac virtual ground.

$$\overline{V_{\text{op amp out}}} = -(19.8\,\mu A_{\text{rms}}\angle 37.6^\circ) \times (33\,\text{k}\Omega\angle 0^\circ)$$

$$= (653.4\,\text{mV}_{\text{rms}}\angle -142^\circ)$$

MATLAB

```
E=250e-3;
Ri=10e3;
Rf=33e3;
Rload=300;
f=440;
Ci=47e-9;
Cout=1e-6;

X_Ci=1/(2*pi*f*Ci);
X_Cout=1/(2*pi*f*Cout);

Zi=(Ri-j*X_Ci);
Ii=E/Zi;
V_Ri=Ii*Ri;

V_op=-Ii*Rf;
V_load=V_op*Rload/(Rload-j*X_Cout);

ans=[abs(V_load) angle(V_load)*57.3]

ans =

      0.41737      -92.096
```

This voltage is applied to the series combination of C_{out} and R_{load}.

$$X_{Cout} = \frac{1}{2\pi \times 440\,\text{Hz} \times 1\,\mu\text{F}} = 362\,\Omega$$

$$\overline{V_{load}} = (653.4\,\text{mV}_{rms}\angle -142°)\frac{(300\,\Omega\angle 0°)}{(300\,\Omega - j362\,\Omega)}$$

$$= 416.9\,\text{mV}_{rms}\angle -91.6°$$

In an ideal circuit this amplifier should have a gain of –3.3. The load voltage should be 825 mV$_{rms}$, and be shifted 180°. Instead, the voltage is only half of this expected value, and the phase is almost 90° off. At 440 Hz, both of the capacitors have significant impedance compared to the resistors they are in series with. They drop a considerable voltage, leaving less for the resistors!

Practice: Calculate the load voltage at $f = 10$ kHz.

Answer: $\overline{V_{load}} = (823.3\,\text{mV}_{rms}\angle -175°)$ There is very little error.

The circuit in Example 4-3 has a 50% error in both magnitude and phase at 440 Hz, but is close at 10 kHz. This error depends on the reactance of the capacitor and the resistance it is in series with. You need to do a little math to develop a general relationship. For the series *RC* coupler, all of e_{in} should appear across the resistor, while none (ideally) is dropped across the capacitor. The voltage divider law indicates

$$\overline{V_R} = \overline{e_{in}} \frac{\overline{Z_R}}{\overline{Z_R} + \overline{Z_C}}$$

$$\overline{V_R} = (e_{in}\angle 0°)\frac{(R\angle 0°)}{(R - jX_C)}$$

To multiply phasors in the numerator, multiply the magnitudes and add the angles.

$$\overline{V_R} = \frac{(e_{in} \times R\angle 0°)}{(R - jX_C)}$$

To divide the phasor of the numerator by the phasor in the denominator, first you must convert the rectangular notation of the denominator to polar.

$$\overline{V_R} = \frac{\left(e_{in} \times R \angle 0^\circ\right)}{\left(\sqrt{R^2 + X_C^2} \angle \arctan \dfrac{-X_C}{R}\right)}$$

To divide phasors, divide the magnitudes and subtract the angles.

$$\overline{V_R} = \frac{e_{in} \times R}{\sqrt{R^2 + X_C^2}} \angle\left(0 - \left(-\arctan\frac{X_C}{R}\right)\right)$$

$$\overline{V_R} = \frac{e_{in} \times R}{\sqrt{R^2 + X_C^2}} \angle \arctan\frac{X_C}{R}$$

Next you must decide just how much error is acceptable. For a 1% drop in magnitude,

$$\left|\overline{V_R}\right| = 0.99 e_{in}$$

Substitute this for the magnitude part of the resistor's voltage.

$$\frac{e_{in} \times R}{\sqrt{R^2 + X_C^2}} = 0.99 e_{in}$$

$$\frac{R}{\sqrt{R^2 + X_C^2}} = 0.99$$

$$R = 0.99\sqrt{R^2 + X_C^2}$$

Square both sides $R^2 = 0.980\left(R^2 + X_C^2\right) = 0.980R^2 + 0.980 X_C^2$

$$0.020 R^2 = X_C^2$$

Take the square root of both sides.

$$X_C = 0.141 R$$

When designing an *RC* coupler, set the capacitor's reactance less than $^1/_7$ the series resistance, at the *lowest* frequency of interest.

$$X_C < \frac{1}{7} R$$

Example 4-4

Select more appropriate values for C_i and C_{out} for the amplifier in Example 4-3 and Figure 4-10.

Solution   For the input capacitor

$$\frac{1}{2\pi \times 440\,Hz \times C_i} < \frac{10\,k\Omega}{7}$$

$$C_i > \frac{1}{2\pi \times 440\,Hz \times \dfrac{10\,k\Omega}{7}} = 253\,nF \qquad C_i = 330\ nF$$

For the output capacitor

$$\frac{1}{2\pi \times 440\,Hz \times C_{out}} < \frac{300\,\Omega}{7}$$

$$C_{out} > \frac{1}{2\pi \times 440\,Hz \times \dfrac{300\,\Omega}{7}} = 8.44\,\mu F \qquad C_{out} = 10\ \mu F$$

Practice: Calculate the voltage across the load with these capacitors.

Answer: $X_{Cin} = 1.10\,k\Omega$, $X_{Cout} = 36.2\,\Omega$, $\overline{I} = 24.85\,\mu A_{rms} \angle 6.3°$,

$\overline{V_{op\,amp\,out}} = 820.1\,mV_{rms} \angle -173.7°$, $\overline{V_{load}} = 814.2\,mV_{rms} \angle -167°$

$\overline{V_{load\ ideal}} = 825\,mV_{rms} \angle -180°$ 99% accurate

4.3 *RCL* Circuits

Capacitors cause the voltage to lag the current (ICE), while inductors have the opposite effect, causing the current to lag the voltage (ELI). The result is that when combined in a series circuit, they tend to cancel each other. How much depends on their reactances, which depend on the frequency. An *RCL* circuit may have radically different behavior, depending on the frequency. That may be very handy. In this section you will see the fundamental phasor calculations using Ohm's law, Kirchhoff's voltage law, and the voltage divider law. These *RCL* cir-cuits are then used to build a cross-over network for the midrange driver

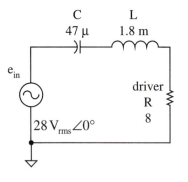

Figure 4-12
RCL series circuit for a midrange cross-over

in your loudspeaker array, and build an amplifier that selects a specific radio station. They also explain why amplifier power supply decoupling may actually cause the amplifier to oscillate.

Simple, Midrange Cross-over Network

Figure 4-12 is the schematic for a simple series *RCL* circuit. These values are appropriate for the cross-over network used to pass signals to the midrange driver in a loudspeaker array. At low frequencies, the capacitor should have a high impedance, there should be little current in the circuit, and the voltage across the resistor should be small.

At $f = 100$ Hz

$$X_C = \frac{1}{2\pi \times 100\,\text{Hz} \times 47\,\mu\text{F}} = 33.9\,\Omega$$

$$X_L = 2\pi \times 100\,\text{Hz} \times 1.8\,\text{mH} = 1.1\,\Omega$$

The circuit's total impedance is

$$\overline{Z}_{\text{total}} = (R + j0) + (0 + jX_L) + (0 - jX_C)$$

$$= [8\,\Omega + j(1.1\,\Omega - 33.9\,\Omega)]$$

$$\overline{Z}_{\text{total}} = (8\,\Omega - j32.8\,\Omega) = (33.8\,\Omega\angle -76.3°)$$

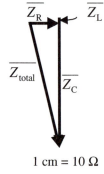

1 cm = 10 Ω

The impedance diagram is shown in Figure 4-13. It starts from the origin with $\overline{Z}_R$, that extends at angle 0° (straight to the right) for 0.8 cm. The impedance of the inductor, $\overline{Z}_L$, starts at the tip of the resistor's phasor, and goes up for only 0.11 cm. This is so small, that the arrow head had to be omitted. The capacitor's impedance, $\overline{Z}_C$, begins where the inductor's impedance ends, and goes down (−90°) for 3.39 cm. Finally, the total impedance, $\overline{Z}_{\text{total}}$, goes from the origin headed downward at a −76.3° angle for 3.38 cm. The tip of the total impedance phasor meets the tip of the last individual component. At this low frequency, the circuit is almost entirely capacitive.

Circuit current is

$$\overline{I}_{\text{total}} = \frac{\overline{E}_{\text{total}}}{\overline{Z}_{\text{total}}}$$

Figure 4-13
Low frequency
RCL impedances

$$= \frac{(28\,\text{V}_{\text{rms}}\angle 0^\circ)}{(33.8\,\Omega\angle - 76.3^\circ)} = (829.3\,\text{mA}_{\text{rms}}\angle 76.3^\circ)$$

Ohm's law gives the voltage across each element.

$$\overline{V_\text{C}} = \overline{I_\text{C}} \times \overline{Z_\text{C}}$$

$$= (829.3\,\text{mA}_{\text{rms}}\angle 76.3^\circ) \times (33.9\,\Omega\angle - 90^\circ)$$

$$= (28.1\,\text{V}_{\text{rms}}\angle - 13.7^\circ)$$

$$\overline{V_\text{L}} = \overline{I_\text{L}} \times \overline{Z_\text{L}}$$

$$= (829.3\,\text{mA}_{\text{rms}}\angle 76.3^\circ) \times (1.1\,\Omega\angle 90^\circ)$$

$$= (0.9\,\text{V}_{\text{rms}}\angle 166.3^\circ)$$

$$\overline{V_\text{R}} = \overline{I_\text{R}} \times \overline{Z_\text{R}}$$

$$= (829.3\,\text{mA}_{\text{rms}}\angle 76.3^\circ) \times (8\,\Omega\angle 0^\circ)$$

$$= (6.6\,\text{V}_{\text{rms}}\angle 76.3^\circ)$$

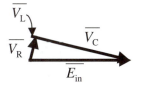

1 cm = 10 V$_{\text{rms}}$

Figure 4-14
Low frequency *RCL*
voltage phasors

The phasor diagram for these four voltages is given in Figure 4-14. Another closed triangle indicates that the voltage drops add up to equal the input signal.

As predicted, most of the voltage appears across the capacitor, with only a little across the resistor, and virtually none across the inductor.

At a low frequency, the impedance is much greater than R, there is little current, and most of the voltage appears across the capacitor. The voltage across the resistor is small. This all changes as the frequency increases. Repeat the calculations at 500 Hz. The results follow. Be sure that you can obtain these numbers yourself.

$$X_\text{C} = 6.77\,\Omega \qquad X_\text{L} = 5.65\,\Omega$$

$$\overline{Z_{\text{total}}} = 8.08\,\Omega\angle - 8.0^\circ$$

$$\overline{I} = 3.47\,\text{A}_{\text{rms}}\angle 8.0^\circ$$

Results at *f* = 500 Hz

$$\overline{V_\text{C}} = 23.49\,\text{V}_{\text{rms}}\angle - 82.0^\circ \qquad \overline{V_\text{L}} = 19.61\,\text{V}_{\text{rms}}\angle 98.0^\circ \qquad \overline{V_\text{R}} = 27.76\,\text{V}_{\text{rms}}\angle 8.0^\circ$$

The capacitor's reactance has dropped and the inductor's has risen until they are almost equal, virtually canceling each other. The result is

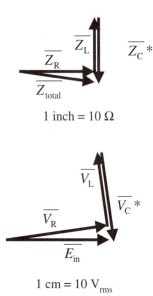

1 inch = 10 Ω

1 cm = 10 V$_{rms}$

* C phasors have been moved to the
right so they can be seen. They
should fall on top of L phasors.

Figure 4-15 *RCL* phasor
diagrams at 500 Hz

a major increase in circuit current and voltage drops across each of the
elements. The voltage across the capacitor and inductor are 180° out of
phase with each other, almost completely canceling each other. This
leaves almost all of the input signal to appear across the resistor. The
impedance diagram and the voltage phasor diagram are shown in Figure
4-15. Notice that the resistor's impedance lies along the *x* axis, the inductor's is at a right angle with it, and the capacitor's impedance falls on
top of the inductor's, going in the opposite direction. The voltage
phasors show a similar shape. The inductor's voltage is 90° ahead of the
resistor's voltage. The capacitor's voltage is 90° behind the resistor's,
on top of and opposite to the inductor's voltage.

At a low frequency, the capacitor hogs all of the voltage. There is
little left for the resistive loudspeaker driver. At this intermediate frequency, the inductor and the capacitor cancel each other, allowing most
of the input to pass to the resistor. Since this cross-over network is supposed to pass *intermediate* frequencies, the high frequencies should be
blocked. The results for 1.5 kHz are shown below.

$$X_C = 2.26\,\Omega \qquad X_L = 16.96\,\Omega$$

$$\overline{Z_{total}} = 16.74\,\Omega\angle 61.4°$$

$$\overline{I} = 1.67\,\text{A}_{rms}\angle -61.4°$$

$$\overline{V_C} = 3.78\,\text{V}_{rms}\angle -151.4°$$

$$\overline{V_L} = 28.32\,\text{V}_{rms}\angle 28.6°$$

$$\overline{V_R} = 13.36\,\text{V}_{rms}\angle -61.4°$$

The phasor diagrams at 1.5 kHz are shown in Figure 4-16. As with
the other impedance diagrams, the resistor's impedance begins at the
origin and extends horizontally to the right. From its tip the inductor's
impedance goes straight up. At this high frequency, it is the dominant
element. The capacitor's impedance is small. It should begin where the
inductor's impedance ends, and extend straight down. It has been offset
slightly to the right so you can see it. At this high frequency, it is very
small. The total impedance is the sum of these phasors, beginning at the
origin and extending to the tip of the last phasor. The triangle closes.

The voltage phasors also form a triangle, with the inductor's voltage
90° ahead of the resistor's, and the capacitor's (very small and offset so
you can see it) is 90° behind. The sum of the voltage drops equals the
applied voltage. This triangle also closes.

Series Tuned Amplifier

The *RCL* circuit was introduced in this chapter as a mid-range cross-over network. It blocked the low tones, and high notes, sending only the middle frequencies to the loudspeaker driver that can respond to them. The same process occurs at radio frequencies when selecting a single station to detect, amplify, and pass on to the loudspeaker or screen. Each station is assigned a specific carrier frequency. After the program (sound, video, or data) is combined with the carrier, the station broadcasts a very narrow range of frequencies. Every other station is also broadcasting a similar narrow range of frequencies, centered around their assigned, different carrier frequency. One of the key jobs of a receiver is to reject all but the one narrow band of frequencies broadcast by the station you want to receive. Those frequencies must be amplified.

The series tuned amplifier shown in Figure 4-17 does just that. The circuit is configured as an inverting amplifier. There is *negative* feedback from the output through R_f to the inverting input. As long as the op amp is not driven into saturation, there is no significant difference in potential between its input pins. There is also no significant current into either input pin. The National Semiconductor LMH6654 is used because it is fast enough to provide the gain needed at the high frequency at which this circuit operates. It is also inexpensive, and works from commonly available ±5 V_{dc} supplies.

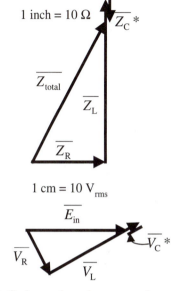

1 inch = 10 Ω

* C phasors have been moved to the right so they can be seen. They should fall on top of L phasors.

Figure 4-16 *RCL* phasor diagrams at 1.5 kHz

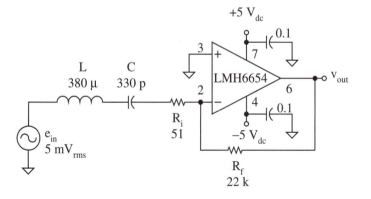

Figure 4-17 Series tuned amplifier

MATLAB

```
E=5e-3;
L=380e-6;
C=330e-12;
Ri=51;
Rf=22e3;
f=[430e3; 450e3; 500e3];

X_L=2×pi×f×L;
X_C=1./(2×pi×f×C);

Z_L=0+j×X_L;
Z_C=0-j×X_C;
Z_R=Ri;
Z_tot=Z_L+Z_C+Z_R;

I=E./Z_tot;
Uout=-I×Rf;
ans=[f abs(Vout) angle(Vout)×57.3]
■

ans =

    4.3e+005      1.0208      -118.26
    4.5e+005      2.1539       177.01
      5e+005     0.46841       102.55
```

Example 4-5

Calculate the output voltage for the circuit in Figure 4-17 at 430 kHz, 450 kHz, and 500 kHz.

Solution

f = 430 kHz

$$X_C = 1.122\,\text{k}\Omega \qquad X_L = 1.027\,\text{k}\Omega$$

$$\overline{Z_{\text{total}}} = (108\,\Omega\angle -61.8°)$$

$$\overline{I} = (46.37\,\mu\text{A}_{\text{rms}}\angle 61.8°)$$

All of this current flows from the *RCL* circuit, through R_f. It produces a voltage drop + to − across R_f.

$$\overline{V_{\text{out}}} = -\overline{I}\times(R_f\angle 0°)$$

$$= -(46.37\,\mu\text{A}_{\text{rms}}\angle 61.8°)\times(22\,\text{k}\Omega\angle 0°)$$

$$\overline{V_{\text{out}}} = (1.020\,\text{V}_{\text{rms}}\angle -118.2°)$$

f = 450 kHz

$$X_C = 1.072\,\text{k}\Omega \qquad X_L = 1.074\,\text{k}\Omega$$

$$\overline{Z_{\text{total}}} = 51\,\Omega\angle 2.2°$$

$$\overline{I} = (97.96\,\mu\text{A}_{\text{rms}}\angle -2.2°)$$

$$\overline{V_{\text{out}}} = (2.155\,\text{V}_{\text{rms}}\angle 177.8°)$$

f = 500 kHz

$$X_C = 965\,\Omega \qquad X_L = 1.194\,\text{k}\Omega$$

$$\overline{Z_{\text{total}}} = 235\,\Omega\angle 77.4°$$

$$\overline{I} = (21.31\,\mu\text{A}_{\text{rms}}\angle -77.4°)$$

$$\overline{V_{\text{out}}} = (468.9\,\text{mV}_{\text{rms}}\angle 102.6°)$$

The signals at 450 kHz are amplified by 431, while lower and higher frequencies receive a lower gain and considerable phase shift.

Practice: Calculate the output voltage at 470 kHz.

Answer: $\overline{V_{out}} = (1.009\,\text{V}_{rms}\angle118.0°)$

Power Supply Bus Decoupling

Power busses run all over most electronics boards. They interconnect the power pins of the ICs and transistor circuits to the power supplies. In doing so, they also connect many different circuits together. *If* the voltage on the bus is truly rock-solid dc, there is no problem. However, every conductor has series inductance. This inductance prevents the power supply, often a meter away, from keeping the supply voltage constant. Variation in the current demanded by the power stage at the end of a chain of amplifiers may cause the voltage on the bus to vary as much as 100 mV$_{rms}$. This variation is passed along the bus, through the series inductance, to the sensitive, high-gain stage at the other end of the amplifier. There, this small variation can easily be coupled through parasitic effects into that high-gain stage. It is amplified, passed to the power stage, and sent back to the beginning along the power supply bus. The whole system breaks into high frequency oscillations.

To prevent this, you are told to place a 0.1 μF decoupling capacitor at the power pins of each analog IC. It is supposed to short out any ac variations on the power supply bus before they can get into the amplifier. This is shown in Figure 4-18.

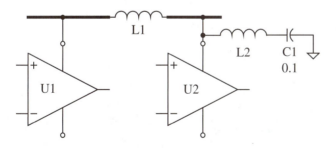

Figure 4-18 Power supply bus

The inductor L1 is the inductance of the power supply bus. In Chapter 3 you learned to calculate the inductance of wire. An inch of 24 AWG wire has about 22 nH of inductance. A 50 mil wide rectangular trace should have even more. So if U1 and U2 are an inch apart L1 will be at least 22 nH. The leads on the decoupling capacitor also have inductance, represented by L2. Leaving each an inch long, L2 = 44 nH.

how big can you dream?™

Example 4-6

Amplifier U1 injects a 100 mV$_{rms}$, 2 MHz signal onto the power bus. Calculate the amplitude of the signal at the power pins of U2. Verify your manual calculations with a simulation.

Solution

The simplified schematic is shown in Figure 4-19. Begin by calculating the reactance and the impedance of each of the three elements.

$$X_{L1} = 2\pi \times 2\,\text{MHz} \times 22\,\text{nH} = 0.28\,\Omega$$

$$X_{L2} = 2\pi \times 2\,\text{MHz} \times 44\,\text{nH} = 0.55\,\Omega$$

$$X_C = \frac{1}{2\pi \times 2\,\text{MHz} \times 0.1\,\mu\text{F}} = 0.80\,\Omega$$

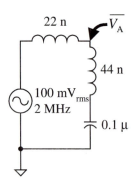

Figure 4-19 Simplified schematic for Example 4-6

Apply the voltage divider law.

$$\overline{V_A} = \overline{E_{in}} \times \frac{\overline{Z_{L2}} + \overline{Z_C}}{\overline{Z_{L1}} + \overline{Z_{L2}} + \overline{Z_C}}$$

$$\overline{V_A} = (100\,\text{mV}_{rms}\angle 0°) \times \frac{(0 + j0.55\,\Omega) + (0 - j0.80\,\Omega)}{(0 + j0.28\,\Omega) + (0 + j0.55\,\Omega) + (0 - j0.80\,\Omega)}$$

$$\overline{V_A} = (833\,\text{mV}_{rms}\angle 180°)$$

Instead of getting rid of the signal, the *LC* circuit has actually made the unwanted signal *much larger*. The simulation shown in Figure 4-20 predicts oscillations.

Practice: Replacing the leaded capacitor with a surface mount capacitor reduces the lead length to 0.01 inches, lowering L2 to 0.44 nF. Calculate the output voltage, with an input of 100 mV$_{rms}$.

Answer: $X_{L2} = 5.53$ mΩ $\overline{V_A} = 154\,\text{mV}_{rms}\angle 0°$

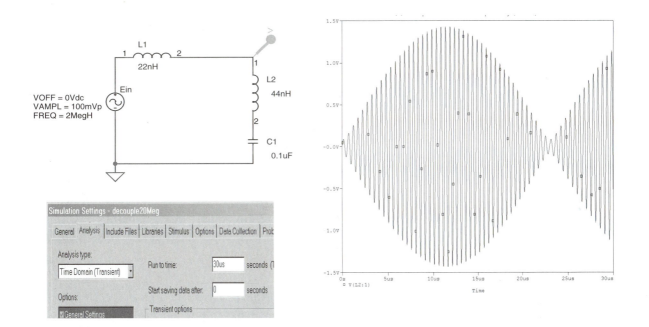

Figure 4-20 Simulation of power bus decoupling

The traditional approach to power supply decoupling may actually increase the interference at a particular frequency. An alternative is shown in Figure 4-21. A *small* resistance resistor is placed in series with the power supply run into each IC, between the bus and the IC power pin. Typically R ≈ 10 Ω. With power supply dc currents of a few milli-amperes, this drops less than 0. 1V_{dc}. However it places a resistive impedance between the noise from U1 and the decoupling capacitor, C1.

Example 4-7

Amplifier U1 injects a 100 mV$_{rms}$, 2 MHz signal onto the power bus shown in Figure 4-21. R = 10 Ω. R and C1 are surface mounted parts. Account for the inductance of the leads to the resistors and the capacitor. Calculate the amplitude of the signal at the power pins of U2. Verify your manual calculations with a simulation.

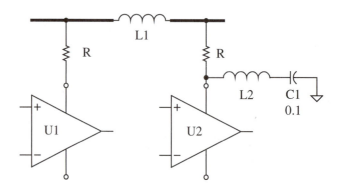

Figure 4-21 *RC* power supply bus decoupling

Solution

The simplified schematic is shown in Figure 4-22. Begin by calculating the reactance and the impedance of each of the elements.

$$X_{L1} = 2\pi \times 2\,\text{MHz} \times 22\,\text{nH} = 0.28\,\Omega$$

$$X_{L2} = 2\pi \times 2\,\text{MHz} \times 0.44\,\text{nH} = 0.006\,\Omega$$

$$X_C = \frac{1}{2\pi \times 2\,\text{MHz} \times 0.1\,\mu\text{F}} = 0.80\,\Omega$$

$$\overline{V_A} = \left(100\,\text{mV}_{\text{rms}}\angle 0°\right) \times$$

$$\frac{\left[0\,\Omega + j\left(0.006\,\Omega - 0.80\,\Omega\right)\right]}{\left[\left(10\,\Omega + 10\,\Omega\right) + j\left(0.006\,\Omega + 0.28\,\Omega + 0.006\,\Omega + 0.006\,\Omega - 0.80\,\Omega\right)\right]}$$

$$\overline{V_A} = \left(3.97\,\text{mV}_{\text{rms}}\angle -89°\right)$$

The MultiSIM simulation is shown in Figure 4-23. Adding the series 10 Ω resistors has solved the *LC* series decoupling problem. There is now less than 4 mV of noise coupled into U2 from U1.

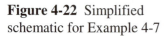

Figure 4-22 Simplified schematic for Example 4-7

Practice: Repeat the calculations using a 2.2 Ω resistor instead of 10 Ω.

Answer: $\overline{V_A} = \left(17.9\,\text{mV}_{\text{rms}}\angle -83°\right)$

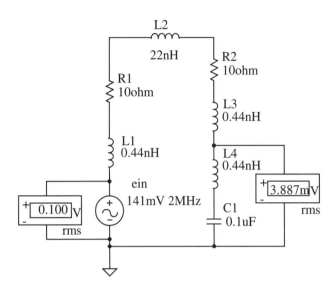

Figure 4-23 Simulation result for Example 4-7

Summary

Components connected end-to-end respond to ac signals just as they do to dc. The only difference is that you must calculate all of the quantities using phasors. This allows you to keep track of the effects on both the magnitude and the phase. The basic laws, restated with phasors are

$$\overline{I} = \overline{I_{E1}} = \overline{I_{E2}} = \overline{I_{Z1}} = \overline{I_{Z2}} = \overline{I_{Z3}} = \overline{I_{Z4}} = ...$$

Kirchhoff's current law

The currents are all the same in a series circuit.

$$\overline{E_1} + \overline{E_2} + ... = \overline{V_1} + \overline{V_2} + \overline{V_3} + \overline{V_4} + ...$$

Kirchhoff's voltage law

The sum of the voltage sources must equal the sum of the voltage drops.

$$\overline{Z_{total}} = \overline{Z_1} + \overline{Z_2} + \overline{Z_3} + \overline{Z_4} + ...$$

Total impedance

Impedances add in a series circuit.

Two versions of Ohm's law

$$\overline{I} = \frac{\overline{E}_{total}}{\overline{Z}_{total}} \qquad\qquad \overline{V}_{Zx} = \overline{I}_{Zx} \times \overline{Z}_x$$

Voltage divider law

$$\overline{V}_{Zx} = \left(\overline{E}_1 + \overline{E}_2 + ...\right) \times \frac{\overline{Z}_x}{\overline{Z}_1 + \overline{Z}_2 + \overline{Z}_3 + \overline{Z}_4 + ...}$$

The phasors of each impedance may be plotted, creating an imped-ance diagram. Using a nose-to-tail technique allows you to begin at the origin and plot impedance after impedance. The total impedance is the phasor drawn from the origin to the tip of the last impedance's phasor. These form a right triangle.

Right triangles are also formed when you plot the phasor voltages dropped across each circuit element. The voltage across an inductor is 90° ahead of the voltage across the resistors. The voltage across a ca-pacitor is 90° behind the resistors' voltages. This means that the induc-tor and capacitor voltages are exactly opposite in direction. Properly drawn, these voltages close with the applied sources to form another right triangle.

In general, the steps in solving a series ac circuit are to calculate the

- individual reactances of the inductors and capacitors
- individual impedances of each element
- total circuit impedance
- impedance phasor diagram
- total circuit current
- voltage across each element using Ohm's law
- voltage phasor diagram

If you only need a few voltages, then the voltage divider law pro-vides a simple combination of several of these steps.

Problems

Review of Fundamental Concepts

4-1 For the series circuit in Figure 4-24, determine each of the following:

a. current through each component

b. applied voltage

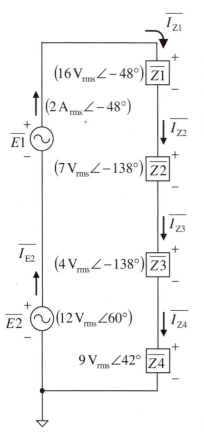

Figure 4-24 Schematic for Problems 4-1 and 4-2

c. total impedance, using Ohm's law

4-2 Repeat Problem 4-1 if the current from $\overline{E1}$ is (5 A$_{rms}$ ∠ −48°).

4-3 A voltage source, (E_{line}∠0°), is applied to a motor with an impedance of (R_{motor} + jX_L) and a series resistor, R_{limit}. Write the equation for the voltage across the motor.

4-4 A voltage source, (E_{out}∠−20°), is applied to a loudspeaker with an impedance of (R_{driver} + jX_L) and a series capacitor with an impedance of (0 − jX_C). Write the equation for the voltage across R_{driver}.

RC and RL Circuits

4-5 A series circuit consists of a voltage source, (40 V$_{rms}$∠0°) at a frequency of 3 kHz, a resistor of 8 Ω, and a 10 μF capacitor.
  a. Draw the schematic.

  b. Calculate the reactance of the capacitor.

  c. Calculate the impedance of the resistor and the capacitor.

  d. Calculate the circuit's total impedance.

  e. Draw the impedance diagram. Be sure to select and indicate an appropriate scale.

  f. Calculate the current through each element.

  g. Calculate the voltage drops using Ohm's law.

  h. Draw the phasor diagram of each voltage, illustrating how the voltage drops add to equal the source voltage

  i. Verify that Kirchhoff's voltage law applies.

4-6 The circuit in Figure 4-25 consists of a motor and a series limiting resistor, R1. Repeat the calculations in Problem 4-5.

4-7 Design a simple cross-over network for a 4 Ω driver that passes the frequencies *below* 500 Hz, and blocks those above.
  a. Draw the schematic.

  b. Set the reactance of the series element at 500 Hz equal to the resistance of the driver.

  c. Calculate the voltage delivered to the 4 Ω resistive driver from a 30 V$_{rms}$ source at the following frequencies:
    100 Hz,   500 Hz,   1000 Hz.

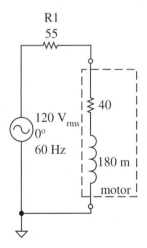

Figure 4-25 Schematic for Problem 4-6

4-8 Design a simple cross-over network for a 16 Ω driver that passes the frequencies *above* 2 kHz, and blocks those below.

a. Draw the schematic.

b. Set the reactance of the series element at 2 kHz equal to the resistance of the driver.

c. Calculate the voltage delivered to the 16 Ω resistive driver from a 30 V_{rms} source at the following frequencies:

    1 kHz,   2 kHz,   10 kHz.

4-9 Calculate the dc and phasor voltages indicated in Figure 4-26.

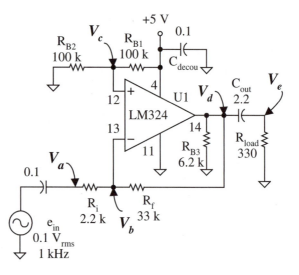

Figure 4-26 Schematic for Problems 4-11 and 4-12

4-10 a. Correct the component values for the amplifier in Figure 4-26 so that there is no more than a 1% drop across the coupling capacitors at 1 kHz.

b. Verify your design by calculating all of the indicated phasor voltages.

RCL Circuits

4-11 A series circuit consists of a voltage, (40 $V_{rms}\angle0°$), 3 kHz, a resistor of 8 Ω, a 58 μF capacitor, and a 270 μH inductor.

   a. Draw the schematic.

   b. Calculate the reactance of the capacitor and the inductor.

   c. Calculate the impedance of each component.

   d. Calculate the circuit's total impedance.

   e. Draw the impedance diagram. Be sure to select and indicate an appropriate scale.

   f. Calculate the current through each element.

   g. Calculate the voltage drops using Ohm's law.

   h. Draw the phasor diagram of each voltage, illustrating how the voltage drops add to equal the source voltage.

4-12 Repeat Problem 4-11 for a frequency of 200 Hz.

4-13 Repeat Problem 4-11 for a frequency of 8 kHz.

4-14 Based on your answers for the voltage across the resistor in the three problems above, explain the purpose of this circuit.

4-15 Calculate the dc and phasor currents and voltages indicated in Figure 4-27 at 500 Hz, 1.5 kHz, and 4 kHz.

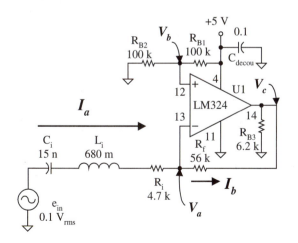

Figure 4-27 Schematic for Problem 4-15

4-16   A 0.3 V_{rms} noise source is 2 inches away from the 0.022 µF decoupling capacitor of the amplifier shown in Figure 4-18. The capacitor's leads are 0.25 inches on each end. Calculate:

    a.   The inductance of the power supply bus.

    b.   The inductance of the capacitor's leads.

    c.   The reactance of each inductance at 5 MHz.

    d.   The phasor voltage at the amplifier's power pin.

4-17   Calculate the phasor voltage at the amplifier's power pin when a 15 Ω resistor is added between the bus and the amplifier.

Series Tuned Amplifier Lab Exercise

A. Biasing

    1.   Build the circuit in Figure 4-28.

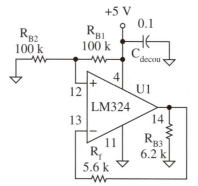

Figure 4-28 Biasing of the amplifier

    2.   Calculate the dc voltage at each input and at the output of the op amp.

    3.   Turn the +5 V_{dc} power supply *on*.

4. With a digital multimeter, verify that these voltages are correct. Do *not* continue until the op amp is properly biased.

B. Low-frequency Calculations

1. Locate C_i, L, and R_i shown in Figure 4-29.

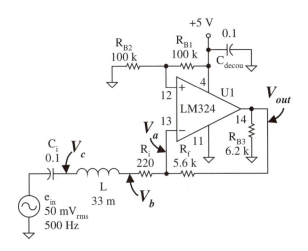

Figure 4-29 Series tuned amplifier

2. Measure each carefully and record their values. Be sure to include the *resistance* of the inductor, as well as its inductance.

3. Using these *measured* values, calculate each of the following quantities (be sure to include the inductor's *resistance*) :

    a. $\overline{Z_{Ci, L, Ri}}$

    b. $\overline{V_a}$ (virtual ground)

    c. $\overline{I_{Ci, L, Ri}}$

    d. $\overline{V_b}$

    e. $\overline{V_c}$ (voltage across R_i *and* L)

    f. $\overline{V_{out}}$ (Watch the phase. This is an *inverting* amplifier.)

C. Low-frequency Measurements
   1. Finish building the circuit in Figure 4-29.

   2. With the digital multimeter, measure the magnitude of the voltage between V_{out} and common. Verify that the measured magnitude corresponds to that calculated in step B3f, above. Do *not* continue until the amplifier is working correctly.

   3. With the oscilloscope, measure the phase of the voltage at the output. Verify that it is correct when compared to the calculations in step B3f, above. Do *not* continue until the amplifier is working correctly.

   4. Use the multimeter to measure the magnitude and the oscilloscope to measure the phase of V_b and V_c.

   5. Construct a table with rows for each of the three voltages, and a column for the calculated magnitude, calculated phase, measured magnitude, measured phase, % magnitude error, and phase error (in degrees).

   6. Add the calculated and measured data to your table.

   7. Calculate the % error of each magnitude measurement and the difference in phase. Explain any magnitude errors over 5% and any phase differences greater than ±4°.

D. Midrange Frequency Calculations and Measurements
   1. Alter the frequency to 2.5 kHz.

   2. Repeat the calculations of section B.

   3. Repeat the measurements and data tabulation of section C.

E. High-frequency Calculations and Measurements
   1. Alter the frequency to 10 kHz.

   2. Repeat the calculations of Section B.

   3. Repeat the measurements and data tabulation of Section C.

5

Parallel Circuits

Introduction

From the previous chapters you learned to analyze circuits in which the elements were connected in series. However, in the *vast* majority of ac circuits, the elements are connected side-by-side, in parallel.

AC parallel circuits obey the same circuit laws as do dc parallel circuits. You just have to use phasors to keep track of the effects on both magnitude and phase as you apply Kirchhoff's current and voltage laws, and Ohm's law. Each branch in a parallel circuit offers another route through which current may flow, lowering the circuit's overall opposition. The more parallel elements there are, the *lower* the impedance. Just as conductance is used in dc circuits to determine the overall parallel resistance, susceptance and admittance allow you to calculate the overall ac opposition in a parallel circuit. Phasor diagrams help you visualize how admittances and currents relate.

Every time you plug in a light or a refrigerator in your home, you are connecting those appliances in parallel. Adjusting parallel industrial power distribution circuits to provide the work needed at the minimum line current is critical. Parallel impedances are also combined with transistors and op amps to customize the amplifiers' frequency responses.

Objectives

Upon completion of this chapter, you will be able to do the following:

- Apply Kirchhoff's current and voltage laws and Ohm's law to parallel ac circuits.

- Define susceptance and admittance, and use them to calculate the total impedance of a parallel ac circuit.

- Draw the phasor diagrams for the admittances and the currents in a parallel circuit.

- Apply parallel circuit analysis techniques to residential and industrial power networks, and to frequency limited amplifiers.

5.1 Review of Fundamental Concepts

A parallel circuit contains a mix of elements all connected side-by-side. For simple ac circuits, these elements are usually voltage or current sources, resistors, capacitors, and inductors. Figure 5-1 shows a generalized ac parallel circuit.

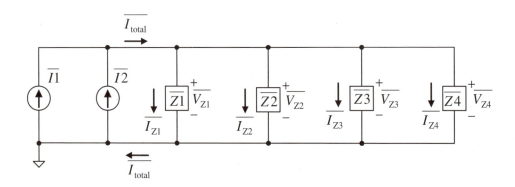

Figure 5-1 General parallel circuit

In a parallel circuit all of the elements are connected between the same two nodes. So, the voltage across each is the same.

Kirchhoff's voltage law

$$\overline{E_{I1}} = \overline{E_{I2}} = \overline{V_{Z1}} = \overline{V_{Z2}} = \overline{V_{Z3}} = \overline{V_{Z4}} = ...$$

The current that flows into the node must equal the current that flows out of the node. It does not build up or leak out.

$$\sum \overline{I_{in}} = \sum \overline{I_{out}}$$

For the circuit in Figure 5-1, this becomes

Kirchhoff's current law

$$\overline{I1} + \overline{I2} = \overline{I_{total}} = \overline{I_{Z1}} + \overline{I_{Z2}} + \overline{I_{Z3}} + \overline{I_{Z4}}$$

This relationship can be used to determine the total opposition of the parallel combination of these impedances.

$$\overline{I_{total}} = \overline{I_{Z1}} + \overline{I_{Z2}} + \overline{I_{Z3}} + \overline{I_{Z4}}$$

According to Ohm's law for ac circuits

$$\overline{I} = \frac{\overline{V}}{\overline{Z}}$$

Apply this to the current summation equation.

$$\frac{\overline{V_{total}}}{\overline{Z_{total}}} = \frac{\overline{V_{Z1}}}{\overline{Z1}} + \frac{\overline{V_{Z2}}}{\overline{Z2}} + \frac{\overline{V_{Z3}}}{\overline{Z3}} + \frac{\overline{V_{Z4}}}{\overline{Z4}}$$

In a parallel circuit all of the voltages are the same.

$$\frac{\overline{V}}{\overline{Z_{total}}} = \frac{\overline{V}}{\overline{Z1}} + \frac{\overline{V}}{\overline{Z2}} + \frac{\overline{V}}{\overline{Z3}} + \frac{\overline{V}}{\overline{Z4}}$$

Divide both sides of the equation by the voltage.

$$\frac{1}{\overline{Z_{total}}} = \frac{1}{\overline{Z1}} + \frac{1}{\overline{Z2}} + \frac{1}{\overline{Z3}} + \frac{1}{\overline{Z4}}$$

$\overline{Z_{total}}$ is the total impedance of the parallel combination of impedances. The reciprocal of the total impedance equals the sum of the reciprocals of each of the individual impedances. Taking the reciprocal of both sides gives an equation for the total impedance.

$$\overline{Z_{total}} = \frac{1}{\dfrac{1}{\overline{Z1}} + \dfrac{1}{\overline{Z2}} + \dfrac{1}{\overline{Z3}} + \dfrac{1}{\overline{Z4}}}$$

Total impedance

This appears a little awkward. A calculator with complex notation provides a reciprocate symbol and allows you to nest parentheses. The result is a more compact notation.

$$\overline{Z_{total}} = \left(\overline{Z1}^{-1} + \overline{Z2}^{-1} + \overline{Z3}^{-1} + \overline{Z4}^{-1} \right)^{-1}$$

Total impedance

The reciprocal of the impedance is called the **admittance**.

Admittance

$$\overline{Y} = \frac{1}{\overline{Z}}$$

The total opposition of a parallel circuit, then, can be rewritten in terms of admittances.

Total admittance

$$\overline{Y_{total}} = \overline{Y1} + \overline{Y2} + \overline{Y3} + \overline{Y4}$$

The admittance of a resistor is

$$\overline{Y}_R = \frac{1}{\overline{Z}_R} = \frac{1\angle 0°}{R\angle 0°}$$

$$= \frac{1}{R}\angle 0°$$

$$\overline{Y}_R = \frac{1}{R}\angle 0° = G\angle 0°$$

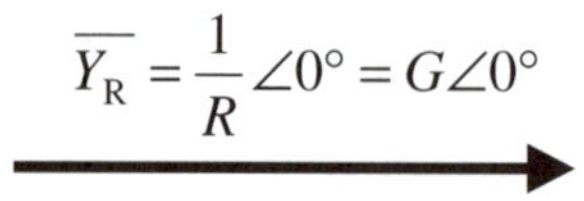

(a) Resistor's admittance

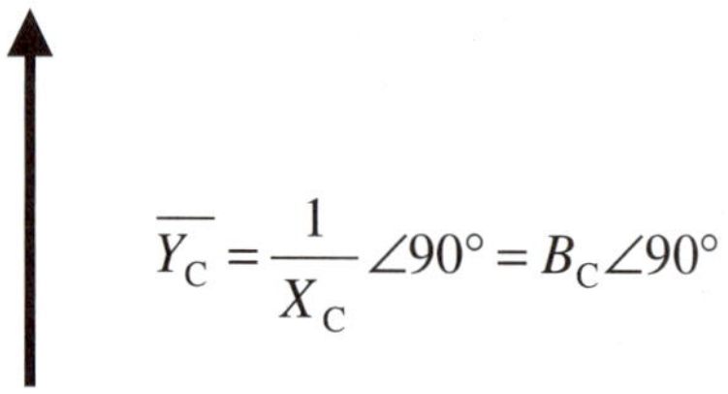

The reciprocal of resistance is **conductance, G**. Its units are siemens, S. Notice this is a capital S. The lower case s is reserved for the time measurement of seconds.

$$\overline{Y}_R = G\angle 0°$$

The admittance of a capacitor is

$$\overline{Y}_C = \frac{1}{\overline{Z}_C} = \frac{1\angle 0°}{X_C \angle -90°}$$

$$\overline{Y}_C = \frac{1}{X_C}\angle 90° = B_C \angle 90°$$

$$= \frac{1}{X_C}\angle 90°$$

(b) Capacitor's admittance

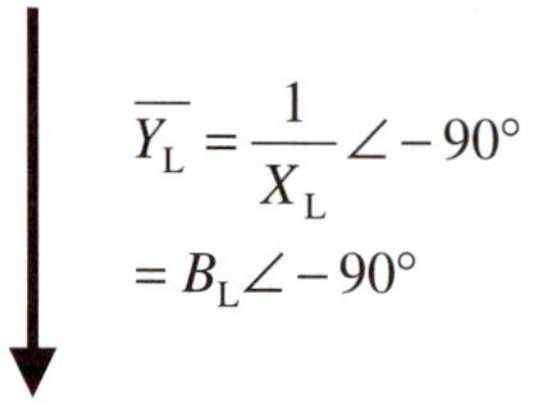

The reciprocal of reactance is susceptance, **B**. Its units are siemens.

$$\overline{Y}_C = B_C \angle 90°$$

The admittance of an inductor is

$$\overline{Y}_L = \frac{1}{\overline{Z}_L} = \frac{1\angle 0°}{X_L \angle 90°}$$

$$\overline{Y}_L = \frac{1}{X_L}\angle -90°$$
$$= B_L \angle -90°$$

$$= \frac{1}{X_L}\angle -90°$$

$$\overline{Y}_L = B_L \angle -90°$$

(c) Inductor's admittance

Figure 5-2
Admitance phasors

Just as you drew an impedance phasor diagram in Chapter 4, you can draw an admittance phasor diagram for parallel circuits. Remember that the magnitudes of the admittances are the reciprocals of the magnitude of the impedances, and that the angles are reversed. This is illustrated in Figure 5-2

5.2 *RC* and *RL* Circuits

Any time resistance, capacitance, and inductance are connected side-by-side a parallel circuit is formed. All of the techniques you have seen so far apply. Be careful, however, not to confuse the relationships within a parallel circuit to those within the series circuits from the previous chapter.

The circuit in Figure 5-3 is a parallel *RC* circuit. The first step is calculating all of the reactances, susceptances, admittances and impedances.

The reactance of the capacitor is

$$X_C = \frac{1}{2\pi f C}$$

$$X_C = \frac{1}{2\pi \times 1\,\text{kHz} \times 33\,\text{nF}} = 4.82\,\text{k}\Omega$$

The susceptance of the capacitor is

$$B_C = \frac{1}{X_C}$$

$$B_C = \frac{1}{4.82\,\text{k}\Omega} = 207\,\mu\text{S}$$

The impedance of the capacitor is

$$\overline{Z_C} = (4.82\,\text{k}\Omega \angle -90°)$$

The admittance of the capacitor is

$$\overline{Y_C} = (207\,\mu\text{S} \angle 90°) = (0 + j207\,\mu\text{S})$$

The conductance of the resistor is

$$G = \frac{1}{R}$$

$$G = \frac{1}{4.7\,\text{k}\Omega} = 213\,\mu\text{S}$$

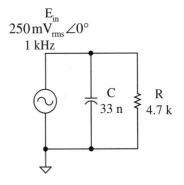

Figure 5-3
RC parallel circuit

The impedance of the resistor is

$$\overline{Z_R} = \left(4.7\,\text{k}\Omega\angle0°\right)$$

The admittance of the resistor is

$$\overline{Y_R} = \left(213\,\mu\text{S}\angle0°\right) = \left(213\,\mu\text{S} + \text{j}0\right)$$

The total admittance is the sum of the individual admittances.

$$\overline{Y_{\text{total}}} = \overline{Y_R} + \overline{Y_C}$$
$$= \left(213\,\mu\text{S} + \text{j}0\right) + \left(0 + \text{j}207\,\mu\text{S}\right)$$
$$\overline{Y_{\text{total}}} = \left(213\,\mu\text{S} + \text{j}207\,\mu\text{S}\right)$$
$$= \left(297\,\mu\text{S}\angle44.2°\right)$$

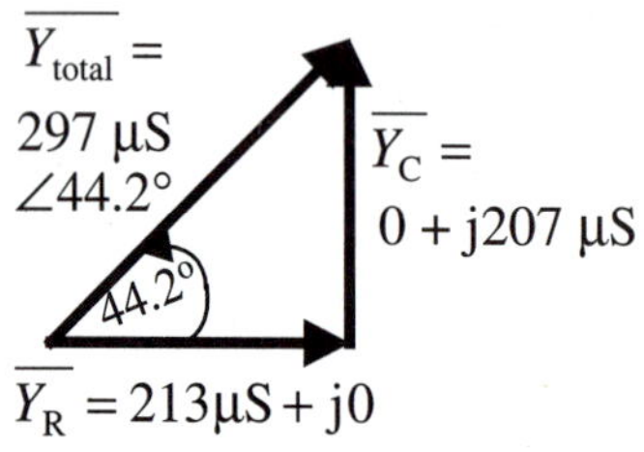

Just as you drew an impedance phasor for a series circuit, the admittance phasor diagram illustrates how the individual admittances add to form the total admittance. Look at Figure 5-4.

The total *impedance* is the reciprocal of the total admittance.

$$\overline{Z_{\text{total}}} = \frac{1}{\overline{Y_{\text{total}}}}$$

$$\overline{Z_{\text{total}}} = \frac{1}{\left(297\,\mu\text{S}\angle44.2°\right)} = 3.37\,\text{k}\Omega\angle-44.2°$$

Taking the reciprocal of each of the individual impedances (to get their admittances), adding them up (to get the total admittance) and then taking the reciprocal of that sum (to get the total impedance) may all be combined into a single step.

$$\overline{Z_{\text{total}}} = \left[\left(4.7\,\text{k}\Omega\angle0°\right)^{-1} + \left(4.82\,\text{k}\Omega\angle-90°\right)^{-1}\right]^{-1} = \left(3.37\,\text{k}\Omega\angle-44.2°\right)$$

Current can be calculated using Ohm's law. For the current through the resistor,

$$\overline{I_R} = \frac{\overline{E_R}}{\overline{Z_R}}$$

$$\overline{I_R} = \frac{\left(250\,\text{mV}_{\text{rms}}\angle0°\right)}{\left(4.7\,\text{k}\Omega\angle0°\right)} = \left(53.2\,\mu\text{A}_{\text{rms}}\angle0°\right)$$

$\overline{Y_{\text{total}}} =$
297 µS
$\angle44.2°$

$\overline{Y_C} =$
$0 + \text{j}207$ µS

$\overline{Y_R} = 213\,\mu\text{S} + \text{j}0$

1 cm = 100 µS

Figure 5-4 Admittance diagram for Figure 5-3

For the current through the capacitor,

$$\overline{I_C} = \frac{\overline{E_C}}{\overline{Z_C}}$$

$$\overline{I_C} = \frac{(250\,\text{mV}_{\text{rms}}\angle 0°)}{(4.82\,\text{k}\Omega\angle -90°)} = (51.9\,\mu\text{A}_{\text{rms}}\angle 90°)$$

Total current is

$$\overline{I_{\text{total}}} = \frac{\overline{E_{\text{total}}}}{\overline{Z_{\text{total}}}}$$

$$\overline{I_{\text{total}}} = \frac{(250\,\text{mV}_{\text{rms}}\angle 0°)}{(3.37\,\text{k}\Omega\angle -44.2°)} = (74.2\,\mu\text{A}_{\text{rms}}\angle 44.2°)$$

According to Kirchhoff's current law,

$$\overline{I_{\text{total}}} = \overline{I_R} + \overline{I_C}$$

$$(74.2\,\mu\text{A}_{\text{rms}}\angle 44.2°)\underset{=}{?}$$

$$(53.2\,\mu\text{A}_{\text{rms}}\angle 0°) + (51.9\,\mu\text{A}_{\text{rms}}\angle 90°)$$

$$(74.2\,\mu\text{A}_{\text{rms}}\angle 44.2°) \cong (74.3\,\mu\text{A}_{\text{rms}}\angle 44.3°)$$

Just as you drew a voltage phasor diagram for a series circuit, the current phasor diagram illustrates how the individual currents add to form the total current. Properly done, the triangle closes. Look at Figure 5-5.

In the next two sections you will see these techniques applied to parallel ac circuits containing a resistor and either an inductor or a capacitor.

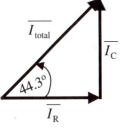

Figure 5-5 Phasor diagram for the currents in Figure 5-3

Residential Circuit

Loads in your home are each connected to the commercial power in parallel. Every time you plug in an appliance, it is connected across the line, in parallel with everything else. Lights and heaters are resistive. Given their power rating, you can determine their steady-state resistance.

$$P = \frac{(V_{\text{rms}})^2}{R}$$

$$R = \frac{\left(V_{rms}\right)^2}{P}$$

Fans, refrigerators, and air conditioners all contain motors. A motor's impedance can be determined from its voltage, current, and power factor ratings. This was presented in detail in Chapter 3.

$$\overline{Z_{motor}} = \frac{V_{rating}\angle 0°}{I_{rating}\angle -\cos^{-1}(power\ factor)}$$

Example 5-1

how big can you dream?™

a. Determine the components for a residential circuit with 100 W of 120 V_{rms} lighting and a small refrigerator motor rated at 120 V_{rms}, 0.8 A_{rms}, with a 0.45 power factor.

b. Draw the schematic.

c. Perform the following:
   Calculate the total admittance.
   Draw the phasor admittance diagram.
   Calculate the current through each element.
   Calculate the total current drawn from the line.
   Draw the current phasor diagram.

d. Verify these currents with a simulation.

Solution

a. The lighting is purely resistive. Once warmed up, it has a resistance of

$$R = \frac{\left(V_{rms}\right)^2}{P}$$

$$R_{lighting} = \frac{\left(120\ V_{rms}\right)^2}{100\ W} = 144\ \Omega$$

Its impedance is

$$\overline{Z_{lighting}} = \left(144\ \Omega\angle 0°\right)$$

The motor in the refrigerator has an impedance of

$$\overline{Z_{motor}} = \frac{V_{rating}\angle 0°}{I_{rating}\angle -\cos^{-1}(power\ factor)}$$

$$\overline{Z_{motor}} = \frac{(120\,V_{rms}\angle 0°)}{0.8\,A_{rms}\angle -\cos^{-1}(0.45)} = (150\,\Omega\angle 63.3°)$$

b. The schematic is shown in Figure 5-6.

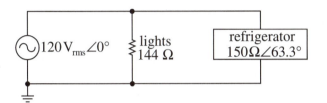

Figure 5-6 Schematic for Example 5-1

c. The admittance of the resistor is

$$\overline{Y_{lighting}} = \frac{1}{(144\,\Omega\angle 0°)} = (6.94\,mS\angle 0°)$$

The admittance of the motor is

$$\overline{Y_{motor}} = \frac{1}{(150\,\Omega\angle 63.3°)} = (6.67\,mS\angle -63.3°)$$

The total admittance is

$$\overline{Y_{total}} = (6.94\,mS\angle 0°) + (6.67\,mS\angle -63.3°)$$
$$= (11.6\,mS\angle -30.9°)$$

Figure 5-7 is the admittance phasor diagram.
This gives a total impedance of

$$\overline{Z_{total}} = \frac{1}{\overline{Y_{total}}}$$

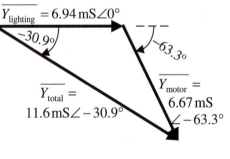

Figure 5-7 Phasor diagram for the admittance in Example 5-1

$$\overline{Z}_{total} = \frac{1}{(11.6\,\text{mS}\angle-30.9°)} = (86.2\,\Omega\angle30.9°)$$

The current through the lighting is

$$\overline{I}_{lighting} = \frac{\overline{E}_{lighting}}{\overline{Z}_{lighting}}$$

$$\overline{I}_{lighting} = \frac{(120\,\text{V}_{rms}\angle0°)}{(144\,\Omega\angle0°)} = (833\,\text{mA}_{rms}\angle0°)$$

The current through the motor is

$$\overline{I}_{motor} = \frac{\overline{E}_{motor}}{\overline{Z}_{motor}}$$

$$\overline{I}_{motor} = \frac{(120\,\text{V}_{rms}\angle0°)}{(150\,\Omega\angle63.3°)} = (800\,\text{mA}_{rms}\angle-63.3°)$$

The total current is

$$\overline{I}_{total} = \frac{\overline{E}_{total}}{\overline{Z}_{total}}$$

$$\overline{I}_{total} = \frac{(120\,\text{V}_{rms}\angle0°)}{(86.2\,\Omega\angle30.9°)} = (1.39\,\text{A}_{rms}\angle-30.9°)$$

The current's phasor diagram is shown in Figure 5-8. Its shape is *identical* to the admittance phasor diagram.

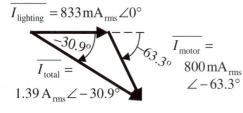

$\overline{I}_{lighting} = 833\,\text{mA}_{rms}\angle0°$

$-30.9°$

$63.3°$ $\overline{I}_{motor} = 800\,\text{mA}_{rms}\angle-63.3°$

$\overline{I}_{total} = 1.39\,\text{A}_{rms}\angle-30.9°$

1 inch = 1 A$_{rms}$

Figure 5-8 Phasor diagram for the currents in Example 5-1

d. To simulate this circuit, you must first determine the *resistance* and *inductance* of the motor. It impedance is

$$\overline{Z}_{motor} = (150\,\Omega\angle63.3°)$$

Convert this polar notation to rectangular.

$$\overline{Z}_{motor} = (67.4\,\Omega + j134.0\,\Omega)$$

This is a *series* combination of a 67.4 Ω resistor with an inductor whose inductive reactance is 134.0 Ω. From the reactance, you can find the inductance.

$$X_{\mathrm{L}} = 2\pi f L$$

$$L_{\mathrm{motor}} = \frac{X_{\mathrm{L}}}{2\pi f}$$

$$L_{\mathrm{motor}} = \frac{134.0\,\Omega}{2\pi \times 60\,\mathrm{Hz}} = 355.4\,\mathrm{mH}$$

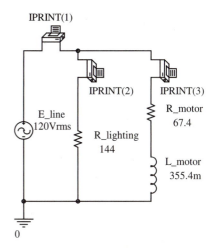

The OrCAD schematic is shown in Figure 5-9. In setting it up, there are several points to remember.

- Select the source as **VAC**.

- Change the name of the ground symbol to **0**.

- To print currents you must insert **IPRINT** symbols in series with each branch.

- Set up each printer by bringing up its dialog box. Then enter **true** in the cells for **AC**, **MAG**, and **PHASE**.

- When you create or edit the **Simulation Profile**, for the **Analysis type**
  - ○ select **AC Sweep**
    - ▪ **AC Sweep Type**
    - ▪ **Start** 60Hz   **End** 60Hz   **points** 1

Figure 5-9
Simulation schematic
for Example 5-1

When you run the analysis, the results appear at the bottom of the **output file.** An edited version is shown in Figure 5-10. The results match the manual calculations.

```
AC ANALYSIS    TEMPERATURE =    27.000 DEG C
*******************************************

 FREQ         IM(V_PRINT1)IP(V_PRINT1)
  6.000E+01    1.391E+00   -3.093E+01

 FREQ         IM(V_PRINT2)IP(V_PRINT2)
  6.000E+01    8.333E-01    0.000E+00

 FREQ         IM(V_PRINT3)IP(V_PRINT3)
  6.000E+01    8.001E-01   -6.330E+01
```

Figure 5-10 Simulation output file for Example 5-1

Practice: Calculate the current from the line if you turn on another 60 W of lighting and a fan with a rating of 1.2 A_{rms} with a 0.6 power factor.

Answer: $\overline{Y_{total}} = (24.5\,\text{mS}\angle -34.8°)$ $\overline{I_{total}} = (2.94\,\text{A}_{rms}\angle -34.8°)$

Amplifier Frequency Limiting

The circuits that you build must operate in a very noisy environment. Radio, television, cell phone, and pager signals wash across everything. Within the circuit itself, logic and microprocessor ICs, power switches, and power supplies generate interference. These high frequencies induce small, high-frequency currents that flow into the op amp's input impedance, producing a voltage. This voltage is amplified right along with the desired signal, often overwhelming it. The result is that the output of a sensitive amplifier may contain more noise than signal, and be completely useless.

Fortunately, the noise is usually at a much higher frequency than the signal that you want to amplify. So the solution is a simple parallel *RC* circuit. Look at Figure 5-11.

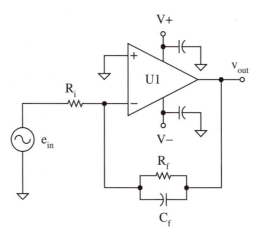

Figure 5-11 Op amp amplifier with frequency limiting

This is a simple modification of the inverting amplifier. For the dc version,

$$V_{out} = -\frac{R_f}{R_i} E_{in}$$

However, R_f and C_f combine to form a complex impedance. This is taken into account by using phasors.

$$\overline{V_{out}} = \frac{\overline{Z_f}}{\overline{Z_i}} \overline{E_{in}} \angle 180°$$

At low frequencies the capacitor's reactance is very large, much greater then R_f's resistance. Since these two are in parallel, the smaller opposition is dominant. The capacitor has little effect at low frequencies. The amplifier operates just as it did at dc, amplifying the signal, with the gain determined by R_f / R_i.

However, at the high frequency of the interference, the capacitive reactance becomes quite small, much smaller than R_f. The feedback impedance falls. The output, at this frequency, also drops drastically. Properly sized, the capacitor in parallel with the feedback resistor enables the amplifier to amplify the desired signal and attenuate the radio frequency noise.

Example 5-2

The amplifier in Figure 5-11 has the following values:

$$\overline{E_{in}} = (250 \, mV_{rms} \angle 0°) \quad \text{At high frequencies, this may be noise.}$$

$$R_i = 10 \, k\Omega \quad R_f = 220 \, k\Omega \quad C_f = 150 \, pF$$

a. Calculate $\overline{V_{out}}$ at 500 Hz, 5 kHz, and 50 kHz.

b. Verify your calculations with a simulation.

Solution

a. At 500 Hz, the reactance of the capacitor is

$$X_{Cf} = \frac{1}{2\pi \times 500 \, Hz \times 150 \, pF} = 2.12 \, M\Omega$$

$$\overline{Z_{Cf}} = (2.12 \, M\Omega \angle -90°)$$

The capacitor's admittance is

multiSIM

MATLAB

```
E=250e-3;
Cf=150e-12;
Ri=10e3;
Rf=220e3;
f=[500; 5e3; 50e3];

X_C=1./(2*pi*f*Cf);

Z_C=0-j*X_C;
Z_R=Rf;
Z_f=1./((1./Z_C)+(1./Z_R));

I=E./Ri;
Uout=-I*Z_f;
ans=[f abs(Uout) angle(Uout)*57.3]
■

ans =

        500      5.4707     174.09
       5000      3.8183     133.98
      50000     0.52807     95.517
```

$$\overline{Y_{Cf}} = \frac{1}{\left(2.12\,M\Omega\angle-90°\right)} = \left(471\,nS\angle 90°\right)$$

The feedback resistor's admittance is

$$\overline{Y_{Rf}} = \frac{1}{\left(220\,k\Omega\angle 0°\right)} = \left(4.55\,\mu S\angle 0°\right)$$

Combine these to determine the total feedback admittance

$$\overline{Y_f} = \left(471\,nS\angle 90°\right) + \left(4.55\,\mu S\angle 0°\right)$$
$$= \left(4.57\,\mu S\angle 5.9°\right)$$

This gives a total feedback impedance of

$$\overline{Z_f} = \frac{1}{\left(4.57\,\mu S\angle 5.9°\right)} = \left(219\,k\Omega\angle-5.9°\right)$$

The output voltage at 500 Hz is

$$\overline{V_{out}} = \frac{\overline{Z_f}}{\overline{Z_i}}\,\overline{E_{in}}\angle 180°$$

$$= \frac{\left(219\,k\Omega\angle-5.9°\right)}{\left(10\,k\Omega\angle 0°\right)}\left(250\,mV_{rms}\angle 180°\right)$$

$$\overline{V_{out\,500Hz}} = \left(5.48\,V_{rms}\angle 174°\right)$$

At 5 kHz, the admittance of the capacitor is

$$\overline{Y_{Cf}} = \frac{1}{\left(212\,k\Omega\angle-90°\right)} = \left(4.72\,\mu S\angle 90°\right)$$

$$\overline{Y_f} = \left(4.72\,\mu S\angle 90°\right) + \left(4.55\,\mu S\angle 0°\right)$$
$$= \left(6.56\,\mu S\angle 46.1°\right)$$

$$\overline{Z_f} = \frac{1}{\left(6.56\,\mu S\angle 46.1°\right)} = \left(153\,k\Omega\angle-46.1°\right)$$

$$\overline{V_{out\,5kHz}} = \frac{\left(153\,k\Omega\angle-46.1°\right)}{\left(10\,k\Omega\angle 0°\right)}\left(250\,mV_{rms}\angle 180°\right)$$

$$\overline{V_{out\,5\,kHz}} = \left(3.83\,V_{rms}\angle 133.9°\right)$$

At 50 kHz, the impedance of the capacitor is

$$\overline{Y_{Cf}} = (47.1\mu S \angle 90°)$$

$$\overline{Z_f} = \frac{1}{(47.4\mu S \angle 84.5°)} = (21.1k\Omega \angle -84.5°)$$

$$\overline{V_{out\ 50\,kHz}} = (0.527\,V_{rms} \angle 95.5°)$$

The high frequency noise signal is amplified *much* less than the lower frequency audio information.

b. The simulation results are shown in Figure 5-12.

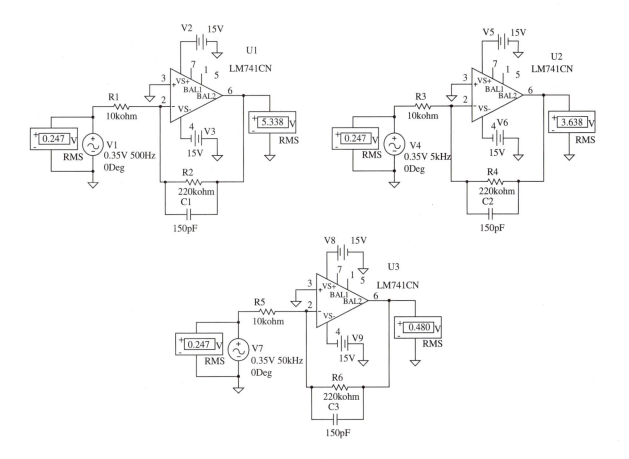

Figure 5-12 Simulation results for Example 5-2

Practice: Calculate the output voltage at 250 Hz and at 100 kHz.

Answer: $\overline{V_{out\ 250\,Hz}} = (5.49\,V_{rms}\angle177°), \overline{V_{out\ 100\,kHz}} = (0.265\,V_{rms}\angle92.8°)$

5.3 *RCL* Circuits

Resistors convert current to voltage, or voltage to current. They dissipate power by producing heat. Incandescent lighting is practically pure resistance, converting power to light and heat. The part of a motor that converts power to shaft motion is also represented as resistance.

Inductors store energy in an electromagnetic field. They oppose a change in current. When a sinusoidal wave is applied, the current through them *lags* the voltage across them by −90°. Relays, solenoids, electromagnets, and motors all have significant inductance caused by their windings.

Capacitors are the complement of inductors. They store energy in an electrostatic field. They oppose a change in voltage. When a sinusoidal wave is applied, the current through them *leads* the voltage across them by +90°.

The current through an inductor *lags* the voltage across it by −90°. The current through a capacitor *leads* the voltage across it by +90°. When connected in parallel, the inductor discharges (sourcing current) at exactly the time that the capacitor charges (sinking current). The capacitor discharges (sourcing current) at exactly the time the inductor charges (sinking current). If the two components are perfectly matched, once the process has begun, significant current circulates between the inductor and the capacitor, back-and-forth, without any current coming from the source. Each appears to be a small, local generator, supplying the current that the other needs.

This complementary nature of parallel inductors and capacitors is used in manufacturing plants running many motors to drastically lower the current required from the utility company. The synchronization of matched charging and discharging occurs at only one frequency. So, an inductor and capacitor connected in parallel may be combined with an amplifier to drastically enhance a signal at one frequency while severely attenuating signals at other frequencies. This is the way radio tuners select a single channel from the broad band of stations received by the antenna.

Power Factor Correction

In Chapter 3, Impedance, power factor was introduced to describe the relative inductive nature of a load. Power factor may be interpreted to indicate the angle between the phasor voltage applied and the phasor current flowing through the load.

$$power\ factor = \cos\theta$$

In a purely resistive load, the voltage and current are in phase, $\theta = 0°$. The power factor is one. All of the current from the utility company is used to produce heat, light, sound, or motion as it is converted to power. This makes the watt-hour meter spin, and electrical power consumption is properly registered.

However, for a purely inductive load, the voltage leads the current by 90°, $\theta = 90°$. The power factor is zero. For half of the cycle, the current from the utility company causes the electromagnetic field to build, *storing* energy. During the other half of the cycle the electromagnetic field collapses. The inductor *generates* current and sends it back to the utility company. No power is dissipated. The watt-hour meter, trying to spin first one direction then the other, ends up not moving at all. No power consumption is registered. *However*, the utility had to generate the current to charge the inductance, and ship it through their lines to the plant. *And* they had to absorb the current returned when the inductor's field collapses, all at no charge to the industrial customer. Left uncorrected, a low power factor requires the generator to produce much more current than is ever converted into useful work. It is just sent back and forth between the generator and the load's inductance.

Example 5-3

The circuit in Figure 5-13 represents a small industrial plant, with about 1000 hp of motors. Calculate the current delivered by the generator.

Solution

The admittance of the resistor is

$$\overline{Y}_R = \frac{1}{(7.5\,\Omega\angle0°)} = (133\,\text{mS}\angle0°)$$

The admittance of the inductance is

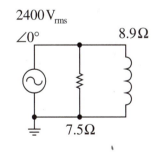

2400 V$_{\text{rms}}$

$\angle0°$ $8.9\,\Omega$

$7.5\,\Omega$

Figure 5-13 Schematic for Example 5-3

$$\overline{Y}_{L} = \frac{1}{(8.9\,\Omega\angle 90°)} = (112\,\text{mS}\angle -90°)$$

The total admittance is

$$\overline{Y}_{\text{total}} = (133\,\text{mS}\angle 0°) + (112\,\text{mS}\angle -90°)$$
$$= (174\,\text{mS}\angle -40.1°)$$

This gives a total impedance of

$$\overline{Z}_{\text{total}} = \frac{1}{(174\,\text{mS}\angle -40.1°)} = (5.75\,\Omega\angle 40.1°)$$

The current from the generator is

$$\overline{I}_{\text{total}} = \frac{(2400\,\text{V}_{\text{rms}}\angle 0°)}{(5.75\,\Omega\angle 40.1°)} = (417\,\text{A}_{\text{rms}}\angle -40.1°)$$

Practice: Calculate the current into the resistor and into the inductor. Verify by Kirchhoff's current law that their sum equals the total current.

Answer: $\overline{I}_{R} = (320\,\text{A}_{\text{rms}}\angle 0°)$, $\overline{I}_{L} = (270\,\text{A}_{\text{rms}}\angle -90°)$
$\overline{I}_{R} + \overline{I}_{L} = (419\,\text{A}_{\text{rms}}\angle -40.2°) \cong (417\,\text{A}_{\text{rms}}\angle -40.1°)$

This example illustrates that there is almost as much current going to charge and discharge the motor's inductance as there is going to its resistance (the part that does the work). All of the inductance's current must be generated, and shipped back and forth to the plant. That current is not converted into shaft rotation. It is just "extra-baggage."

Placing a capacitor in parallel with the inductance solves this problem. The capacitor charges when the inductor's field collapses, generating current. The energy is stored in the capacitor, rather than being shipped back to the generator. The capacitor discharges at precisely the time that the inductance needs current to reestablish its field. Normally this current would have come from the generator. The capacitor acts as a local reservoir, absorbing and releasing charge as the inductance needs it.

Example 5-4

Calculate the current delivered by the generator for the circuit in Figure 5-14.

Solution

The admittance of the capacitor is

$$\overline{Y}_C = \frac{1}{(12\,\Omega\angle -90°)} = (83\,\text{mS}\angle 90°)$$

From Example 5-3 the admittance of R and L are

$$\overline{Y}_R = \frac{1}{(7.5\,\Omega\angle 0°)} = (133\,\text{mS}\angle 0°)$$

$$\overline{Y}_L = \frac{1}{(8.9\,\Omega\angle 90°)} = (112\,\text{mS}\angle -90°)$$

$$\overline{Y}_{\text{total}} = (133\,\text{mS}\angle 0°) + (112\,\text{mS}\angle -90°) + (83\,\text{mS}\angle 90°)$$
$$= (136\,\text{mS}\angle -12.3°)$$

$$\overline{Z}_{\text{total}} = \frac{1}{(136\,\text{mS}\angle -12.3°)} = (7.35\,\Omega\angle 12.3°)$$

$$\overline{I}_{\text{total}} = \frac{(2400\,\text{V}_{\text{rms}}\angle 0°)}{(7.35\,\Omega\angle 12.3°)} = (327\,\text{A}_{\text{rms}}\angle -12.3°)$$

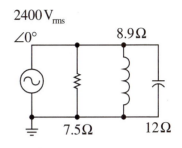

$2400\,\text{V}_{\text{rms}}$
$\angle 0°$
$8.9\,\Omega$
$7.5\,\Omega$ $12\,\Omega$

Figure 5-14 Schematic for Example 5-4

Practice: Calculate the current into the resistor, the inductor, and the capacitor. Verify that their sum equals the total current.

Answer: $\overline{I}_R = (320\,\text{A}_{\text{rms}}\angle 0°)$, $\overline{I}_L = (270\,\text{A}_{\text{rms}}\angle -90°)$,
$\overline{I}_C = (200\,\text{A}_{\text{rms}}\angle 90°)$

$\overline{I}_R + \overline{I}_L + \overline{I}_C = (328\,\text{A}_{\text{rms}}\angle -12.3°) = \overline{I}_{\text{total}}$

The capacitor has provided practically all of the "excess" current needed by the motors' inductance. The magnitudes of their currents are close, and they are 180° out of phase (one charging while the other discharges). The current from the generator is barely more than that needed by the resistor.

Summary

Parallel ac circuits obey the same laws as dc circuits, with phasors.

Kirchhoff's voltage law

$$\overline{E_{I1}} = \overline{E_{I2}} = \overline{V_{Z1}} = \overline{V_{Z2}} = \overline{V_{Z3}} = \overline{V_{Z4}} = ...$$

Kirchhoff's current law

$$\overline{I1} + \overline{I2} = \overline{I_{total}} = \overline{I_{Z1}} + \overline{I_{Z2}} + \overline{I_{Z3}} + \overline{I_{Z4}}$$

Total impedance

$$\overline{Z_{total}} = \frac{1}{\frac{1}{\overline{Z1}} + \frac{1}{\overline{Z2}} + \frac{1}{\overline{Z3}} + \frac{1}{\overline{Z4}}}$$

Admittance is the reciprocal of impedance. So the total admittance is the sum of the individual parallel element admittances.

Admittances

$$\overline{Y} = \frac{1}{\overline{Z}} \qquad \overline{Y_{total}} = \overline{Y1} + \overline{Y2} + \overline{Y3} + \overline{Y4}$$

Susceptance, B, is the reciprocal of reactance.

Susceptance

$$B = \frac{1}{X}$$

In solving parallel ac circuits, the elements' individual impedances are found, the total impedance is calculated, and then the total current or voltage. This allows you to determine the current through each element and to draw a current phasor diagram. This diagram (like the voltage phasor diagram for series circuits) shows the relationship of each of the currents and how they combine to equal the total (applied) current.

A capacitor in parallel with the feedback resistor of an op amp-based amplifier reduces the amplifier's gain at high frequencies. This makes it less susceptible to high-frequency noise.

In a residential setting, most loads are resistive (lights and heaters) or a combination of R and L (motors in fans, refrigerators, pumps). This inductance causes excessive current to be demanded and returned to the generator as it charges and discharges. Placing a capacitor in parallel with the inductive load (*RCL*) allows this charging and discharging current to be stored locally, lowering the current sent back and forth to the generator. This is referred to as power factor correction.

Problems

Review of Fundamental Concepts

5-1 For the parallel circuit in Figure 5-15, and the following values, determine each of the parameters below

$$\overline{I1} = (3\,\text{A}_{\text{rms}}\angle 120°), \quad \overline{V1} = (220\,\text{V}_{\text{rms}}\angle 60°),$$

$$\overline{I}_{Z1} = (0.9\,\text{A}_{\text{rms}}\angle 60°), \quad \overline{I}_{Z2} = (6\,\text{A}_{\text{rms}}\angle 150°),$$

$$\overline{Z3} = (45\,\Omega\angle -30°), \quad \overline{Z4} = (40\,\Omega\angle 60°)$$

   a. Calculate the voltage across each component.

   b. Calculate the current through each component.

   c. Calculate the current from the sources and the total current.

   d. Draw the phasor diagram of the currents.

   e. Calculate the total impedance, using Ohm's law.

   f. Calculate the total admittance.

   g. Calculate the admittance of each component.

   h. Draw the phasor diagram for the admittances.

   i. Verify that the admittances add to the total admittance.

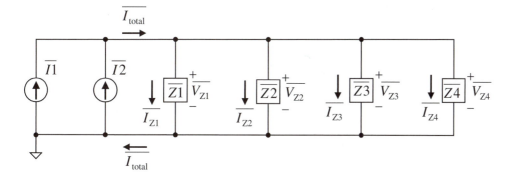

Figure 5-15 Schematic for Problems 5-1 and 5-2

5-2 Repeat Problem 5-1 with the following values.

$$\overline{I1} = (200\mu A_{rms} \angle 90°), \quad \overline{V1} = (3 V_{rms} \angle 45°),$$
$$\overline{I_{Z2}} = (136\mu A_{rms} \angle 45°), \quad \overline{I_{Z4}} = (100\mu A_{rms} \angle -45°),$$
$$\overline{Z1} = (47 k\Omega \angle 0°), \qquad \overline{Z3} = (40 k\Omega \angle 90°)$$

RC and *RL* Circuits

5-3 A parallel circuit consists of a voltage source, $(12.6 V_{rms} \angle 0°)$ at a frequency of 120 Hz, a resistor of 5 Ω, and a 150 μF capacitor.

    a. Draw the schematic.

    b. Calculate the reactance of the capacitor.

    c. Calculate the impedance of the resistor and the capacitor.

    d. Calculate the admittance of the resistor and the capacitor.

    e. Calculate the circuit's total admittance

    f. Calculate the circuit's total impedance.

    g. Draw the admittance diagram

    h. Calculate the current through each element and the source.

    i. Draw the phasor diagram of each current.

5-4 Repeat Problem 5-3 for a voltage source of $(3.5 V_{rms} \angle 125°)$ at a frequency of 100 MHz, a resistor of 51 Ω, and a 52 pF capacitor.

5-5 A 120 V_{rms} residential circuit has 300 W of lighting, a refrigerator rated at 120 V_{rms}, 2 A_{rms} with a 0.6 power factor, and a whole-house ventilation fan rated at 120 V_{rms}, 1.3 A_{rms}, with a power factor of 0.5.

    a. Draw the schematic.

    b. Calculate the impedance and admittance of each element.

    c. Calculate the total admittance.

    d. Draw the admittance diagram.

    e. Calculate the current through each element.

    f. Calculate the total current drawn from the line.

    g. Draw the current phasor diagram.

5-6   Repeat Problem 5-5 if the lighting load drops to 100 W, the refrigerator turns *off*, and the fan switches to low (0.9 A_{rms}).

5-7   The amplifier in Figure 5-16 has the following values:

$$\overline{E_{in}} = (10\,mV_{rms}\angle 0°), \quad R_i = 1\;k\Omega, \quad R_f = 47\;k\Omega, \quad C_f = 330\;pF$$
$$f = 10\;kHz$$

Calculate each of the following:
a.   voltage at the op amp's inverting input

b.   current through R_i

c.   current through the parallel feedback elements

d.   reactance of C_f

e.   admittance of C_f

f.   impedance of R_f in parallel with C_f

g.   voltage dropped across these parallel elements

h.   output voltage

5-8   Repeat Problem 5-7 for a frequency of 6 kHz.

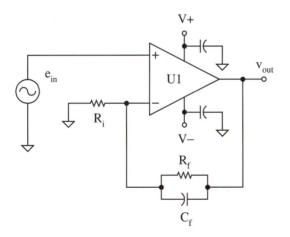

Figure 5-16 Schematic for Problems 5-7 and 5-8

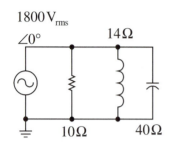

Figure 5-17 Schematic for Problems 5-9 and 5-10

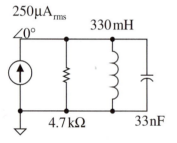

Figure 5-18 Schematic for Problems 5-11 and 5-12

RLC Circuits

5-9   Calculate the current delivered by the generator for Figure 5-17.

5-10   **a.**   Repeat Problem 5-9 with the capacitor's reactance at 15 Ω.

   b.   Compare the results of Problem 5-9 and 5-10 and suggest an optimum value for the capacitor.

5-11   For the circuit in Figure 5-18, at $f = 1.4$ kHz:
   a.   Calculate the voltage across the circuit.

   b.   Calculate the current through each element.

   c.   Draw the phasor diagram of these currents.

5-12   **a.**   Repeat Problem 5-11 at $f = 700$ Hz.

   b.   Repeat Problem 5-11 at $f = 2.8$ kHz.

Parallel Circuits Lab Exercise

A.   Passive Circuit
   1.   Find, measure, and record the components needed to build the circuit in Figure 5-19.

   2.   Build the circuit shown in Figure 5-19.

   3.   Use the oscilloscope and the digital multimeter to set the source to 4 V_{rms} at 15 kHz.

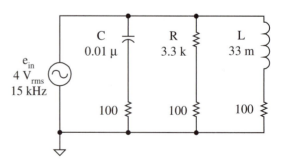

Figure 5-19 Passive parallel *RCL* circuit

4. Measure and record in a spreadsheet the magnitude and the phase shift of the voltage across each of the 100 Ω resistors. Measure the magnitude with the digital multimeter, and the phase, with respect to the source, with the oscilloscope.

5. Using these measured voltages, calculate and record the current (magnitude and phase) through each of the 100 Ω resistors in your spreadsheet. These are the *measured* branch currents.

6. Add these three measured branch currents to determine the *measured* total current. Remember to use phasors. Record this measured current in your spreadsheet.

7. Calculate and record all *theoretical* phasor currents using the actual component values.

8. Calculate and record the *measured* value of the total impedance.

$$\overline{Z_{total\ measured}} = \frac{\left(V_{rms\ source\ measured}\angle 0°\right)}{\overline{I_{total\ measured}}}$$

9. Compare the measured and the theoretical values. There should be less than ±5% magnitude error and ±4° phase error.

B. Frequency Limited Amplifier: Low-frequency Calculations
   1. Locate the components shown in Figure 5-20.

   2. Measure each carefully and record their values.

   3. Using these *measured* values, calculate each of the following quantities (be sure to include the inductor's *resistance*):
      a. $\overline{Z_{Ri}}$ b. $\overline{V_a}$ (virtual ground)   c. $\overline{I_{Ri}}$
      d. $\overline{Z_{Cf,\ Rf}}$ f. $\overline{V_{out}}$ (Watch the phase)

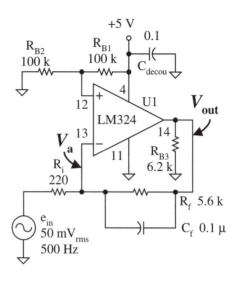

Figure 5-20 Frequency limited amplifier

D. Frequency Limited Amplifier: Low-frequency Measurements

1. Finish building the circuit in Figure 5-20.

2. With the digital multimeter, measure the magnitude of the voltage between V_{out} and common. Verify that the measured magnitude corresponds to that calculated in step C3f, above. Do *not* continue until the amplifier is working correctly.

3. With the oscilloscope, measure the phase of the voltage at the output. Verify that it is correct when compared to the calculations in step C3f, above. Do *not* continue until the amplifier is working correctly.

4. Construct a table with rows for V_{out} and $\overline{I_{in}}$, and a column for the calculated magnitude, calculated phase, measured magnitude, measured phase, % magnitude error, and phase error (in degrees).

5. Add the calculated and measured data to your table.

6. Calculate the % error of each magnitude measurement and the difference in phase. Explain any magnitude errors over 5% and any phase differences greater than ±4°.

E. Frequency Limited Amplifier: Midrange Performance
   1. Alter the frequency to 2.5 kHz.

   2. Repeat the calculations of Section C.

   3. Repeat the measurements and data tabulation of Section D.

F. Frequency Limited Amplifier: High-frequency Performance
   1. Alter the frequency to 10 kHz.

   2. Repeat the calculations of Section C.

   2. Repeat the measurements and data tabulation of Section D.

Filter Applications

The purpose and use of an electrical signal is often strongly influenced by its frequency. In audio, low-frequency tones are sent to the woofer while high pitches must be separated and sent to the tweeter. Each button you push on a telephone creates its unique set of tones. Radio signals, whether from a cellular telephone or a commercial television station, are each assigned to a specific carrier frequency. Measurements from your bathroom scale and from the transducers in the engine compartment of your car often are dc but survive in an electrical environment that is filled with signals of many higher frequencies.

These signals exist in a continuum, each carrying unique information that must be separated from all of the other signals, based on *frequency*. The opposition of inductors and capacitors depends on the *frequency* of the signal. So, they can be combined with resistors and amplifiers to build circuits that pass some signals, while blocking others, depending on the signals' frequencies. These circuits are called **filters**.

Filters are specified, analyzed, and designed using a frequency response plot. Its shape and parameters define how the filter responds to different frequencies. Chapter 6 provides the fundamentals of the plot and the definitions of the parameters.

The values of the resistors, inductors, and capacitors used to configure a filter determine its key parameters. In Chapters 7 and 8 you will learn to quickly evaluate a filter's performance just by looking at its schematic. A full derivation then verifies this with the mathematical relationship of the circuit's components to the filter parameters.

Proper amplifier performance demands that high-frequency noise be rejected. This assures stable operation at low frequencies. The op amp's gain bandwidth product and slew rate are key specifications that determine that IC's frequency response. In Chapter 9 you will learn how to select an op amp for a given application, determine the limitations of an op amp, and customize an op amp's frequency response.

Resonance is the property that allows a system to operate over a very narrow range of frequencies. You have seen this when playing on a swing (that swings back and forth at one rate) or playing a musical instrument (that resonates at the pitch of the note you play). Electrical circuits, both series and parallel, resonate as well. Their performance and a variety of applications that allow you to tune to a specific radio channel and run a motor efficiently are covered in Chapter 10.

6

Filter Terminology

Introduction

Filters are circuits that pass certain signals while rejecting others, depending on the signals' frequencies. These circuits find wide applications in communications, sensor processing (your bathroom scale, furnace thermostat, car's engine), and industrial power distribution.

Each filter amplifies or attenuates a signal's voltage, as set by its components' impedances, like the series and parallel circuits you have already studied. But the *results* of manipulating these voltages, currents, and impedances are specified by a set of terms that are unique to filters.

These parameters all center around a plot of the gain (magnitude and phase) of a circuit on the vertical axis, versus the signal's frequency on the horizontal axis. This is the **frequency response plot**. In the first several sections of this chapter, you will learn to identify filter types and performance from that plot, then to scale its horizontal axis logarithmically and its vertical axis in decibels.

How steeply the filter transitions from passing one frequency to blocking another is defined by the **roll-off rate**, which in turn is established by the filter's **order**. The **pass band gain** is the key vertical axis specification, while **critical frequency** and the **half-power point** scale the horizontal axis. These are defined in the chapter's later sections.

Objectives

Upon completion of this chapter, you will be able to do the following:

- Identify low-pass, high-pass, and band-pass responses.

- Properly draw and logarithmically scale a frequency response plot.

- Convert between ratio and a variety of decibel parameters.

- Define and locate on a frequency response plot:

     Pass band gain, critical frequency, half-power point,

     filter order, roll-off rate

6.1 The Frequency Response Plot and Filter Types

circuit

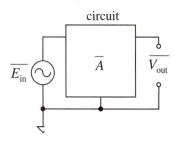

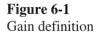

Figure 6-1
Gain definition

The **gain** of a circuit is the ratio of the output parameter to the input parameter. Often these parameters are voltage. Look at Figure 6-1.

$$\overline{A} = \frac{\overline{V}_{out}}{\overline{E}_{in}}$$

There are two important observations. First, all quantities are *phasors*. The gain indicates how much larger (amplification) or smaller (attenuation) the signal becomes as it passes through the circuit. It also shows how much the signal's phase is shifted, lead or lag. Secondly, when the parameters are input and output voltage, the magnitude of the gain is unitless, that is volts/volts.

Filters provide different gain to signals of different frequencies. Some frequency signals are passed with amplification and little phase shift, while other frequency signals may be severely attenuated with significant phase shift. It is common to illustrate this with a frequency response plot, usually with gain magnitude on the vertical axis and frequency on the horizontal. Look at Figure 6-2.

These are the responses of a low-pass filter. The ideal response,

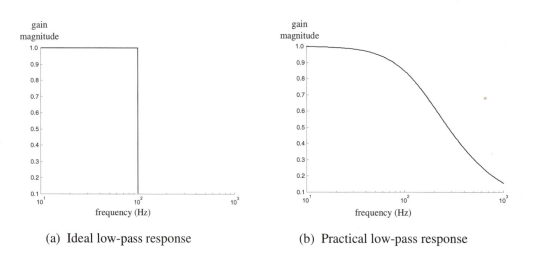

(a) Ideal low-pass response

(b) Practical low-pass response

Figure 6-2 Low-pass filter frequency response

Figure 6-2 (a), indicates that signals at or below about 10^2 Hz are passed with a gain magnitude of 1.0. That is, the output is the same as the input. Above 10^2 Hz, none of the input is passed to the output.

A more practical response is shown in Figure 6-2 (b). As the frequency goes up, the gain falls. Low-frequency signals are passed with little loss. But higher frequency signals are severely attenuated.

Low-pass filters are used to pass the bass signals to woofer loudspeakers, while attenuating the high pitches (to which the woofer cannot respond). They are also used extensively when measuring physical parameters, such as weight, pressure, or temperature. These signals vary at the rate of the physical world that produces them. It takes several seconds for the oil pressure or engine temperature in your car to change noticeably. However, these signals must exist next to the spark plug pulse of hundreds of volts that occurs in a few microseconds. Low-pass filters assure that interference from the high-frequency spark plugs pulses, or radio, or microprocessors are attenuated, while the slow variations from the sensors are passed on to the instrumentation and control cluster.

Figure 6-3 shows the frequency response of a high-pass filter. Low frequencies are attenuated while signals at higher frequencies are passed. This filter passes the high pitches of an audio signal to the tweeter loudspeaker, while blocking the low tones that should go only to the woofer. You may have also seen these filters as the coupler between amplifier

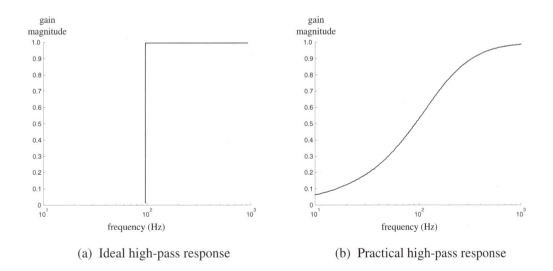

(a) Ideal high-pass response   (b) Practical high-pass response

Figure 6-3 High-pass filter frequency response

stages, or whenever you ac couple the input channel of an oscilloscope. The dc bias voltage (0 Hz) is blocked, while the ac signal passes.

Properly configured, the low-pass and high-pass filters combine to form the band-pass filter shown in Figure 6-4. Signals below or above the selected band are rejected. Only those within the chosen band are passed with little attenuation. The band-pass filter used in a communications receiver is very sharp, selecting one signal at 100 MHz, while rejecting another adjacent station only 0.3 MHz away. However, the midrange audio band-pass filter may pass signals from 200 Hz to 2 kHz.

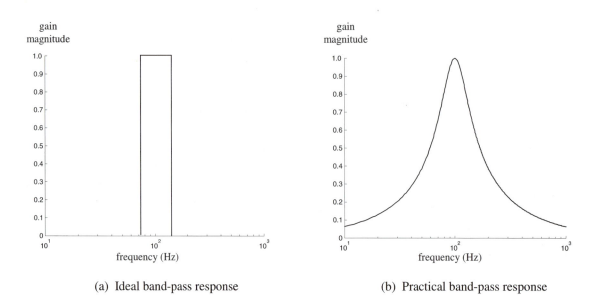

(a) Ideal band-pass response   (b) Practical band-pass response

Figure 6-4 Bandpass filter frequency response

6.2 The Frequency Response Plot's Horizontal Axis

Look carefully at the horizontal axis of Figure 6-4. The scale is *not* linear. It is logarithmic. Equal intervals are assigned to decade increases in frequency. On a linear scale the divisions would have gone up by integers, 1, 2, 3. However, on this logarithmic scale, the divisions are decades, 10^1, 10^2, 10^3; that is 10, 100, 1000.

Scaling the horizontal axis of a frequency response plot logarithmically allows a wide range of signals to be displayed without giving the higher tones a disproportionately large part of the axis.

Logarithmically scaled plots have three key characteristics. Look at Figure 6-5. First, equal distances advance the frequency by a factor of 10. This is called a **decade**. In Figure 6-5, the major divisions go from 10, to 100, to 1000, to 10,000, to 100,000. Secondly, there is no zero. That is because the log 0 is undefined. (There is no exponent of 10 that will result in 0.) If you move a major division further to the left the frequency drops by 0.1, from 10 to 1, then to 0.1, then to 0.01, and so on.

Finally, the minor divisions are *not* linearly placed. They bunch-up as you get closer to the next decade. You will see some graphs that only have every other minor division. Between 10 and 100 there may be only 4 ticks. These are at 20, 40, 60, and 80, even though the spacing may seem unusual.

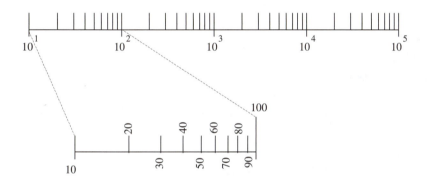

Figure 6-5 Logarithmic horizontal scale

Example 6-1

The low-pass filter frequency response shown in Figure 6-2 was generated from the set of equations : $R = 1\,\mathrm{k}\Omega$ $C = 1\,\mu\mathrm{F}$

$$X_C = \frac{1}{2\pi f C} \qquad |A| = \frac{X_C}{\sqrt{R^2 + X_C^2}}$$

Plot this gain magnitude versus frequency, from 10 Hz to 1 kHz, with the horizontal axis scaled logarithmically. Use a spreadsheet and MATLAB.

	A	B	C
1			= =1/(2*PI()*A6*$B$3)
2	R (ohms)	1.00E+0	
3	C (Farads)	1.00E-06	
4			= =B6/SQRT($B$2^2+B6^2)
5	frequency (Hz)	Xc (ohms)	gain mag
6	10	15915	0.998
7	20	7958	0.992
8	30	5305	0.983
9	40	3979	0.970
10	50	3183	0.954
11	60	2653	0.936
12	70	2274	0.915
13	80	1989	0.893
14	90	1768	0.870
15	100	1592	0.847
16	200	796	0.623
17	300	531	0.469
18	400	398	0.370
19	500	318	0.303
20	600	265	0.256
21	700	227	0.222
22	800	199	0.195
23	900	177	0.174
24	1000	159	0.157

Figure 6-6
Spreadsheet for
Example 6-1

Solution

The spreadsheet is shown in Figure 6-6. The frequency cells increase in steps of 10 to 100, then in steps of 100 to 1000.

The values for the resistor and for the capacitor are placed in their own cells at the top of the sheet. This way, if you want to see the effect of changing the resistor or capacitor, you change one cell, and the sheet and graph are automatically updated.

The capacitive reactance is calculated in the second column. The equation is shown in the insert. Once this cell is correctly calculated, just pull it down to fill the rest of the column.

The last column calculates the gain magnitude. The formula is also shown in the insert. The resulting calculation is then pulled down to fill the third column. The series of steps needed to create the logarithmic plot are shown in Figure 6-7.

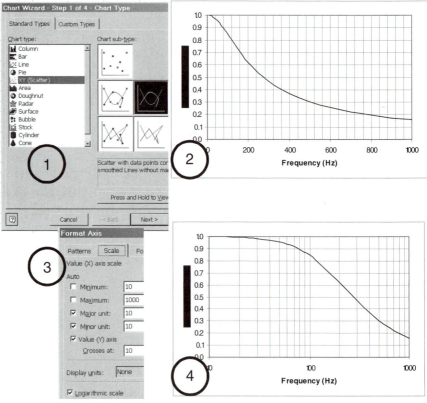

Figure 6-7 A spread-sheet-based logarithmically scaled plot

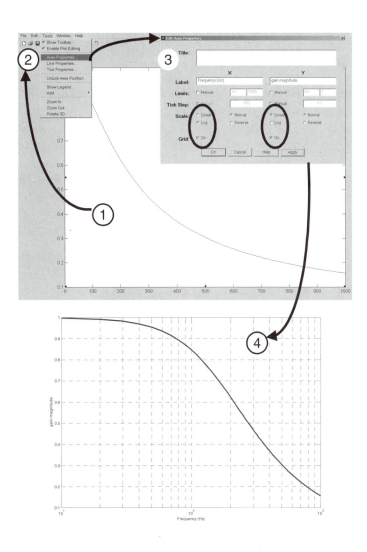

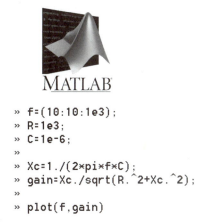

```
» f=(10:10:1e3);
» R=1e3;
» C=1e-6;
»
» Xc=1./(2*pi*f*C);
» gain=Xc./sqrt(R.^2+Xc.^2);
»
» plot(f,gain)
```

1. When the plot command is exe-
   cuted, the linear plot is created in a
   separate window.

2. From the top menu bar, select
   <u>Tools</u>.

   Then select <u>Axis Properties</u>.

3. In the Axis Properties dialog box,

   enter appropriate labels
   check the X axis Scale Log button
   check the Grid On box

4. The *X* axis is scaled logarithmically

Figure 6-8 MATLAB-based logarithmically scaled plot

Practice: Plot the frequency response for the following filter with
10 Hz $\leq f \leq$ 1 kHz. Use both a spreadsheet and MATLAB.

$$R = 1 \ k\Omega, \qquad C = 1 \ \mu F, \qquad |A| = \frac{R}{\sqrt{R^2 + X_C^2}}$$

Answers: See Figure 6-3.

6.3 The Frequency Response Plot's Vertical Axis

The frequency response plot's horizontal axis is scaled logarithmically to display a wide range of frequencies, and to accommodate the way we hear. For these same reasons, special units scale the *vertical* axis too.

There has been considerable study into the relationship between the amplitude of the signal sent to a telephone's loudspeaker and the perceived volume. As with many other human senses, it is *logarithmic*, and depends on the *power* sent to the loudspeaker, not its voltage. For that reason, the **bel** (in honor of Alexander Graham Bell) was defined.

$$bel = B = \log \frac{P_{out}}{P_{in}}$$

Notice that the bel is a measure of power *gain*. It is the log of a ratio.

It is common for a circuit to produce a power gain that is between 0.2 and 5. Over this region of operation, the resulting gain in bels is a fraction. Especially before the arrival of pocket calculators, this was inconvenient. So, the decibel became the traditional unit.

$$dB = 10 \log \frac{P_{out}}{P_{in}}$$

Key points you should notice about the decibel include:

- A decade (factor of 10) increase or decrease in the ratio produces a linear increase or decrease in dB.

- A ratio gain of 1 means the input and output power are the same. The dB gain is 0. Nothing is added or subtracted from the gain.

- Attenuation, where the output power is smaller than the input power, is expressed as a *negative* dB gain.

- Increasing the gain by 100 adds 20 dB, while attenuating the gain by $\frac{1}{100}$ subtracts 20 dB (i.e., –20 dB).

- Doubling the output power is expressed as a 3 dB gain, while cutting it in half results in –3 dB. You will see this particular point on the frequency response plot repeatedly.

Voltage provided at the input and delivered to the load is much easier to measure than power. Gain in dB can be expressed in terms of input and output *voltage*.

$$dB = 10\log\frac{P_{out}}{P_{in}} \qquad P_{out} = \frac{(V_{load})^2}{R_{load}} \qquad P_{in} = \frac{(E_{in})^2}{R_{in}}$$

$$dB = 10\log\frac{\dfrac{(V_{load})^2}{R_{load}}}{\dfrac{(E_{in})^2}{R_{in}}}$$

$$dB = 10\log\left(\frac{(V_{load})^2}{R_{load}} \times \frac{R_{in}}{(E_{in})^2}\right)$$

$$dB = 10\log\left(\left(\frac{V_{load}}{E_{in}}\right)^2 \times \frac{R_{in}}{R_{load}}\right)$$

In balanced systems, $R_{in} = R_{load}$. Communications systems impedances are forced to 50 Ω or 75 Ω, while the original telephone and audio systems set all impedances to 600 Ω. In these two cases the resistances cancel, and the ratio becomes considerably simpler. Even for most amplifiers and filters, it is common to ignore the ratio of the resistances.

$$dB = 10\log\left(\left(\frac{V_{load}}{E_{in}}\right)^2\right)$$

$$\log(x^2) = 2\log x$$

$$dB = 20\log\left(\frac{V_{load}}{E_{in}}\right)$$

dB gain in terms of voltages

Example 6-2

Complete the following calculations.

a. Given: $P_{in} = 100$ mW, $P_{out} = 25$ W
   Calculate: gain dB.

b. Given: $V_{out} = 50$ mV$_{rms}$, gain $= -45$ dB

Calculate: E_{in}.

Solution

a.
$$dB = 10\log\left(\frac{P_{out}}{P_{in}}\right)$$

$$dB = 10\log\left(\frac{25\,\text{W}}{100\,\text{mW}}\right) = 23.98\,\text{dB}$$

b. Voltage calculation requires that the equation be rearranged.

$$dB = 20\log\left(\frac{V_{out}}{E_{in}}\right)$$

$$\frac{dB}{20} = \log\left(\frac{V_{out}}{E_{in}}\right)$$

$$10^{\frac{dB}{20}} = \frac{V_{out}}{E_{in}}$$

$$E_{in} = \frac{V_{out}}{10^{\frac{dB}{20}}} = \frac{50\,\text{mV}_{rms}}{10^{\frac{-45\,\text{dB}}{20}}} = 8.89\,\text{V}_{rms}$$

MATLAB

The proper function is

log10

```
» dB=20×log10(50e-3/8.89)
dB =

    -44.999
```

Practice: **a.** $E_{in} = 3\ \text{V}_{rms}$, $V_{out} = 62\ \text{mV}_{rms}$ Calculate the dB gain.
b. $E_{in} = 1.5\ \text{V}_{rms}$, $dB = -28\ \text{dB}$ Calculate V_{out}.

Answers: **a.** $dB = -33.69\ \text{dB}$ **b.** $V_{out} = 59.72\ \text{mV}_{rms}$.

The decibel is a measure of *gain*, the log of the *ratio* of output compared to input. The dB gain tells you the output level only if you know the level of the input. However, it is convenient to express level in dB. To do this, a reference (input) must be specified.

dBV: voltage level
1 V_{rms} reference

$$dBV = 20\log\left(\frac{V}{1\,\text{V}_{rms}}\right)$$

Often audio systems describe signal level in dBu.

dBu: voltage level
0.775 V_{rms} reference

$$dBu = 20\log\left(\frac{V}{0.775\,\text{V}_{rms}}\right)$$

This reference level, 0.775 V_{rms}, was established because that is the voltage necessary to deliver 1 mW to a 600 Ω load.

$$P = \frac{(V_{rms})^2}{R}$$

$$P = \frac{(0.775\,V_{rms})^2}{600\,\Omega} = 1.00\,mW$$

When expressing power levels, rather than rms voltage, dB can be used as well. The dBm compares the measured power to 1 mW into a 600 Ω load. This is usually used for signal processing equipment.

$$dBm = 10\log\left(\frac{P}{1\,mW}\right)\bigg|_{R=600\,\Omega}$$

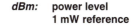

dBm: **power level**
         1 mW reference

The dBW uses 1 W as the reference, more appropriate for amplifiers.

$$dBW = 10\log\left(\frac{P}{1W}\right)$$

dBW: **power level**
         1 W reference

The frequency response plots shown in the previous figures have scaled their vertical axis in *ratio* as shown again in Figure 6-9 (a). It is the standard, however, to scale the vertical axis of a frequency response plot in *dB*. This is shown in Figure 6-9 (b).

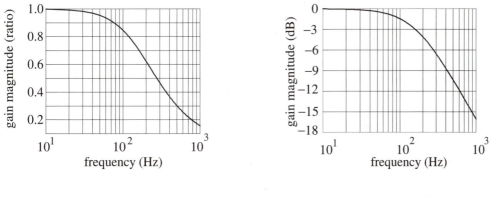

(a) Y axis: ratio   (b) Y axis: decibels

Figure 6-9 Frequency response plots: vertical axis comparison

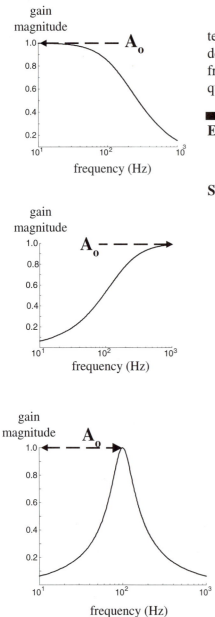

The **pass band gain**, A_o, is the largest gain at the extreme of the filter's operation. For a low-pass filter, the pass band gain is determined at dc. For a high-pass filter, it is measured at the circuit's highest operating frequency. The band-pass filter peaks somewhere in the middle frequencies. That is where A_o is measured. Figure 6-10 illustrates these.

Example 6-3

Use simulation software to plot the frequency response of the filter in Figure 6-11 and to determine its pass band gain.

Solution

Once you have drawn the schematic, next verify that the dc bias voltages are correct. The dc voltages on the noninverting input pin (pin 12), the inverting input pin (pin 13) and the output pin (pin 14) should all be 2.5 V_{dc}.

When the bias is correct, verify that the waveform at the output is correct. At 500 Hz, the output should be a sine wave with a 2.5 mV_p amplitude, riding on a 2.5 V_{dc} level.

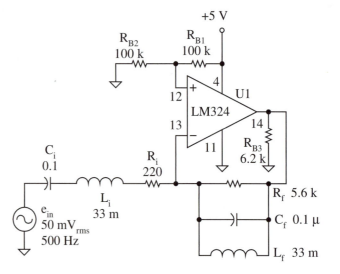

Figure 6-10
Pass band gain for several filters

Figure 6-11 Filter schematic for Example 6-3

The frequency response simulation is shown in Figure 6-12. Look carefully at the **Bode Plotter's** connections, **F** and **I** blocks

The cursor may be moved across the display. The gain magnitude and frequency at the intersection of the cursor and the frequency response plot are indicated in the lower right window of the **Bode Plotter**. In this example, $A_o = 28.1$ dB at 2.75 kHz.

Practice: Determine A_o if $R_i = 120$ kΩ, and $L_i = L_f = 470$ mH.

Answer: $A_o = -26.6$ dB at 724 Hz

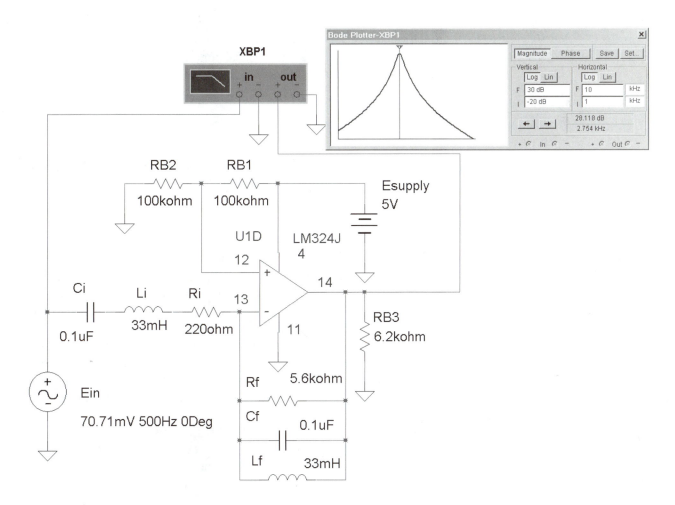

Figure 6-12 Frequency response simulation for Example 6-3

6.4 Roll-off Rate and Filter Order

You can build a filter with a resistor and a single capacitor or a single inductor. The circuit is simple and performs all of the functions of a filter. Adding capacitors and/or inductors increases the filter's complexity and cost. These added energy storage elements may also significantly improve the filter's performance. Look at Figure 6-13.

The number of inductors and/or capacitors in the circuit sets the filter's order. Each of these energy storage elements produces an increase of the derivative order needed to describe the circuit's behavior.

$$v_{\mathrm{L}} = L\frac{di}{dt} \quad \text{and} \quad i_C = C\frac{dv}{dt}$$

A circuit with resistors and two capacitors or two inductors or one capacitor and one inductor is described by a differential equation containing a second order derivative.

$$\frac{d^2}{dt^2}$$

A resistor, two capacitors and an inductor form a *third* order circuit. Its equation contains a third order derivative.

$$\frac{d^3}{dt^3}$$

Fortunately, the behavior for the more useful circuits has been fully investigated and the results tabulated. The circuits and the frequency responses in Figure 6-13 are for **Butterworth** filters. The design of both low- and high-pass, higher order filters using the Butterworth tables are covered in the following chapters.

The pass band gain and the frequency characteristics have been set to be equal. This makes comparisons easier. The major difference, then, among the various order filters is how steeply the gain falls as the frequency increases. This is called the **roll-off rate**. The higher the filter's order, the higher the roll-off rate. Since the purpose of a filter is to pass certain frequency signals and reject others, the higher the order, the more ideal the performance. Remember, however, that this improved performance is bought with increased circuit complexity and expense. Proper behavior often requires nonstandard component values, making the hardware even more difficult to actually build. It is usually a good idea to select the lowest order filter that works.

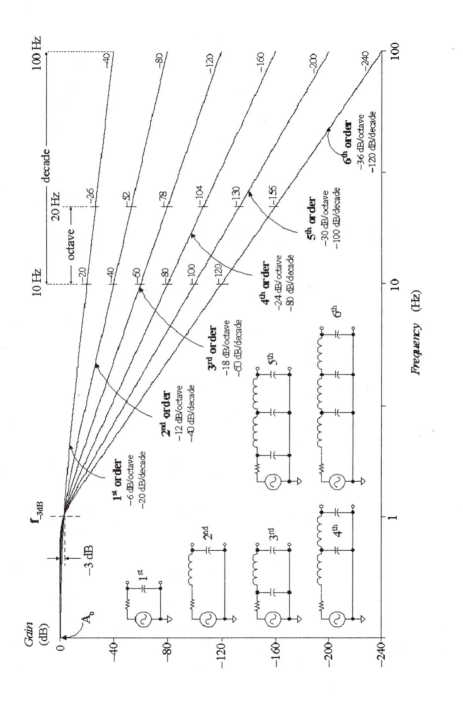

Figure 6-13 Frequency responses of first through sixth order low-pass filters

Decade:
    x 10 or x $^1/_{10}$

Octave:
    x 2 or x $^1/_2$

Roll-off rate depends on the filter's order.

Roll-off rate is a slope, expressing how rapidly the gain *changes* with a change in frequency. The gain is measured in dB. The change in frequency is measured in either decades or octaves. A **decade** change in frequency is an increase or decrease by a **factor of ten**. In Figure 6-13, a decade change is shown between 10 Hz and 100 Hz. An **octave** change is a change by **two**. In Figure 6-13, the frequency changes an octave between 10 Hz and 20 Hz. Audio filters usually have their roll-off rates defined in dB/octave because musical notes are also arranged in octaves. This allows finer resolution. There are approximately 10 octaves between 20 Hz and 20 kHz, but only three decades.

The higher the order, the steeper the roll-off is. When the gain first begins to fall, it falls *nonlinearly*. Within a couple of octaves, the slope becomes *linear*. The roll-off becomes constant, depending on order, at

$$-20 \text{ dB/decade/order} \qquad -6 \text{ dB/octave/order}$$

The polarity is negative since the gain falls as the frequency increases. For high-pass filters the slope is positive.

6.5 Cut-off Frequencies

The key gain parameter (vertical axis) is the pass band gain, A_o. This tells you the filter's maximum gain.

When plotting the gain *magnitude*, the key *frequency* parameter (horizontal axis) is the **half-power point frequency**, f_{-3dB}. The purpose of a filter is to pass power to the load from signals of certain frequencies, and to block power to the load for signals of other frequencies. At the half-power point frequency, the power passed to the load is one-half of the power passed to the load at the filter's maximum gain (A_o).

$$P_{\text{half-power}} = \frac{1}{2} P_{\text{max}}$$

$$\frac{P_{\text{half-power}}}{P_{\text{max}}} = \frac{1}{2}$$

$$dB_{\text{half-power}} = 10\log\left(\frac{P_{\text{half-power}}}{P_{\text{max}}}\right) = 10\log\left(\frac{1}{2}\right) = -3\text{dB}$$

At the half-power point, the power delivered to the load has fallen −3 dB below the power delivered when the filter is passing its maximum output.

In terms of voltage gain,

$$\frac{\left(V_{\text{half-power}}\right)^2}{R_{\text{load}}} = \frac{1}{2}\frac{\left(V_{\text{max}}\right)^2}{R_{\text{load}}}$$

$$\left(V_{\text{half-power}}\right)^2 = \frac{1}{2}\left(V_{\text{max}}\right)^2$$

$$V_{\text{half-power}} = \frac{1}{\sqrt{2}}V_{\text{max}} = 0.707\,V_{\text{max}}$$

At the half-power point, the output voltage has fallen to 0.707 of its maximum value. That's a 30% drop in voltage, producing a 50% drop in power delivered to the load.

To determine the effect on the voltage *gain* at the half-power point, divide each side of the equation by E_{in}.

$$\frac{V_{\text{half-power}}}{E_{\text{in}}} = \frac{1}{\sqrt{2}}\times\frac{V_{\text{max}}}{E_{\text{in}}}$$

$$A_{\text{half-power}} = \frac{1}{\sqrt{2}}\times A_o$$

$$A_{\text{half-power dB}} = 20\log\!\left(\frac{1}{\sqrt{2}}\times A_o\right)$$

The log of the product is the sum of the log of each number.

$$A_{\text{half-power dB}} = 20\log\!\left(A_o\right) + 20\log\!\left(\frac{1}{\sqrt{2}}\right)$$

$$A_{\text{half-power dB}} = A_{o\,\text{dB}} - 3\,\text{dB}$$

At $f_{\text{-3dB}}$:

$P = 0.5\,P_{\text{max}}$

$V = 0.707\,V_{\text{max}}$

$A_{\text{dB}} = A_{o\,\text{dB}} - 3\,\text{dB}$

Example 6-4

Run a simulation for the circuit in Figure 6-14.

a. Produce a frequency response plot and determine $f_{\text{-3dB}}$.

b. Determine the rms voltage and power delivered to the resistor at $10\,f_{\text{-3dB}}$ and at $f_{\text{-3dB}}$.

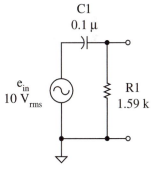

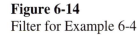

Figure 6-14
Filter for Example 6-4

Solution

The frequency response plot is given in Figure 6-15. This is a high-pass filter with $A_o = 0$ dB and $f_{-3dB} = 1$ kHz.

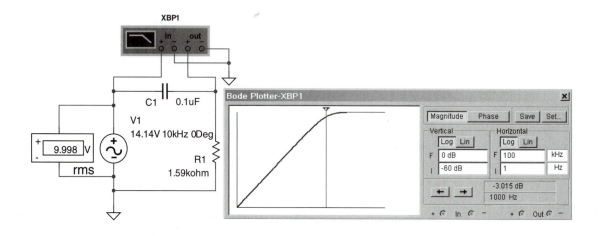

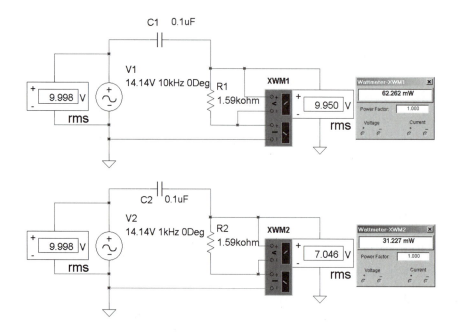

Figure 6-15 Simulation results for Example 6-4

Practice: Replace the 0.1 µF capacitor in Figure 6-14 with a 25.3 mH inductor and repeat the simulation.

Answer: This is a low-pass filter. $A_o = 0$ dB, $f_{-3dB} = 10$ kHz, $P_{R @ 1kHz} = 62$ mW, $V_{R @ 1 kHz} = 10$ V$_{rms}$, $P_{R @ 10 kHz} = 31$ mW, $V_{R @ 10 kHz} = 7.0$ V$_{rms}$

The frequency response plots shown so far have been for gain *magnitude* versus frequency. However, a change in frequency also shifts the phase of a filter's output voltage. In audio systems, a phase shift between two loudspeakers makes the sound seem to move, without fading. The phase of the carrier signal of one form of communications system is shifted to encode the information being transmitted (rather than shifting the amplitude, AM, or the frequency, FM). When many signals of different frequencies and phases all arrive at a filter at the same time, the shape of the composite output signal strongly depends on the phase shift each of the different sine waves receives. The phase shift produced by the filter has a critical effect on the output voltage.

For a first order, *RC*, low-pass filter, the phase shift is

$$\Theta_{low\ pass} = -90° + \arctan\left(\frac{X_C}{R}\right)$$

Configured as a first order high-pass filter the phase shift is

$$\Theta_{high\ pass} = \arctan\left(\frac{X_C}{R}\right)$$

Adding an inductor produces a band-pass filter with a phase shift of

$$\Theta = -\arctan\left(\frac{X_L - X_C}{R}\right)$$

These three basic filters' phase frequency responses are plotted in Figure 6-16. The MATLAB commands used are also included. The low-pass filter's phase frequency response begins near zero, and shifts more and more negative as the frequency increases, headed toward, but never reaching −90°. Low-pass filters cause the phase of the output to **lag.**

The high-pass filter is the complement of the low-pass. At high frequencies, where the filter is passing the input signal, the phase is shifted very little. As the frequency falls, and the filter begins to reject the signal, the phase is *advanced*, toward +90°. High-pass filters cause the phase of the output to **lead** the input signal.

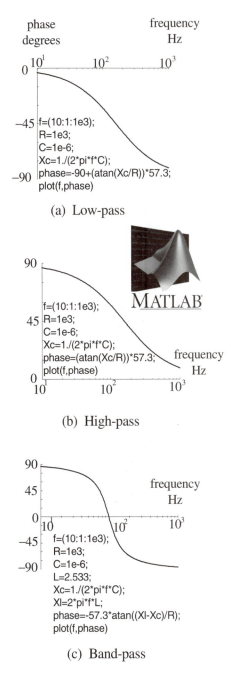

(a) Low-pass

(b) High-pass

(c) Band-pass

Figure 6-16 Phase shift frequency responses with MATLAB commands

For the band-pass the low frequencies are rejected. This is like a high-pass filter. The phase at low frequencies also behaves like it does in a high-pass filter. It leads, being shifted nearly +90° at very low frequencies, and dropping to 0°. The bandpass filter rejects high frequencies too, just as a low-pass filter does. The upper frequency output phase lags, starting at 0° near the center, and shifting toward −90°.

The **critical frequency**, f_0 is that frequency at which the phase has been shifted ±45°/order. Low-pass filters cause the phase to lag (−45°/order). High-pass filters cause the phase to lead (+45°/order). Increasing the order of the filter increases the phase shift at f_0. Figure 6-17 is a plot of the phase shift of low-pass filters, first to sixth order.

These are the same filters whose magnitude frequency response, and schematics, are shown in Figure 6-13. For certain classes of filters these two key frequency parameters occur at the same frequency. For others, the half-power point frequency is related to the critical frequency. Remember, these two parameters define two different phenomena. At the half-power point frequency (f_{-3dB}), the power to the load has fallen by 50% and the voltage gain's *magnitude* has dropped to 0.707 A_o. At the critical frequency, the phase has been shifted some multiple of 45°.

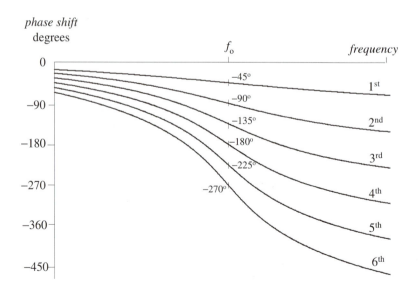

Figure 6-17 Phase shift of first through sixth order low-pass filters

Summary

Gain is the *phasor* ratio of output to input. Both magnitude and phase are calculated. Filters provide different gain for different frequency signals, allowing certain frequency signals to pass while blocking others. Low-pass filters are used to drive bass loudspeakers, and to remove high frequency. High-pass filters pass the notes to the tweeter loudspeaker, and couple signals between amplifier stages while blocking their dc bias. Band-pass filters drive the mid-range loudspeakers.

$$dB = 10 \times \log\frac{P_{out}}{P_{in}}$$

$$= 20 \times \log\frac{V_{out}}{E_{in}}$$

The horizontal axis of a filter's frequency response plot is scaled logarithmically. Both spreadsheets and MATLAB allow graphs to have their horizontal axis scaled logarithmically.

Gain magnitude often is expressed in decibels. The dB may indicate the *power* gain or the *voltage* gain. There are two key dB levels. A gain of 0 dB indicates that the output power or voltage equals the input power or voltage. A gain of −3 dB indicates that the output is below the input. The output power is half of the input power. The output voltage has fallen to 0.707 of the input voltage. Since dB is a measure of *gain*, it must be a ratio. However, the dB may also be used to express a level if a reference is specified, such as dBV, dBu, dBm, and dBW.

$$dBV = 20 \times \log\frac{V_{out}}{1V_{rms}}$$

$$dBu = 20 \times \log\frac{V_{out}}{0.775\,V_{rms}}$$

$$dBm = 10 \times \log\frac{P_{out}}{1mW}$$

$$dBW = 10 \times \log\frac{P_{out}}{1W}$$

A filter's order is equal to the number of inductors and capacitors in the circuit. The higher the filter's order is, the more steeply the gain magnitude rolls-off. Roll-off rate may be expressed in dB/decade or dB/octave. Low-pass filters have a negative slope, while the slope for high-pass filters is positive.

Roll-off rate:
   20 dB/decade/order

   6 dB/octave/order

The pass band gain, A_o, is the largest gain at the extreme of the filter's operation. For a low-pass filter, the pass band gain is determined at dc. For a high-pass filter, it is measured at the circuit's highest operating frequency. The band-pass filter's gain peaks somewhere in the middle.

The key frequency parameter on a gain magnitude plot is the half-power point frequency, f_{-3dB}. At that frequency, the dB gain has fallen 3 dB below A_o. Power to the load is 50% of what it is at the filter's maximum output. The voltage gain falls to 0.707 A_o.

The critical frequency f_o is the main frequency parameter of the gain phase frequency response. At that frequency the phase has shifted by ±45°/order. Low-pass filters lag, sending the phase negative. High-pass filters lead, producing a positive phase shift. The higher the filter's order is, the greater the phase shift at the critical frequency.

For many filters, the half-power point frequency f_{-3dB} and the critical frequency f_o occur at the same frequency. For other types of filters, these two parameters are related, but occur at different frequencies.

Problems

The Frequency Response Plot

6-1   **a.**   Plot the frequency response of the following gain magnitude between the frequencies of 1 kHz and 1 MHz. A spreadsheet or MATLAB will help to automate the calculations and plotting. See Example 6-1 if you need help.

$$|A| = \frac{8R}{\sqrt{R^2 + (X_L - X_C)^2}}$$

$$R = 2.2 \text{ k}\Omega, \; L = 2.2 \text{ mH}, \; C = 1 \text{ nF}$$

b.   Identify the *type* of filter.

c.   Explain one application for this filter that is not described in the text.

6-2   Repeat the three parts of Problem 6-1 for the following.

$$|A| = \frac{3X_L}{\sqrt{R^2 + (X_L - X_C)^2}}$$

$$R = 2.2 \text{ k}\Omega, \; L = 2.2 \text{ mH}, \; C = 1 \text{ nF}$$

The Horizontal Axis

6-3   Plot the gain magnitude for the circuit in Problem 6-1 with its horizontal axis scaled logarithmically.

6-4   Plot the gain magnitude for the circuit in Problem 6-2 with its horizontal axis scaled logarithmically.

The Vertical Axis

6-5   Calculate the *dB* for each of the following:
   a. $P_{in} = 200$ W, $P_{out} = 20$ mW

   b. $V_{in} = 100$ mV$_{rms}$, $V_{out} = 70.7$ mV$_{rms}$

6-6   Calculate the *dB* for each of the following:
   a. $P_{in} = 1.2$ W, $P_{out} = 300$ m W

   b. $V_{in} = 28.3$ V$_{rms}$, $V_{out} = 250$ mV$_{rms}$

6-7   A filter has a gain of −8 dB and an input of 12 mW. Calculate the power delivered to the load.

6-8   A filter has a gain of 12 dB and an input of 300 mV$_{rms}$. Calculate the output voltage.

6-9   What input power is required to deliver 1.4 W to a load from a filter with a gain of −24 dB?

6-10   What input voltage is required to deliver 4.6 V$_{rms}$ to a load from a filter with a gain of 41 dB?

6-11   Convert the following:
    a.   50 mW to dBm   **b.**   25 W to dBW

    c.   350 mV$_{rms}$ to dBu   **d.**   7.3V$_{rms}$ to dBV

6-12   Convert the following:
    a.   230 mW to dBm   **b.**   75 W to dBW

    a.   230 mV$_{rms}$ to dBu   **d.**   8.9V$_{rms}$ to dBV

6-13   **a.**   Plot the gain magnitude from Problem 6-1. Express the gain magnitude in dB. Scale the horizontal axis logarithmically.

    b.   Determine the pass band gain, A_o.

6-14   **a.**   Plot the gain magnitude from Problem 6-2. Express the gain magnitude in dB. Scale the horizontal axis logarithmically.

    b.   Determine the pass band gain, A_o.

Roll-off Rate and Filter Order
6-15   Determine the roll-off rate for each of the following filters.
    a.   Third order low-pass. Answer in dB/decade.

    b.   Second order high-pass. Answer in dB/octave.

    c.   Sixth order low-pass. Answer in dB/decade.

    d.   Fifth order high-pass. Answer in dB/octave.

6-16   Determine the order needed for each of the following filters.
    a.   At 3 kHz the gain is −3dB. At 6 kHz the gain is −15 dB.

    b.   At 100 kHz the gain is −40 dB. At 10 kHz the gain is 0 dB.

    c.   At 100 Hz the gain is −3dB. At 400 Hz the gain is −39 dB.

    d.   At 1 MHz the gain is −90 dB. At 10 kHz the gain is +70 dB.

Cut-off Frequencies

6-17   Determine f_{-3dB} for the filter in Problem 6-1. There is both a low frequency and a high frequency half-power point frequency.

6-18   Determine f_{-3dB} for the filter in Problem 6-2. There is both a low-frequency and a high-frequency half-power point frequency.

6-19   If the vertical axis of the frequency response plot is scaled in ratio, and $A_o = 23$, where does f_{-3dB} occur?

Frequency Response Lab Exercise

A. Passive High-pass Filter

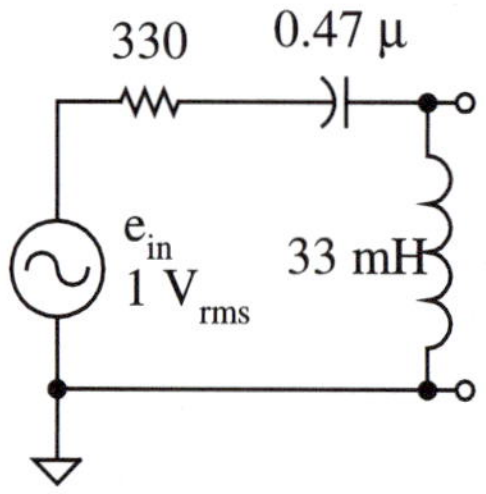

1. Build the circuit in Figure 6-18. Keep the leads from the signal generator and the leads to each component as short as practical.

2. Set the signal generator's amplitude to 1 V_{rms}, and its frequency to 100 Hz.

3. Verify the pass band gain (A_o) of your circuit by connecting the oscilloscope and digital multimeter between the output and common. Then, raise the frequency to 10 kHz. V_{out} should be a maximum at 10 kHz. Record this gain ratio.

4. Lower the frequency until the output voltage has fallen to 0.707 of its value in step A3. This frequency is f_{-3dB}. Record it.

5. Measure and record the critical frequency, f_o (+90° phase shift).

6. Lower the frequency to 200 Hz. Record the gain in dB.

7. Lower the frequency to 100 Hz. Record the gain in dB again.

8. Calculate the dB/octave roll-off rate.

330   0.47 μ

e_{in}
1 V_{rms} 33 mH

Figure 6-18
High pass filter

$A_{@\ 200\ Hz}$ = _____ dB

$A_{@\ 100\ Hz}$ = _____ dB

Roll-off rate = _____ dB/octave

9. Construct a spreadsheet with columns for *frequency* (kHz), V_{out} (V_{rms}), *gain* (dB), *phase* (degrees). Provide a row for the column titles and then 19 rows for data. Enter the frequency data, one in each row (0.1, 0.2, 0.3, 0.4, 0.5, 0.6, 0.7, 0.8, 0.9, 1.0, 2.0, 3.0, 4.0, 5.0, 6.0, 7.0, 8.0, 9.0, 10.0).

10. Create a frequency response graph of *frequency* (log scale) on the *x* axis versus *gain* (dB) on the *y* axis. Do this before you begin taking data.

11. Create a frequency response graph of *frequency* (log scale) on the *x* axis versus *phase* on the *y* axis. Do this before you begin taking data.

12. Set the input to 100 Hz, 1 V_{rms}. Measure the output rms voltage magnitude and phase. Allow the spreadsheet to calculate the dB.

13. Change the frequency to that indicated in the next row. Adjust the input voltage magnitude to 1 V_{rms}. Measure the output magnitude and phase.

14. Compete the table and the plots.

B. Active Low-pass Filter
1. Build the circuit in Figure 6-19. Keep the leads from the signal generator and the leads to each component as short as practical.

2. Set the signal generator's amplitude to 150 mV_{rms}, and its frequency to 10 kHz.

3. Verify the pass band gain (A_o) of your circuit by connecting the oscilloscope and digital multimeter between the output and common. Then, lower the frequency to 100 Hz. The output should be a maximum at 100 Hz. Record this gain ratio.

4. Raise the frequency until the output voltage has fallen to 0.707 of its value in step B3. This frequency is f_{-3dB}. Record it.

5. Measure and record the critical frequency (f_o).

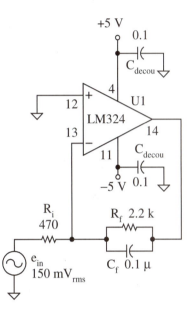

Figure 6-19
Active low-pass filter

$A_o =$ _____

$f_{-3dB} =$ _____

$f_o =$ _____

A@ 5 kHz = _____ **dB**

A@ 10 **kHz** = _____ **dB**

Roll-off rate = _____**dB/octave**

6. Raise the frequency to 5 kHz. Record the gain in dB.

7. Raise the frequency to 10 kHz. Record the gain in dB again.

8. Calculate the dB/octave roll-off rate.

9. Construct a spreadsheet with columns for *frequency* (kHz), V_{out} (V_{rms}), *gain* (dB), *phase* (degrees). Provide a row for the column titles and then 19 rows for data. Enter the frequency data, one in each row (0.1, 0.2, 0.3, 0.4, 0.5, 0.6, 0.7, 0.8, 0.9, 1.0, 2.0, 3.0, 4.0, 5.0, 6.0, 7.0, 8.0, 9.0, 10.0).

10. Create a frequency response graph of *frequency* (log scale) on the *x* axis versus *gain* (dB) on the *y* axis. Do this before you begin taking data.

11. Create a frequency response graph of *frequency* (log scale) on the *x* axis versus *phase* on the *y* axis. Do this before you begin taking data.

12. Set the input to 100 Hz, 150 mV_{rms}. Measure the output rms voltage magnitude and phase. Allow the spreadsheet to calculate the dB.

13. Change the frequency to that indicated in the next row. Maintain the input voltage magnitude at 150 mV_{rms}. Measure the output magnitude and phase.

14. Compete the table and the plots.

7

Low-pass Filters

Introduction

Low-pass filters are among the most commonly used filters. They remove high frequency noise created by microprocessors, spark plugs, radio communications, and even the utility power lines while passing dc and the more slowly varying signals that contain the information about how much you weigh to the scale, or how fast you are going to the speedometer display, or the dc power to your amplifier. They are used to pass the low tones to the woofer part of your loudspeaker array. Low-pass filters extract the audio and video information carried by rf communications signals, while rejecting the higher frequency carrier signals.

Resistors, capacitors, inductors, and even op amps may be combined to build a low-pass filter with the pass band gain, half-power point frequency, and roll-off rate that you need. Which configuration you choose depends on these parameters as well as considerations of load resistance, space, cost, system support, and part availability. In this chapter, you will investigate four of the more common low-pass filter arrangements.

The same treatment is used for each filter. A quick-look technique is applied first to determine the filter's type, order, and pass band gain. Trigonometry is then used to derive the equations for the cut-off frequencies. Finally an analysis or design example is presented to allow you to see this circuit in a practical application.

Objectives

Upon completion of this chapter, you will be able to do the following for *RC*, *LR*, op amp based, and *RLC* higher order low-pass filters:

- Draw the schematic.

- Determine the filter type, pass band gain, order, and roll-off rate.

- Derive the cut-off frequencies.

- Analyze the filter to plot its frequency response.

- Design each of these filters to meet given specifications.

163

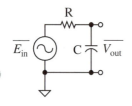

Figure 7-1
RC low-pass filter

7.1 The *RC* Low-pass Filter

The *RC* filter shown in Figure 7-1 is very commonly used. It is spread throughout most measurement and communications systems to prevent adjacent high-frequency signals from contaminating the lower frequency signal that contains the information you are trying to process. In fact, every power supply and every circuit with a decoupling capacitor at its power supply input are also *RC* low-pass filters.

Quick-look

A **quick-look** analysis allows you to determine the type of filter (low-pass, high-pass, band-pass), the filter's pass band gain, the order, and roll-off rate by inspection or with a few simple dc calculations. Always perform a quick-look before starting derivations. Start by determining the filter's response at dc. At dc, capacitors look like an open.

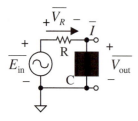

Figure 7-2
Quick-look at dc

$$X_{C@dc} = \frac{1}{2\pi \times 0 \times C} \rightarrow \infty\,\Omega$$

At dc, inductors look like a short.

$$X_{L@dc} = 2\pi \times 0 \times L = 0\,\Omega$$

The dc version of the *RC* low-pass filter is shown in Figure 7-2. Since C is an open, there is no current flow. The resistor drops no voltage. So, all of the voltage provided by the source arrives at the output.

$$\overline{I_{@dc}} = 0\,A$$

$$\overline{V_{R@dc}} = \overline{I} \times (R\angle 0°) = 0\,V$$

$$\overline{V_{out@dc}} = \overline{E_{in}} - \overline{V_{R@dc}} = \overline{E_{in}}$$

At dc, the output equals the input.

$$A_{dc} = 1$$

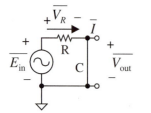

Figure 7-3
Quick-look as $f \rightarrow \infty$

Next, determine the performance of the circuit at a very high frequency. As $f \rightarrow \infty$, the capacitor becomes a short. This is shown in Figure 7-3. The voltage dropped by an impedance of $0\,\Omega$ is $0\,V$. So the output voltage falls to zero. The source voltage is across the resistor.

$$X_{C @ f \to \infty} = \frac{1}{2\pi \times \infty \times C} \to 0\,\Omega$$

$$\overline{V_{out @ f \to \infty}} = \overline{I} \times 0\,\Omega = 0\,\text{V}$$

$$A_{f \to \infty} = 0$$

Gain Derivation

Determining the detailed performance uses a little trigonometry. The output voltage equation can be written with the voltage divider law.

$$\overline{V_{out}} = \overline{E_{in}} \times \frac{\overline{Z_C}}{\overline{Z_R} + \overline{Z_C}}$$

The gain is needed in order to plot the frequency response.

$$\overline{A} = \frac{\overline{V_{out}}}{\overline{E_{in}}} = \frac{\overline{Z_C}}{\overline{Z_R} + \overline{Z_C}}$$

Express the impedances in terms of R and X_C.

$$\overline{A} = \frac{(X_C \angle -90°)}{(R \angle 0°) + (X_C \angle -90°)}$$

To complete the addition in the denominator, the phasors must be placed in rectangular form.

$$\overline{A} = \frac{(X_C \angle -90°)}{(R + j0) + (0 - jX_C)}$$

$$\overline{A} = \frac{(X_C \angle -90°)}{(R - jX_C)}$$

Division of the numerator by the denominator requires that both quantities be expressed in their polar form. Figure 7-4 shows the phasor diagram. The length of the hypotenuse is

$$c = \sqrt{a^2 + b^2}$$

$$|R - jX_C| = \sqrt{R^2 + X_C^2}$$

The angle comes from the tangent.

RC **filter quick-look results:**
$A_{dc} = 1$
$A_{f \to \infty} = 0$
∴ $A_o = 1 = 0$ dB
   type = low-pass

one capacitor
∴ **first order**
   –6 dB/octave
   $\theta_{fo} = -45°$

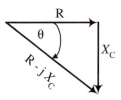

Figure 7-4
Rectangular-polar conversion

$$\tan(\theta) = \frac{opposite}{adjacent} = \frac{-X_C}{R}$$

The complementary operation of the tangent is the arctangent.

$$\theta = \arctan\left(-\frac{X_C}{R}\right)$$

$$\arctan(-\phi) = -\arctan(\phi)$$

$$\theta = -\arctan\left(\frac{X_C}{R}\right)$$

Combining these operations converts the denominator to polar.

$$(R - jX_C) = \left(\sqrt{R^2 + X_C^2} \angle -\arctan\left(\frac{X_C}{R}\right)\right)$$

Substitute this into the denominator of the gain equation.

$$\overline{A} = \frac{(X_C \angle -90°)}{\left(\sqrt{R^2 + X_C^2} \angle -\arctan\left(\frac{X_C}{R}\right)\right)}$$

When dividing phasors, divide the magnitudes and subtract the angles.

$$\overline{A} = \frac{X_C}{\sqrt{R^2 + X_C^2}} \angle \left(-90° - \left(-\arctan\frac{X_C}{R}\right)\right)$$

The *magnitude* of the gain is

Gain *magnitude*

$$|\overline{A}| = \frac{X_C}{\sqrt{R^2 + X_C^2}}$$

At low frequency, X_C is very large, much greater than R. The denominator is just a little more than X_C. So the gain magnitude goes to 1. At very high frequencies, X_C falls to 0, as does the gain.

The gain's phase shift is

Gain *phase*

$$\Theta = -90° + \arctan\left(\frac{X_C}{R}\right)$$

At low frequencies, X_C is very large. The arctangent of a very large angle approaches 90°. So the phase shift at low frequencies is almost 0°.

When frequency goes up, the reactance falls. The arctangent adds less to −90°, shifting the phase negative, eventually to −90°. This is a first order low-pass filter with a phase shift at f_o of −45°.

Example 7-1

Plot the frequency response for the *RC* low-pass filter with

R = 1 kΩ, and C = 1 μF.

Sweep the frequency logarithmically from 10 Hz to 10 kHz.

Plot both the gain magnitude (in dB) and the phase.

Identify each of the following on the graph:

A_o, f_{-3dB}, f_o

Solution

The spreadsheet is shown in Figure 7-5. Calculations for dB are performed in column D. Phase is calculated in column E.

The gain magnitude and phase plots are given in Figure 7-6. To plot *both* dB gain and phase versus frequency, highlight cells A6–A24, then press the control key while highlighting cells D6–D24 and E6–E24.

To create a secondary y axis on the right, for phase, double click on the phase line. Under the **Axis** tab in the dialog box, select **Secondary axis**.

The pass band gain, A_o, is identified on the graph at 0 dB. This is a gain ratio of 1, which matches the theoretical quick-look value. The half-power point frequency occurs when the gain in dB has fallen −3dB below A_o. To help locate that point on the graph, the dB axis divisions are set in steps of 3 dB.

The critical frequency, f_o, for this first order low-pass filter occurs when the phase has shifted −45°. So the major divisions on the phase axis are set to steps of 15°. These two key frequency specifications occur at the same point.

=20*LOG(C6)

=B6/SQRT($B$2^2+B6^2)   =−90+57.3*ATAN(B6/$B$2)

	A	B	C	D	E
2	R (ohms)	1.00E+03			
3	C (Farads)	1.00E-06			
4					
5	frequency (Hz)	Xc (ohms)	gain mag	dB	phase
6	10	15915	0.998	-0.02	-4
7	20	7958	0.992	-0.07	-7
8	30	5305	0.983	-0.15	-11
9	40	3979	0.970	-0.27	-14
10	50	3183	0.954	-0.41	-17
11	60	2653	0.936	-0.58	-21
12	70	2274	0.915	-0.77	-24
13	80	1989	0.893	-0.98	-27
14	90	1768	0.870	-1.20	-29
15	100	1592	0.847	-1.45	-32
16	200	796	0.623	-4.11	-51
17	300	531	0.469	-6.58	-62
18	400	398	0.370	-8.64	-68
19	500	318	0.303	-10.36	-72
20	600	265	0.256	-11.82	-75
21	700	227	0.222	-13.08	-77
22	800	199	0.195	-14.19	-79
23	900	177	0.174	-15.18	-80
24	1000	159	0.157	-16.07	-81

Figure 7-5
Spreadsheet for Example 7-1

Practice: Change the capacitor to 22 nF and repeat Example 7-1. Sweep the frequency from 1 kHz to 100 kHz. Locate and determine values for A_o, f_{-3dB}, and f_o.

Answer: A_o = 0 dB, f_{-3dB} = 7.2 kHz, f_o = 7.2 kHz

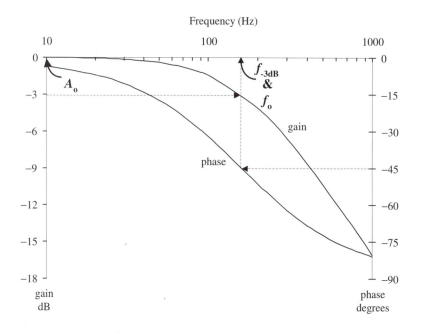

Figure 7-6 Frequency response plot for Example 7-1

Half-power Point Frequency, f_{-3dB}

Completing an entire frequency response plot to determine a circuit's cut-off frequencies is cumbersome. The gain equations can be used to *derive* equations for the cut-off frequencies in terms of R and C.

At the half-power point frequency, the gain ratio has fallen below A_o.

$$\left|A_{@\,f-3db}\right| = \frac{A_o}{\sqrt{2}}$$

For this *RC* low-pass filter, $A_o = 1$. Substitute this into the general gain magnitude equation.

$$\frac{A_o}{\sqrt{2}} = \frac{1}{\sqrt{2}} = \frac{X_C}{\sqrt{R^2 + X_C^2}}$$

Clear the fractions.

$$X_C \times \sqrt{2} = \sqrt{R^2 + X_C^2}$$

Square both sides.

$$2X_C^2 = R^2 + X_C^2$$

$$X_C^2 = R^2$$

$$X_C = R$$

At f_{-3dB} $X_C = R$

Substitute the equation for the capacitive reactance.

$$R = \frac{1}{2\pi f C}$$

Rearrange the equation to isolate f on the left. This is the half-power point frequency.

$$f_{-3dB} = \frac{1}{2\pi RC}$$

For Example 7-1, in Figure 7-6, f_{-3dB} is between 100 Hz and 200 Hz.

$$f_{-3dB} = \frac{1}{2\pi \times 1k\Omega \times 1\mu F} = 159\,Hz$$

Above the half-power point frequency, the gain magnitude falls linearly at about −6 dB/octave, −20 dB/decade. However, below f_{-3dB}, the gain climbs *nonlinearly* toward A_o. At $f_{5\%}$, the gain is 5% below A_o. This occurs at $\frac{1}{3} f_{-3dB}$. You can verify this by repeating the preceding derivation, beginning with

$$|A| = 0.95 = \frac{X_C}{\sqrt{R^2 + X_C^2}}$$

$$f_{5\%} = \frac{f_{-3dB}}{3}$$

If a 5% drop is too large, at $f_{1\%}$, the gain is 1% below A_o. This occurs at $\frac{1}{7} f_{-3dB}$. That derivation begins with

$$|A| = 0.99 = \frac{X_C}{\sqrt{R^2 + X_C^2}}$$

$$f_{1\%} = \frac{f_{-3dB}}{7}$$

Critical Frequency, f_o

At the critical frequency, the phase shift for a first order, low-pass filter is −45°. Substitute this into the equation for the gain phase shift.

$$\Theta = -90° + \arctan\left(\frac{X_C}{R}\right)$$

$$-45° = -90° + \arctan\left(\frac{X_C}{R}\right)$$

$$45° = \arctan\left(\frac{X_C}{R}\right)$$

Take the tangent of each side.

$$1 = \frac{X_C}{R}$$

$$R = X_C$$

This is just like the half-power point frequency. At the critical frequency, the reactance equals the resistance.

$$R = \frac{1}{2\pi f_o C}$$

Critical frequency
f_o

$$f_o = \frac{1}{2\pi RC}$$

Figure 7-6 indicates that the phase is shifted −45° at the same frequency (f_o) that the gain falls −3dB below A_o. This matches the derivations.

7.2 The *LR* Low-pass Filter

The *RC* low-pass filter in Section 7.1 places the output voltage across a capacitor. If you want to pass the voltage to a resistive load, you must place R_{load} in parallel with the capacitor. That loads down the filter, lowering the pass band gain and raising the cut-off frequencies.

The *LR* low-pass filter places the output voltage across the resistor. If your load is resistive, then R_{load} is R_{filter}. The schematic of a first order, low-pass, *LR* filter is given in Figure 7-7.

Quick-look

To begin the quick-look analysis, consider the circuit's performance at dc. The reactance of the inductor is

$$X_L = 2\pi \times 0 \times L = 0\,\Omega$$

Replacing the inductor with a short means that the input voltage all appears at the output.

$$\overline{V}_{out} = \overline{E}_{in}\frac{(R\angle 0°)}{(R\angle 0°)+(X_L\angle 90°)}$$

$$\overline{A}_{@\,dc} = \frac{\overline{V}_{out\,@\,dc}}{\overline{E}_{in}} = \frac{(R\angle 0°)}{(R\angle 0°)+(0\angle 90°)}$$

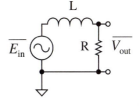

At very high frequencies, the inductor's impedance becomes an open. None of the input voltage arrives at the output.

$$\overline{A}_{f\to\infty} = \frac{\overline{V}_{out\,f\to\infty}}{\overline{E}_{in}} = \frac{(R\angle 0°)}{(R\angle 0°)+(\infty\angle 90°)} = 0$$

Since there is only one inductor and no capacitors, the filter is first order. The look at dc and as $f\to\infty$ indicates that this is low-pass, $A_o = 1$.

Figure 7-7
LR low-pass filter

LR filter quick-look results:
$A_{dc} = 1$
$A_{f\to\infty} = 0$
∴ $A_o = 1 = 0$ dB
   type = low-pass

one inductor
∴   first order
   –6 dB/octave
   $\theta_{fo} = -45°$

Gain Derivation

As with the *RC* filter of Section 7.1, the equation for the gain can be written using the voltage divider law.

$$\overline{A} = \frac{(R\angle 0°)}{(R\angle 0°)+(X_L\angle 90°)}$$

The phasors must be placed in rectangular form

$$\overline{A} = \frac{(R\angle 0°)}{(R+j0)+(0+jX_L)}$$

$$\overline{A} = \frac{(R\angle 0°)}{(R+jX_L)}$$

Division of the numerator by the denominator requires that both quantities be expressed in their polar form

$$\left| R + jX_L \right| = \sqrt{R^2 + X_L^{\,2}}$$

The angle comes from the tangent.

$$\theta = \arctan\left(\frac{X_L}{R}\right)$$

$$\left(R + jX_L \right) = \left(\sqrt{R^2 + X_L^{\,2}} \angle \arctan\left(\frac{X_L}{R}\right) \right)$$

Substitute this into the denominator of the gain equation.

$$\overline{A} = \frac{\left(R \angle 0^\circ \right)}{\left(\sqrt{R^2 + X_L^{\,2}} \angle \arctan\left(\frac{X_L}{R}\right) \right)}$$

When dividing phasors, divide the magnitudes and subtract the angles.

$$\overline{A} = \frac{R}{\sqrt{R^2 + X_L^{\,2}}} \angle \left(0^\circ - \arctan\frac{X_L}{R} \right)$$

The *magnitude* of the gain is

$$\left| \overline{A} \right| = \frac{R}{\sqrt{R^2 + X_L^{\,2}}}$$

At low frequency, X_L is small, much smaller than R. The denominator is just a little more than R. So the gain magnitude goes to 1. At very large frequencies, X_L becomes very large, as does the denominator. The gain falls toward zero. This matches the results from the quick-look.

The gain's phase shift is

$$\varTheta = -\arctan\left(\frac{X_L}{R}\right)$$

At low frequencies, X_L is very small. The arctangent of a very small angle approaches 0°. So the phase shift at low frequencies is almost 0. When frequency goes up, the reactance rises. The arctangent increases toward 90°, so the filter's phase *lags* (–) more and more as the frequency increases. This also matches the quick-look, that indicated this is a first order low-pass filter with a phase shift at f_o of -45°.

Half-power Point Frequency, f_{-3dB}

At the half-power point frequency, the gain ratio has fallen below A_o.

$$\left|A_{@ f-3db}\right| = \frac{A_o}{\sqrt{2}}$$

For this *LR* low-pass filter, $A_o = 1$. Substitute this into the general gain magnitude equation.

$$\frac{A_o}{\sqrt{2}} = \frac{1}{\sqrt{2}} = \frac{R}{\sqrt{R^2 + X_L^{\,2}}}$$

Clear the fractions.

$$R \times \sqrt{2} = \sqrt{R^2 + X_L^{\,2}}$$

$$2R^2 = R^2 + X_L^{\,2}$$

$$R^2 = X_L^{\,2}$$

$$R = X_L \qquad\qquad\qquad \text{At } f_{-3dB} \quad X_L = R$$

Substitute the equation for the inductive reactance.

$$R = 2\pi f_{-3dB} L$$

Rearrange the equation to isolate f on the left.

$$f_{-3dB} = \frac{R}{2\pi L}$$

Critical Frequency, f_o

At the critical frequency, the phase shift for a first order, low-pass filter is $-45°$. Substitute this into the equation for the gain phase shift.

$$\Theta = -\arctan\left(\frac{X_L}{R}\right)$$

$$-45° = -\arctan\left(\frac{X_L}{R}\right)$$

$$45° = \arctan\left(\frac{X_L}{R}\right)$$

Take the tangent of each side.

$$1 = \frac{X_L}{R}$$

$$R = X_L$$

This is just like the half-power point frequency. At the critical frequency, the reactance equals the resistance.

$$R = 2\pi f_o L$$

Critical frequency
f_o

$$f_o = \frac{R}{2\pi L}$$

As with the *RC* low-pass filter, the critical frequency and the half-power point frequency occur at the same frequency.

Example 7-2

    a. Design an *LR* low-pass filter. The resistance is the 8 Ω resistance of a loudspeaker. Set f_{-3dB} = 300 Hz.

    b. Calculate the frequency where A = 0.95.

how big can you dream?™

    c. Use simulation to confirm A_o, f_{-3dB}, $f_{5\%}$, the dB/octave and dB/decade roll-off rates.

Solution

    a. $$f_{-3dB} = \frac{R}{2\pi L}$$

$$L = \frac{R}{2\pi f_{-3dB}} = \frac{8\Omega}{2\pi \times 300\,\text{Hz}} = 4.3\,\text{mH}$$

    b. $$f_{5\%} = \frac{f_{-3dB}}{3} = \frac{300\,\text{Hz}}{3} = 100\,\text{Hz}$$

    c. The schematic and probe output are shown in Figure 7-8.

Practice: Alter the design to provide f_{-3dB} = 200 Hz. Find f_o.

Answers: L = 6.4 mH, f_o = 200 Hz

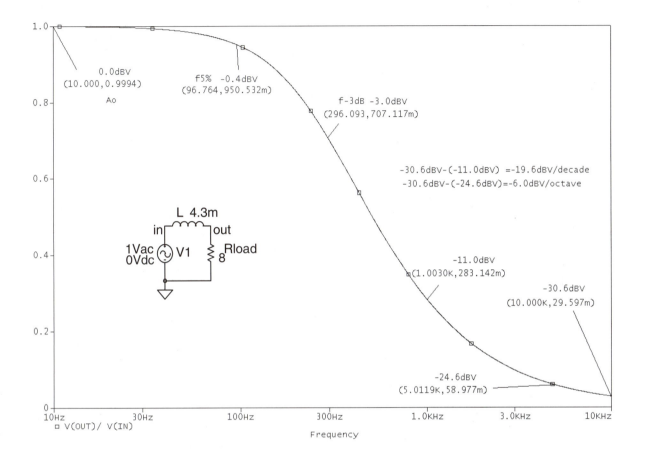

Figure 7-8 Simulation results for Example 7-2

7.3 Simple Active Low-pass Filter

Both the *RC* and *LR* filters have a pass band gain of 1. If you need a higher pass band gain, you could just follow each filter with an op amp based amplifier. However, Figure 7-9 shows a simpler solution. This is an inverting op amp based amplifier with a capacitor, C_f, placed in parallel across the feedback resistor, R_f.

Quick-look

Start the quick-look analysis by determining the circuit's response at dc (0 Hz). At dc, capacitors look like an open. The circuit, then, is just the inverting op amp amplifier you have seen before.

$$A_{dc} = -\frac{R_f}{R_i}$$

or

$$|A_{dc}| = \frac{R_f}{R_i}$$

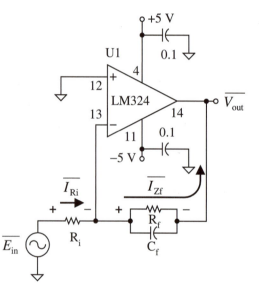

Figure 7-9 First order *active* low-pass filter

Next, determine the performance of the circuit at a very high frequency. As $f \to \infty$, the capacitor becomes a short. It shorts out R_f. The voltage dropped by an impedance of 0 Ω is 0 V. So, as the frequency becomes higher and higher, the output voltage falls to zero.

$$\overline{V_{\text{out} @ f \to \infty}} = \overline{I} \times 0\,\Omega = 0\,\text{V}$$

At low frequencies the gain is R_f/R_i. As frequency goes up the gain falls, eventually to 0. That is the performance of a low-pass filter.

There is only one energy storage element, the capacitor. So this is a first order low-pass filter. Its roll-off rate is −6 dB/octave. Also, as a first order low-pass filter, the phase shift at the critical frequency should be −45°. However, the inverting configuration of the op amp adds 180° of phase shift. So the overall phase shift at the critical frequency, is

$$\theta_{@ \text{fo}} = -45° + 180° = 135°$$

Active filter quick-look results:
$A_{dc} = R_f/R_i$
$A_{f \to \infty} = 0$
$\therefore \quad A_o = R_f/R_i$
    type = low-pass

one capacitor
$\therefore$ **first order**
    −6 dB/octave
    $\theta_{\text{fo}} = 135°$

Gain Derivation

The gain for a simple inverting op amp amplifier is

$$A = -\frac{R_f}{R_i}$$

Since the feedback resistor is paralleled by a capacitor, this becomes

$$\overline{A} = \frac{\overline{Z_f}}{\overline{Z_i}} \angle 180°$$

$$\overline{Z_i} = (R_i \angle 0°)$$

The feedback impedance is the *parallel* combination of R_f and C_f.

$$\overline{Z_f} = \left[\frac{1}{(R_f \angle 0°)} + \frac{1}{(X_C \angle -90°)} \right]^{-1}$$

Combine these two fractions by obtaining a common denominator.

$$\overline{Z_f} = \left[\frac{(X_C \angle -90°)}{(R_f \angle 0°)(X_C \angle -90°)} + \frac{(R_f \angle 0°)}{(R_f \angle 0°)(X_C \angle -90°)} \right]^{-1}$$

$$\overline{Z_f} = \left[\frac{(X_C \angle -90°) + (R_f \angle 0°)}{(R_f \angle 0°)(X_C \angle -90°)} \right]^{-1}$$

You can take the reciprocal of a fraction just by inverting it.

$$\overline{Z}_f = \frac{(R_f \angle 0°)(X_C \angle -90°)}{(R_f \angle 0°) + (X_C \angle -90°)}$$

$$\overline{Z}_f = \frac{(R_f \angle 0°)(X_C \angle -90°)}{(R_f - jX_C)}$$

This is the numerator of the gain equation.

$$\overline{A} = \frac{\overline{Z}_f}{\overline{Z}_i} \angle 180°$$

$$\overline{A} = \frac{\dfrac{(R_f \angle 0°)(X_C \angle -90°)}{(R_f - jX_C)}}{(R_i \angle 0°)} \angle 180°$$

Clear this fraction by inverting the denominator and multiplying.

$$\overline{A} = \frac{(R_f \angle 0°)(X_C \angle -90°)}{(R_i \angle 0°)(R_f - jX_C)} \angle 180°$$

$$\overline{A} = \left(\frac{R_f}{R_i}\right) \frac{(X_C \angle -90°)}{(R_f - jX_C)} \angle 180°$$

To complete the division, the denominator must be converted into its polar notation, just as has been done for the *RC* and *LR* filters.

$$\overline{A} = \frac{\left(\dfrac{R_f}{R_i}\right)(X_C \angle -90°)}{\left(\sqrt{R_f^2 + X_C^2} \angle \arctan\left(\dfrac{-X_C}{R_f}\right)\right)} \angle 180°$$

Divide the magnitudes and subtract the angles.

$$\overline{A} = \left(\frac{R_f}{R_i} \frac{X_C}{\sqrt{R_f^2 + X_C^2}} \angle -90° + 180° - \arctan\left(\frac{-X_C}{R_f}\right)\right)$$

Simplify the angles.

$$\overline{A} = \left(\frac{R_f}{R_i} \frac{X_C}{\sqrt{R_f^2 + X_C^2}} \angle 90° + \arctan\left(\frac{X_C}{R_f}\right) \right)$$

$$|\overline{A}| = \frac{R_f}{R_i} \frac{X_C}{\sqrt{R_f^2 + X_C^2}}$$

$$\Theta = 90° + \arctan\frac{X_C}{R_f}$$

First order, active, low-pass filter's gain

Look back at the *RC* low-pass filter's gain derivation. The gain equation for the op amp version is the same, combined with the pass band gain set by R_f/R_i and the additional 180° phase shift.

Half-power Point Frequency, f_{-3dB}

At the half-power point frequency, the gain ratio has fallen below A_o.

$$|\overline{A}_{@\,f-3db}| = \frac{A_o}{\sqrt{2}}$$

For this active low-pass filter, $A_o = R_f / R_i$

$$\frac{A_o}{\sqrt{2}} = \frac{\dfrac{R_f}{R_i}}{\sqrt{2}} = \frac{R_f}{R_i} \frac{X_C}{\sqrt{R_f^2 + X_C^2}}$$

The R_f/R_i can be divided from both sides.

$$\frac{1}{\sqrt{2}} = \frac{X_C}{\sqrt{R_f^2 + X_C^2}}$$

$$X_C \times \sqrt{2} = \sqrt{R_f^2 + X_C^2}$$

$$2X_C^2 = R_f^2 + X_C^2$$

$$X_C^2 = R_f^2$$

$$X_C = R_f$$

Substitute the equation for the capacitive reactance.

$$R_f = \frac{1}{2\pi f C_f}$$

Rearrange the equation to isolate f on the left.

Half-power point frequency, f_{-3dB}

$$f_{-3dB} = \frac{1}{2\pi R_f C_f}$$

This is just like the *RC* low-pass filter. Be sure to remember to use the *feedback* resistor.

Critical Frequency, f_o

The critical frequency can be derived just as the half-power point frequency. At the critical frequency, the phase shift for a first order, low-pass filter is −45°. However, since there is an additional inversion from the op amp, at the critical frequency, $\theta_{@ fo} = -45° + 180° = 135°$

Substitute this into the equation for the gain phase shift.

$$\Theta = 90° + \arctan\left(\frac{X_C}{R_f}\right)$$

$$135° = 90° + \arctan\left(\frac{X_C}{R_f}\right)$$

$$45° = \arctan\left(\frac{X_C}{R_f}\right)$$

Take the tangent of each side.

$$1 = \frac{X_C}{R_f}$$

$$R_f = X_C$$

$$R_f = \frac{1}{2\pi f_o C_f}$$

Critical frequency f_o

$$f_o = \frac{1}{2\pi R_f C_f}$$

Example 7-3

A signal from a microphone (1 kHz, 0.1 V_p) is passed to an analog-to-digital converter. Before it reaches the converter, it is contaminated by a 0.5 V_p, 200 kHz square wave from the digital processing circuitry.

multiSIM

a. Design an active filter that will amplify the 1 kHz signal by -10, and will severely reject the digital noise.

b. Confirm your circuit's performance by simulation.

Solution

a. The schematic is shown in Figure 7-9. The feedback resistor is selected first, set to as large a value as is practical. Much greater than 470 kΩ may cause nonideal dc errors with the op amp and may make the amplifier sensitive to small noise currents induced by stray electromagnetic fields.

The input resistor is calculated from the pass band gain.

$$A_o = 10 = \frac{R_f}{R_i}$$

$$R_i = \frac{R_f}{10} = \frac{470\,k\Omega}{10} = 47\,k\Omega$$

This is reasonable. It sets the input impedance, and is large enough not to load professional quality microphones.

The half-power point frequency must be chosen to be above 1 kHz (the signal's frequency) but considerably below the frequency of the noise. For a 5% error in the gain,

$$f_{5\%} = \frac{f_{-3dB}}{3} = 1\,kHz$$

$$f_{-3dB} = 3\,kHz$$

The feedback capacitor can now be calculated.

$$f_{-3dB} = \frac{1}{2\pi R_f C_f} \qquad C_f = \frac{1}{2\pi R_f f_{-3dB}}$$

$$C_f = \frac{1}{2\pi \times 470\,k\Omega \times 3\,kHz} = 113\,pF \quad \text{Pick 100 pF}$$

    b. The simulation is shown in Figure 7-10. The blurred upper trace on the oscilloscope is the high frequency square wave noise overwhelming the much smaller 1 kHz sine wave. The lower trace is the recovered and amplified 1 kHz sine wave. The negative dc offset at the output is the result of amplifying (and inverting) the offset from the square wave noise generator. The resulting sine wave amplitude is shown by the cursors as 2 V_{pp}.

Practice: What are the effects of altering the amplifier to noninverting?

Answer: The pass band gain goes to 11, f_{-3dB} shifts to 3.3 kHz, and there is much more noise at the output because the entire noise input signal is passed from the noninverting input pin to the negative feedback pin, becoming part of the output.

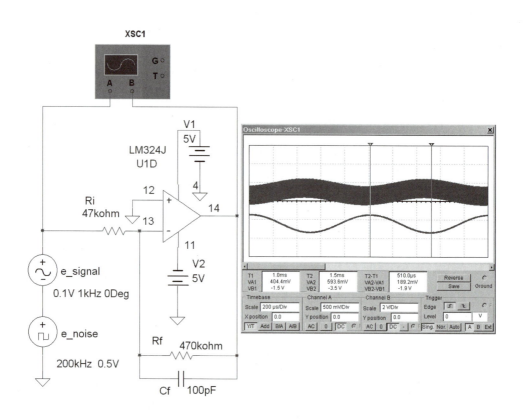

Figure 7-10 Simulation results for Example 7-3

7.4 Higher Order Low-pass Filters

As the order increases, the solution of the gain equation and the selection of damping becomes *much* more complicated. These circuits have been the subjects of extensive study. They are grouped into the following categories, each with their own performance, damping, and set of component values.

Butterworth
Chebyshev
Bessel
Linear phase
Elliptic

Of these, the Butterworth filters have the flattest response in the pass band. Other filters either ripple up and down, or sag early. The Butterworth filters also are stable, showing little tendency to break into oscillations. Finally, they are more tolerant of variation in component value than the other filters.

Beyond the half-power point frequency, Butterworth filters roll-off linearly. In terms of dB the gain at any frequency, f_x, depends on f_{-3dB} and on the order, n.

$$A_{dB} = 10\log\left[1 + \left(\frac{f_x}{f_{-3dB}}\right)^{2n}\right]$$

If you know the gain at two frequencies, as shown in Figure 7-11, then the required order for the filter may be calculated by

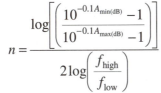

$$n = \frac{\log\left[\left(\frac{10^{-0.1A_{min(dB)}} - 1}{10^{-0.1A_{max(dB)}} - 1}\right)\right]}{2\log\left(\frac{f_{high}}{f_{low}}\right)}$$

There are a variety of ways that a resistor, capacitors, and inductors can be arranged to form Butterworth low-pass filters. Figure 7-12 shows eight orders. Table 7-1 gives the required component relationships.

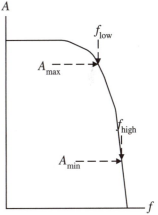

Figure 7-11
Order selection

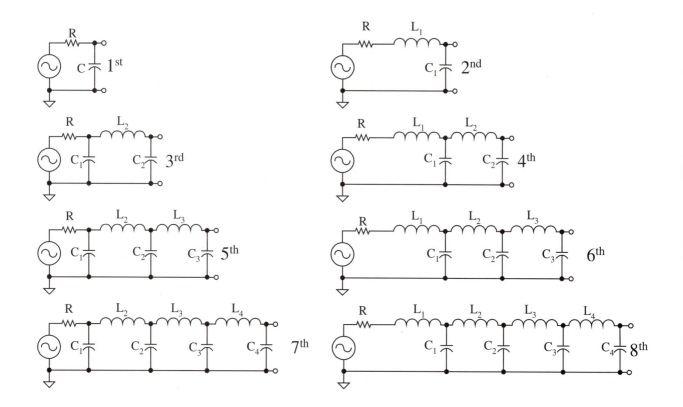

Figure 7-12 First through eighth order passive low-pass filters

Table 7-1 Butterworth component relationships

order	L1	C1	L2	C2	L3	C3	L4	C4
2	0.7071	1.414						
3		0.500	1.333	1.500				
4	0.3827	1.082	1.577	1.531				
5		0.309	0.894	1.382	1.694	1.545		
6	0.2588	0.758	1.202	1.553	1.759	1.533		
7		0.223	0.656	1.054	1.397	1.659	1.799	1.588
8	0.1951	0.578	0.937	1.259	1.528	1.729	1.824	1.561

A.B. Williams, F.J. Taylor, *Electronic Filter Design Handbook* 3rd ed, McGraw Hill, pages 11.5 - 11.11

The values in Table 7-1 are *normalized*. They assume that R = 1 Ω and ω_{-3dB} = 1 rad/s. To convert these into values for an actual circuit,

$$L_{actual} = \frac{R_{actual} \times L_{table}}{2\pi f_{-3dB}}$$

and

$$C_{actual} = \frac{C_{table}}{2\pi f_{-3dB} \times R_{actual}}$$

To design a circuit, then, you first determine which order filter is needed. Then repeatedly try standard values of R_{actual} until you find a set of capacitors and inductors that are close to standard values. This could be a tedious process. However, Table 7-1 has been combined with these two equations in a spreadsheet, which is included on the cd that accompanies this text. You enter f_{-3dB} and then a standard value for R_{actual}. Examine the resulting inductors and capacitors displayed by the spreadsheet.

Example 7-4

    a. Design a filter using standard value components, with f_{-3dB} = 3 kHz, and gain = −40 dB at 10 kHz.

    b. Confirm the performance at 3 kHz and 10 kHz

Solution

    a. A_{max} = −3dB, f_{low} = 3 kHz; A_{min} = −40 dB, f_{high} = 10 kHz

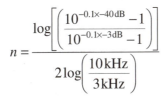

$$n = \frac{\log\left[\left(\dfrac{10^{-0.1 \times -40\,dB} - 1}{10^{-0.1 \times -3\,dB} - 1}\right)\right]}{2\log\left(\dfrac{10\,kHz}{3\,kHz}\right)}$$

$$n = \frac{\log\left(\dfrac{9999}{0.995}\right)}{1.0458} = 3.83$$

    Select a fourth order filter. The gain at 10 kHz should be a little more negative than the −40 dB specified. By trial and error or with the spreadsheet, pick R = 560 Ω.

$$L_1 = \frac{560\,\Omega \times 0.3827}{2\pi \times 3\,\text{kHz}} = 11.37\,\text{mH} \qquad \text{Pick 11 mH}$$

$$C_1 = \frac{1.082}{2\pi \times 3\,\text{kHz} \times 560\,\Omega} = 0.1025\,\mu\text{F} \quad \text{Pick 0.1 μF}$$

$$L_2 = 46.9\,\text{mH} \qquad\qquad\qquad\qquad\qquad \text{Pick 47 mH}$$

$$C_2 = 0.145\,\mu\text{F} \qquad\qquad\qquad\qquad\qquad \text{Pick 0.15 μF}$$

b. The simulation results are shown in Figure 7-13.

Practice: Design a filter $f_{-3\text{dB}} = 200$ Hz, $A_{800\,\text{Hz}} = -60$ dB, R = 150 Ω

Answer: n = 5, C_1 = 1.6 μF, L_2 = 110 mH, C_2 = 7.5 μF, L_3 = 200 mH, C_3 = 8.2 μF

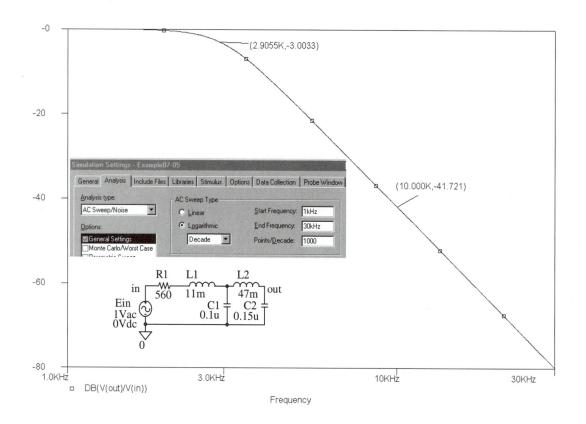

Figure 7-13 Simulation results for Example 7-4

Summary

Begin analyzing a filter with a quick-look. First consider the filter's performance at dc, with all inductors replaced with shorts and all capacitors replaced with opens. Determine the gain for this simplified circuit. Then determine the filter's gain at extremely high frequency. Replace the inductors with opens and the capacitors with shorts. From this brief review you can determine the type of filter, its pass band gain, its order, its roll-off rate, and the phase shift at the critical frequency.

You derive the gain equation by calculating the output voltage in terms of the input signal, the resistors, capacitive reactances, and inductive reactances in the circuit. Remember to *add* phasors in their rectangular form, and to convert them to polar for *division*. The final gain equation can be separated into a magnitude and a phase.

The equation for the half-power point frequency is determined by setting the gain magnitude equation equal to $A_o/\sqrt{2}$. The frequency is encoded in the reactances. The equation for the critical frequency comes from setting the gain phase equation equal to $-45°$ for first order filters or $-90°$ for second order filters, and solving for the frequency.

These steps were applied to several filters: RC, LR, and a simple active filter, and a RLC second order.

RC	LR	active
$f_{-3dB} = \dfrac{1}{2\pi RC}$	$f_{-3dB} = \dfrac{R}{2\pi L}$	$f_{-3dB} = \dfrac{1}{2\pi R_f C_f}$
$f_0 = \dfrac{1}{2\pi RC}$	$f_0 = \dfrac{R}{2\pi L}$	$f_0 = \dfrac{1}{2\pi R_f C_f}$

Higher order filters can be built by configuring a resistor with series inductors and capacitors to common. The Butterworth filter is a good, general purpose filter. First calculate the order needed from the specified gains at two frequencies. Then scale the coefficients given in Table 7-1 for f_{-3dB} and for the R you want. You will have to repeat this process until you find a set of component values that are close to standard values.

Problems

The *RC* Low-pass Filter

7-1   For a simple *RC* low-pass filter with R = 1 kΩ and C = 56 nF, calculate:

   A_o
   filter order
   roll-off rate

   f_{-3dB}
   f_o
   gain magnitude at f_{-3dB}
   phase shift at f_o

7-2   Repeat Problem 7-1 if R = 50 Ω, and C = 33 pF

7-3   Design an *RC* low-pass filter with a minimum input impedance of 10 kΩ, and f_{-3dB} = 280 Hz.

7-4   Design an *RC* low-pass filter with a minimum input impedance of 75 Ω, and f_{-3dB} = 540 kHz.

7-5   For the circuit in Figure 7-14:
   a.   Complete a quick-look analysis.

   b.   Derive the gain magnitude and gain phase equations.

   c.   Derive the equation for f_{-3dB}.

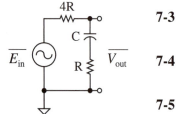

Figure 7-14
Schematic for
Problem 7-5

The *LR* Low-pass Filter

7-6   For a simple *LR* low-pass filter with R = 1 kΩ and L = 470 mH, calculate:

   A_o
   filter order
   roll-off rate

   f_{-3dB}
   f_o
   gain magnitude at f_{-3dB}
   phase shift at f_o

7-7 Repeat Problem 7-6 if R = 50 Ω, and L = 75 nH.

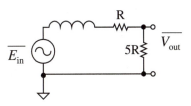

7-8 Design an *LR* low-pass filter for a cross-over network using a 4 Ω loudspeaker, with $f_{-3dB} = 5$ kHz.

7-9 Design an *LR* low-pass filter with a load resistance of 50 Ω, with $f_{-3dB} = 540$ kHz

7-10 For the circuit in Figure 7-15:
   a. Complete a quick-look analysis.
   b. Derive the gain magnitude and gain phase equations
   c. Derive the equation for f_{-3dB}.
   d. Derive the equation for f_o.

Figure 7-15
Schematic for Problem 7-10

The Simple Active Low-pass Filter

7-11 For a simple active low-pass filter with $R_i = 1$ kΩ, $R_f = 22$ kΩ, and $C_f = 1.5$ nF, calculate:
    A_o
    filter order
    roll-off rate
    f_{-3dB}
    f_o
    gain magnitude at f_{-3dB}
    phase shift at f_o

7-12 Repeat Problem 7-11 if $R_i = 51$ Ω, $R_f = 150$ Ω and $C_f = 100$ pF.

7-13 Design a simple active low-pass filter with an input impedance of 1 kΩ, $A_o = 30.4$ dB, and $f_{-3dB} = 1.5$ kHz.

7-14 Design a simple active low-pass filter with an input impedance of 75 Ω, $A_o = 45$ dB, and $f_{-3dB} = 540$ kHz.

7-15 For the circuit in Figure 7-16,
   a. Complete a quick-look analysis.
   b. Derive the gain magnitude and gain phase equations.

Higher Order Low-pass Filters

7-16 Design a Butterworth filter with $A_o = -3$ dB, $f_{-3dB} = 3$ kHz, and $A_{10 \text{ kHz}} < -16$ dB. Confirm your design's performance by simulation.

7-17 Design a Butterworth filter with $A_o = -3$ dB, $f_{-3dB} = 500$ kHz, and $A_{2 \text{ MHz}} < -65$ dB. Confirm your design's performance by simulation.

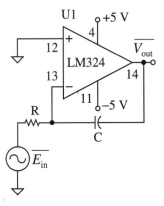

Figure 7-16
Schematic for Problem 7-15

Low-pass Filters Lab Exercise

A. Second Order Low-pass Filter

1. Build a second order low-pass filter with

$$R = 100\ \Omega, \quad L = 33\ \text{mH}, \quad C = 5.6\ \mu\text{F}$$

Keep the leads from the signal generator and the leads to each component as short as practical.

2. Set the signal generator's amplitude to 1 V_{rms} and its frequency to 50 Hz.

A_o
measured = _____   theory = _____

3. Verify the pass band gain (A_o) of your circuit by connecting the oscilloscope and digital multimeter between the output and common. V_{out} should be a maximum at 50 Hz. Record this gain ratio, and compare it to the theoretical pass band gain.

f_{-3dB}
measured = _____   theory = _____

f_o
measured= _____   theory = _____

4. Raise the frequency until the output voltage has fallen to 0.707 of its value in step A3. This frequency is f_{-3dB}. Record it and compare it to the theoretical half-power point frequency.

5. Measure, record, and compare the critical frequency (f_o).

6. Raise the frequency to 2 kHz. Record the gain in dB.

7. Raise the frequency to 4 kHz. Record the gain in dB again.

$A_{@\ 2\ kHz} =$ _____ dB

8. Calculate the dB/octave roll-off rate. Compare that to the theoretical roll-off rate.

$A_{@\ 4\ kHz} =$ _____ dB

Roll-off rate
   measured = _____dB/octave

   theory = _____ dB/octave

9. Construct a spreadsheet with columns for *frequency* (kHz), V_{out} (V_{rms}), *gain* (dB), and *phase* (degrees). Provide a row for the column titles and then 19 rows for data. Enter the frequency data, one in each row (0.1, 0.2, 0.3, 0.4, 0.5, 0.6, 0.7, 0.8, 0.9, 1.0, 2.0, 3.0, 4.0, 5.0, 6.0, 7.0, 8.0, 9.0, 10.0).

10. Create a frequency response graph of *frequency* (log scale) on the *x* axis versus *gain* (dB) on the *y* axis. Do this before you begin taking data.

11. Create a frequency response graph of *frequency* (log scale) on the *x* axis versus *phase* on the *y* axis. Do this before you begin taking data.

12. Set the input to 100 Hz, 1 V_{rms}. Measure the output rms voltage magnitude and phase. Allow the spreadsheet to calculate the dB and begin plotting the graph.

13. Change the frequency to that indicated in the next row. Adjust the input voltage magnitude to 1 V_{rms}. Measure the output magnitude and phase. Check the spreadsheet's plots.

14. Complete the table and the plots.

B. Simple Active Low-pass Filter

Repeat all of the steps from Section A for the simple active low-pass filter that you designed in Problem 7-13.

- Use an LM324 op amp.

- Set its supplies to ±5 V_{dc}.

- Set the input signal's amplitude to 50 mV_{rms}.

8

High-pass Filters

Introduction

Two of the most pervasive signals are dc and the 50 Hz to 60 Hz of the utility power. They are so commonly used that too often they also contaminate the signals you are trying to process, amplify, detect, decode, or encode. These *desired* signals are usually at much higher frequencies. The solution for removing the dc and power line frequencies from the higher frequency signals is to use a high-pass filter.

Passing the high notes to the tweeter driver in your loudspeaker array requires a high-pass filter. Passing the rf signals between stages of a radio without disturbing each stage's bias requires a high-pass filter.

Resistors, capacitors, inductors, and even op amps may be combined to build a high-pass filter with the pass band gain, half-power point frequency, and roll-off rate that you need. Which configuration you choose depends on these parameters as well as considerations of load resistance, space, cost, system support, and part availability. In this chapter, you will investigate four of the more common high-pass filter arrangements.

The same treatment is used for each filter. A quick-look technique is applied first to determine the filter's type, order, and pass band gain. Trigonometry is then used to derive the equations for the cut-off frequencies. Finally an analysis or design example is presented to allow you to see this circuit in a practical application.

Objectives

Upon completion of this chapter, you will be able to do the following for *CR*, *RL*, op amp based, and *RCL* higher order high-pass filters:

- Draw the schematic.

- Determine the filter type, pass band gain, order, and roll-off rate.

- Derive the cut-off frequencies.

- Analyze the filter to plot its frequency response.

- Design each of these filters to meet given specifications.

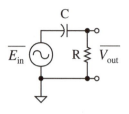

Figure 8-1
CR high-pass filter

8.1 The *CR* High-pass Filter

The *CR* filter shown in Figure 8-1 is common. It is used to couple high frequency energy from one source to a resistive load. Low frequencies, particularly dc, are blocked. You will see this circuit used between signal processing stages in most circuits biased from a single polarity supply. The signal passes through the capacitor to the next stage, but the bias voltages are kept completely separate. This circuit is also a simple form of cross-over network for the tweeter driver in your loudspeaker.

Quick-look

Before beginning the complex gain derivation, always perform a quick-look. Start by determining the filter's response at dc (0 Hz). At dc, capacitors look like an open.

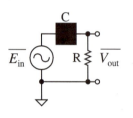

Figure 8-2
Quick-look at dc

$$X_{C@dc} = \frac{1}{2\pi \times 0 \times C} \to \infty\,\Omega$$

The dc version of the *CR* high-pass filter is shown in Figure 8-2. Since C is an open, none of the input signal reaches the resistor.

$$\overline{I}_{@dc} = 0\,A$$

$$\overline{V}_{out@dc} = \overline{I} \times (R\angle 0°) = 0\,V$$

Next, determine the performance of the circuit at a very high frequency. As $f \to \infty$, the capacitor becomes a short. This is shown in Figure 8-3. All of the source voltage is dropped across the resistor.

$$X_{C@f\to\infty} = \frac{1}{2\pi \times \infty \times C} \to 0\,\Omega$$

$$\overline{V}_{out@f\to\infty} = \overline{E}_{in}$$

$$A_{f\to\infty} = 1$$

At low frequencies the gain is 0. As frequency goes up the gain rises, eventually to 1. That is the performance of a high-pass filter, with $A_o = 1 = 0$ dB.

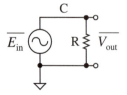

Figure 8-3
Quick-look as $f \to \infty$

There is only one energy storage element, the capacitor. So this is a first order high-pass filter. Its roll-off rate is 6 dB/octave or 20

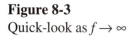

dB/decade. The polarity is positive, because as the frequency goes up, so does the gain. Also, as a first order high-pass filter, the phase shift at the critical frequency, $\theta_{@\,fo} = 45°$. It is a *lead* filter. The phase is advanced, not retarded as it was by low-pass filters.

Gain Derivation

The equation for the output voltage comes from the voltage divider law.

$$\overline{V_{out}} = \overline{E_{in}} \times \frac{\overline{Z_R}}{\overline{Z_R} + \overline{Z_C}}$$

The gain is needed in order to plot the frequency response.

$$\overline{A} = \frac{\overline{V_{out}}}{E_{in}} = \frac{\overline{Z_R}}{\overline{Z_R} + \overline{Z_C}}$$

Express the impedances in terms of R and X_C.

$$\overline{A} = \frac{(R\angle 0°)}{(R\angle 0°) + (X_C\angle -90°)}$$

Place the impedances in rectangular form.

$$\overline{A} = \frac{(R\angle 0°)}{(R + j0) + (0 - jX_C)}$$

$$\overline{A} = \frac{(R\angle 0°)}{(R - jX_C)}$$

Division of the numerator by the denominator requires that both quantities be expressed in their polar form

$$|R - jX_C| = \sqrt{R^2 + X_C^2}$$

The angle comes from the tangent.

$$\tan(\theta) = \frac{opposite}{adjacent} = \frac{-X_C}{R}$$

Do *not* forget the negative sign, since X_C extends vertically *down*. The complementary operation of the tangent is the arctangent.

$$\theta = \arctan\left(-\frac{X_C}{R}\right)$$

A trigonometric identity states that

$$\arctan(-\phi) = -\arctan(\phi)$$

$$\theta = -\arctan\left(\frac{X_C}{R}\right)$$

These operations convert the denominator from rectangular to polar.

$$(R - jX_C) = (c\angle\theta)$$

$$(R - jX_C) = \left(\sqrt{R^2 + X_C^2}\angle -\arctan\left(\frac{X_C}{R}\right)\right)$$

Substitute this into the denominator of the gain equation.

$$\overline{A} = \frac{(R\angle 0°)}{\left(\sqrt{R^2 + X_C^2}\angle -\arctan\left(\frac{X_C}{R}\right)\right)}$$

When dividing phasors, divide the magnitudes and subtract the angles.

$$\overline{A} = \frac{R}{\sqrt{R^2 + X_C^2}}\angle\left(0° - \left(-\arctan\frac{X_C}{R}\right)\right)$$

The *magnitude* of the gain is

Gain *magnitude*

$$\left|\overline{A}\right| = \frac{R}{\sqrt{R^2 + X_C^2}}$$

At low frequency, X_C is very large, much greater than R. So, the gain magnitude falls toward 0. At very large frequencies, X_C becomes very small. The gain is 1. This matches the results from the quick-look.

The gain's phase shift is

Gain *phase*

$$\Theta = \arctan\left(\frac{X_C}{R}\right)$$

At low frequencies, X_C if very large. The arctangent of a very large angle approaches 90°. So the phase shift at low frequencies is almost 90°. When frequency goes up, the reactance falls. The arctangent falls toward 0°. This also matches the quick-look, which indicated this is a first order high-pass filter with a phase shift at f_o of 45°.

Half-power Point Frequency, f_{-3dB}

At the half-power point frequency, the gain ratio has fallen below A_o.

$$\left| A_{@\,f-3db} \right| = \frac{A_o}{\sqrt{2}}$$

For this *CR* high-pass filter, $A_o = 1$.
Substitute this into the general gain magnitude equation.

$$\frac{A_o}{\sqrt{2}} = \frac{1}{\sqrt{2}} = \frac{R}{\sqrt{R^2 + X_C^{\,2}}}$$

$$R \times \sqrt{2} = \sqrt{R^2 + X_C^{\,2}}$$

$$2R^2 = R^2 + X_C^{\,2}$$

$$R^2 = X_C^{\,2}$$

$$R = X_C$$

$$R = \frac{1}{2\pi f C}$$

Rearrange the equation to isolate f on the left. This is the half-power point frequency.

$$f_{-3dB} = \frac{1}{2\pi RC}$$

This is just like the *RC* low-pass filter.

Below the half-power point frequency, the gain magnitude falls linearly at about 6 dB/octave, 20 dB/decade. However, above f_{-3dB}, the gain climbs *nonlinearly* toward A_o. At $f_{5\%}$, the gain is 5% below A_o. This occurs at $3\,f_{-3dB}$. You can verify this by repeating the preceding derivation, beginning with

$$|A| = 0.95 = \frac{R}{\sqrt{R^2 + X_C^{\,2}}}$$

$f_{5\%} = 3\,f_{-3dB}$

If a 5% drop is too large, at $f_{1\%}$, the gain is 1% below A_o. This occurs at $7\,f_{-3dB}$. That derivation begins with

$$f_{1\%} = 7\,f_{-3dB}$$

$$|A| = 0.99 = \frac{R}{\sqrt{R^2 + X_C^2}}$$

Critical Frequency, f_o

At the critical frequency, the phase shift for a first order, high-pass filter is 45°. Substitute this into the equation for the gain phase shift.

$$\Theta = \arctan\left(\frac{X_C}{R}\right)$$

$$45° = \arctan\left(\frac{X_C}{R}\right)$$

Take the tangent of each side.

$$1 = \frac{X_C}{R}$$

$$R = X_C$$

At the critical frequency, the reactance equals the resistance.

$$R = \frac{1}{2\pi f_o C}$$

Critical frequency
f_o

$$f_o = \frac{1}{2\pi RC}$$

The critical frequency for a CR high-pass filter occurs at the same frequency as the half-power point, and at the same point as it does when the same components are used to form an RC low-pass filter.

Example 8-1

a. Design a cross-over network (high-pass filter) for an 8 Ω tweeter driver. Set the cut-off frequencies at 3 kHz.

b. Verify the pass band gain, f_{-3dB}, f_o, and roll-off rate by simulation.

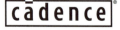

how big can you dream?™

Solution

a.
$$C = \frac{1}{2\pi R \times f_{-3dB}} = \frac{1}{2\pi \times 8\Omega \times 3kHz} = 6.6\mu F$$

This should be made with high-quality, nonpolarized capacitors that exhibit stable characteristics up to at least 20 kHz.

b.   The simulation schematic and results are shown in Figure 8-4. Both the gain in dB and the phase are plotted. The pass band gain is 0 dB. Both cut-off frequencies are 3 kHz. The roll-off rates are 6 dB/octave, 20dB/decade. These are correct for a first order filter.

Practice: Alter the circuit for a 4 Ω driver and a 2 kHz cut-off.

Answer: C = 20 μF

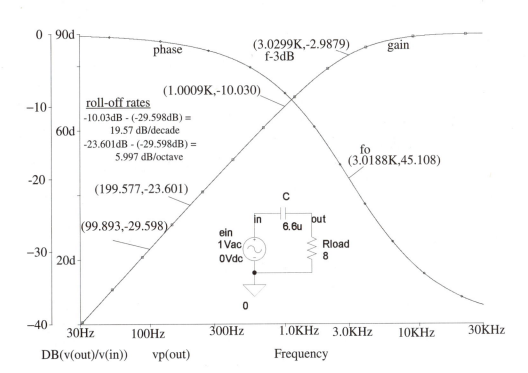

Figure 8-4 Simulation results for Example 8-1

8.2 The *RL* High-pass Filter

Many loads are inductive, not resistive. Motors, relays, solenoids, loud-speaker arrays (to some degree), even the bare wires and printed circuit board traces used to interconnect components have inductance. When combined with series resistance, the effect is a high-pass filter. The schematic of a first order, high-pass, *RL* filter is given in Figure 8-5.

Quick-look

Consider the circuit's dc performance. The reactance of the inductor is

$$X_L = 2\pi \times 0 \times L = 0\,\Omega$$

Replacing the inductor with a short means that the output voltage is zero.

At very high frequencies, the inductor's impedance becomes much greater than the resistor's. All of the input voltage arrives at the output.

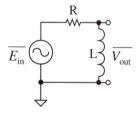

Figure 8-5
RL high-pass filter

$$\overline{A_{f\to\infty}} = \frac{\overline{V_{out\,f\to\infty}}}{\overline{E_{in}}} = \frac{(\infty\angle 90°)}{(R\angle 0°)+(\infty\angle 90°)} = 1$$

Since there is only one inductor and no capacitors, the filter is first order. The quick-look indicates that this is high-pass, $A_o = 1$.

RL filter quick-look results:
  $A_{dc} = 0$
  $A_{f\to\infty} = 1$
  ∴ $A_o = 1 = 0$ dB
      type = high-pass

one inductor
  ∴   **first order**
      6 dB/octave
      $\theta_{fo} = 45°$

Gain Derivation

Using the voltage divider law, express the impedance of the resistor and inductor in terms of R and X_L.

$$\overline{A} = \frac{(X_L\angle 90°)}{(R\angle 0°)+(X_L\angle 90°)}$$

The phasors must be placed in rectangular form

$$\overline{A} = \frac{(X_L\angle 90°)}{(R + jX_L)}$$

Division of the numerator by the denominator requires that both quantities be expressed in their polar form

$$\left|R + jX_L\right| = \sqrt{R^2 + X_L{}^2}$$

The angle comes from the tangent.

$$\theta = \arctan\left(\frac{X_L}{R}\right)$$

Convert the denominator from rectangular to polar.

$$(R + jX_L) = \left(\sqrt{R^2 + X_L^2} \angle \arctan\left(\frac{X_L}{R}\right)\right)$$

Substitute this into the denominator of the gain equation.

$$\overline{A} = \frac{(X_L \angle 90°)}{\left(\sqrt{R^2 + X_L^2} \angle \arctan\left(\frac{X_L}{R}\right)\right)}$$

When dividing phasors, divide the magnitudes and subtract the angles.

$$\overline{A} = \frac{X_L}{\sqrt{R^2 + X_L^2}} \angle \left(90° - \arctan\frac{X_L}{R}\right)$$

The *magnitude* of the gain is

$$|\overline{A}| = \frac{X_L}{\sqrt{R^2 + X_L^2}}$$

Gain *magnitude*

At low frequency, X_L is very small, much smaller than R. The gain magnitude is 0. At very large frequencies, X_L becomes much larger than R. The gain becomes 1. This matches the results from the quick-look.

The gain's phase shift is

$$\Theta = 90° - \arctan\left(\frac{X_L}{R}\right)$$

Gain *phase shift*

Half-power Point Frequency, f_{-3dB}

At the half-power point frequency, the gain ratio has fallen below A_o.

$$\left|\overline{A}_{@\,f-3db}\right| = \frac{A_o}{\sqrt{2}}$$

$A_o = 1$. Substitute this into the general gain magnitude equation.

$$\frac{A_o}{\sqrt{2}} = \frac{1}{\sqrt{2}} = \frac{X_L}{\sqrt{R^2 + X_L^{\,2}}}$$

$$X_L \times \sqrt{2} = \sqrt{R^2 + X_L^{\,2}}$$

$$2X_L^{\,2} = R^2 + X_L^{\,2}$$

$$R^2 = X_L^{\,2}$$

At f_{-3dB} $X_L = R$ $$R = X_L$$

Substitute the equation for the inductive reactance.

$$R = 2\pi f_{-3dB} L$$

Solve for f. This is the half-power point frequency.

f_{-3dB} $$f_{-3dB} = \frac{R}{2\pi L}$$

This is just like the *LR* low-pass filter.

Critical Frequency, f_o

At the critical frequency, the phase shift for a first order, high-pass filter is 45°. Substitute this into the equation for the gain phase shift.

$$\Theta = 90° - \arctan\left(\frac{X_L}{R}\right)$$

$$45° = 90° - \arctan\left(\frac{X_L}{R}\right)$$

$$-45° = -\arctan\left(\frac{X_L}{R}\right)$$

$$45° = \arctan\left(\frac{X_L}{R}\right)$$

Take the tangent of each side.

$$1 = \frac{X_L}{R}$$

$$R = X_L$$

At the critical frequency, the reactance equals the resistance.

$$R = 2\pi f_o L$$

$$f_o = \frac{R}{2\pi L}$$

Critical frequency
f_o

As with the *CR* high-pass filter, the critical frequency and the half-power point frequency occur at the same point.

8.3 The Simple Active High-pass Filter

Both the *CR* and *RL* filters have a pass band gain of 1. Figure 8-6 shows a way to add amplification. This is an inverting op amp based amplifier with a capacitor, C_i, placed in series across the input resistor, R_i.

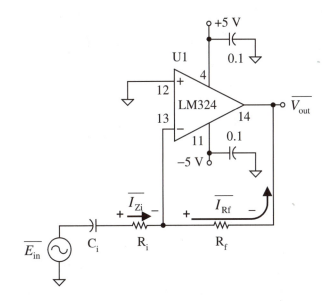

Figure 8-6 First order *active* high-pass filter

Quick-look

The gain for this amplifier is

$$\overline{A} = -\frac{R_f \angle 0°}{\overline{Z}_i}$$

At very low frequencies, the impedance of the capacitor becomes very large. There is little input current, and the gain goes to zero.

$$\overline{A_{dc}} = -\frac{R_f \angle 0°}{\infty} \rightarrow 0$$

As $f \rightarrow \infty$, the capacitor becomes a short. This just leaves the input resistor between the voltage source and the op amp, and sets A_o.

$$\left| A_{f \rightarrow \infty} \right| = A_o = \frac{R_f}{R_i}$$

At low frequencies the gain is 0. As frequency goes up the gain climbs to R_f/R_i. That is the performance of a high-pass filter.

There is only one energy storage element, the capacitor. So this is a first order high-pass filter. Its roll-off rate is 6 dB/octave. As a first order high-pass filter, the phase shift at the critical frequency should be 45°. However, the inverting configuration of the op amp subtracts 180° of phase shift. So the overall phase shift at the critical frequency, is

$$\theta_{@\,fo} = 45° - 180° = -135°$$

Active filter quick-look results:
$A_{dc} = 0$
$A_{f \rightarrow \infty} = R_f/R_i$
∴ $A_o = R_f/R_i$
    type = high-pass

one capacitor
∴   **first order**
    6 dB/octave
    $\theta_{fo} = -135°$

Gain Derivation

The gain for a simple inverting op amp amplifier is

$$A = -\frac{R_f}{R_i}$$

Since the input resistor is in series with a capacitor, this becomes

$$\overline{A} = \frac{\overline{Z_f}}{\overline{Z}_i} \angle -180°$$

where

$$\overline{Z}_f = \left(R_f \angle 0° \right)$$

and $\qquad \overline{Z}_i = \left(R_i \angle 0° \right) + \left(X_C \angle -90° \right)$

In rectangular form the input impedance is

$$\overline{Z}_i = \left(R_i + j X_C \right)$$

$$\overline{A} = \frac{\left(R_f \angle 0° \right)}{\left(R_i + j X_C \right)} \angle -180°$$

Convert the denominator into its polar form.

$$\overline{A} = \frac{\left(R_f \angle 0° \right)}{\left(\sqrt{R_i^2 + X_C^2} \angle \arctan \left(\dfrac{-X_C}{R_i} \right) \right)} \angle -180°$$

Divide the magnitudes and subtract the angles.

$$\overline{A} = \left(\frac{R_f}{\sqrt{R_i^2 + X_C^2}} \angle -180° - \arctan \left(\frac{-X_C}{R_i} \right) \right)$$

Simplify the angles.

$$\overline{A} = \left(\frac{R_f}{\sqrt{R_i^2 + X_C^2}} \angle -180° + \arctan \left(\frac{X_C}{R_i} \right) \right)$$

$$\left| \overline{A} \right| = \frac{R_f}{\sqrt{R_i^2 + X_C^2}}$$

First order, active, high-pass filter's gain

$$\varTheta = -180° + \arctan \frac{X_C}{R_i}$$

Half-power Point Frequency, f_{-3dB}

At the half-power point frequency, the gain ratio has fallen below A_o.

$$\left| A_{@f-3db} \right| = \frac{A_o}{\sqrt{2}}$$

For this active high-pass filter, $A_o = R_f/R_i$. Substitute this into the general gain magnitude equation.

$$\frac{A_o}{\sqrt{2}} = \frac{\frac{R_f}{R_i}}{\sqrt{2}} = \frac{R_f}{\sqrt{R_i^{\,2} + X_C^{\,2}}}$$

$$R_f \times \sqrt{2} = \frac{R_f}{R_i}\sqrt{R_i^{\,2} + X_C^{\,2}}$$

$$2R_f^{\,2} = \frac{R_f^{\,2}}{R_i^{\,2}}\left(R_i^{\,2} + X_C^{\,2}\right)$$

$$2 = \frac{R_i^{\,2} + X_C^{\,2}}{R_i^{\,2}}$$

$$2R_i^{\,2} = R_i^{\,2} + X_C^{\,2}$$

$$R_i^{\,2} = X_C^{\,2}$$

At f_{-3dB} $X_C = R_i$
$$R_i = X_C$$

Substitute the equation for the capacitive reactance.

$$R_i = \frac{1}{2\pi f C_i}$$

Solve for f. This is the half-power point frequency.

Half-power point frequency, f_{-3dB}
$$f_{-3dB} = \frac{1}{2\pi R_i C_i}$$

This is just like the *CR* high-pass filter, and similar in form to the *RC* low-pass and active low-pass filters. Be sure to use the correct resistor.

Critical Frequency, f_o

The critical frequency can be derived just as the half-power point frequency. At the critical frequency, the phase shift for a first order, high-pass filter is 45°. However, since there is an additional inversion from the op amp, at the critical frequency, $\theta = 45° - 180° = -135°$.

Substitute this into the equation for the gain phase shift.

$$\Theta = -180° + \arctan\frac{X_C}{R_i}$$

$$-135° = -180° + \arctan\left(\frac{X_C}{R_i}\right)$$

$$45° = \arctan\left(\frac{X_C}{R_i}\right)$$

Take the tangent of each side.

$$1 = \frac{X_C}{R_i}$$

$$R_i = X_C$$

At the critical frequency, the reactance equals the resistance.

$$R_i = \frac{1}{2\pi f_o C_i}$$

$$f_o = \frac{1}{2\pi R_i C_i}$$

Critical frequency
f_o

Example 8-2

The single-supply op amp from Example 4-4 and Figure 4-10 is shown again in Figure 8-7.

a. Calculate its pass band gain, half-power point frequency, and the frequency at which the gain has fallen 5%.

b. Confirm your circuit's performance by simulation.

Solution

a. The pass band gain is

$$|A_o| = \frac{R_f}{R_i} = \frac{33\,\text{k}\Omega}{10\,\text{k}\Omega} = 10.4\,\text{dB}$$

The half-power point frequency is

$$f_{-3\text{dB}} = \frac{1}{2\pi R_i C_i}$$

$$f_{-3\text{dB}} = \frac{1}{2\pi \times 10\,\text{k}\Omega \times 47\,\text{nF}} = 339\,\text{Hz}$$

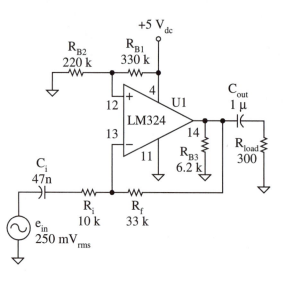

Figure 8-7 Single-supply high-pass filter for Example 8-2

At this frequency the gain drops to

$$A_{-3dB} = 10.4\,dB - 3\,dB = 7.4\,dB$$

In terms of *ratio* gain

$$A_{-3dB} = \frac{3.3}{\sqrt{2}} = 2.3$$

The frequency at which the gain has only fallen 5% is $f_{5\%}$.

$$f_{5\%} = 3f_{-3dB} = 3 \times 339\,Hz = 1016\,Hz$$

At this frequency the gain will have dropped to

$$A_{5\%} = 0.95 \times 3.3 = 3.14$$

b. The simulation is shown in Figure 8-8. The input is set to 353 mV$_p$ = 250 mV$_{rms}$ at 10 kHz. The output is 824 mV$_{rms}$. This is a pass band gain of 3.3. The upper Bode plotter has a log vertical scale. Its cursor indicates 7.4 dB of gain at 347 Hz. This is f_{-3dB}. The lower Bode plotter has a linear scale. Its cursor is at $f_{5\%}$, and indicates a gain of 3.14 at 1047 Hz. All of these measurements confirm the manual calculations.

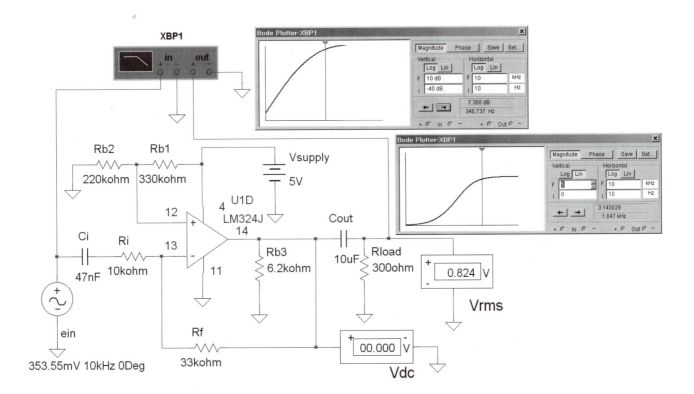

Figure 8-8 Simulation results for Example 8-2

Practice: What are the effects of changing R_i to 4.7 kΩ?

Answer: $A_o = 7 = 16.9$ dB, $f_{-3dB} = 720$ Hz, and $f_{5\%} = 2.2$ kHz.

8.4 Higher Order High-pass Filters

Section 7.4 explained the higher order Butterworth low-pass filters built by combining inductors and capacitors. Just replace each of the capacitors with an inductor and each of the inductors with a capacitor, and you have a comparable *high-pass* filter, shown in Figure 8-9.

Order selection is calculated just as it was for the higher order low-pass filters in section 7.4. However, the graph has flipped.

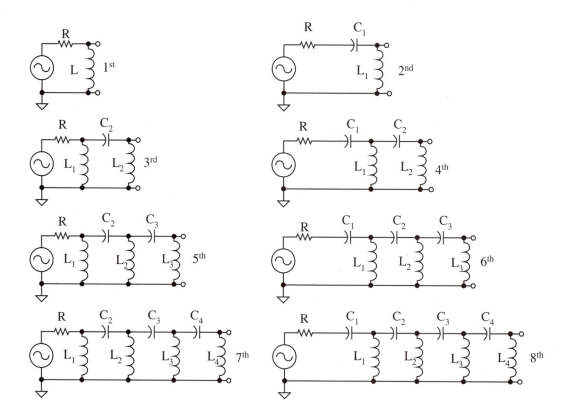

Figure 8-9 First through eighth order passive high-pass filters

Look at Figure 8-10.

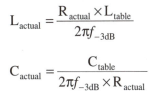

$$n = \frac{\log\left[\left(\dfrac{10^{-0.1A_{min(dB)}} - 1}{10^{-0.1A_{max(dB)}} - 1}\right)\right]}{2\log\left(\dfrac{f_{high}}{f_{low}}\right)}$$

In Table 8-1, the inductors and capacitors have been swapped from Table 7-1, and their values have been reciprocated. The values must be adjusted since they assume that R = 1 Ω and ω_{-3dB} = 1 rad/s. To convert these into values you can use in a circuit,

$$L_{actual} = \frac{R_{actual} \times L_{table}}{2\pi f_{-3dB}}$$

and

$$C_{actual} = \frac{C_{table}}{2\pi f_{-3dB} \times R_{actual}}$$

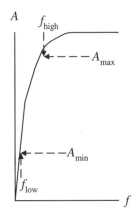
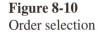

Figure 8-10
Order selection

To design a circuit, then, you first determine which order filter is needed. Then repeatedly try standard values of R_{actual} until you find a set of capacitors and inductors that are close to standard values. This could be a tedious process. However, Table 8-1 has been combined with these two equations in a spreadsheet, which is included on the cd that accompanies this text. You enter f_{-3dB} and then a standard value for R_{actual}. Examine the resulting inductors and capacitors displayed by the spreadsheet. If these values are inconvenient, try a different R_{actual}.

Table 8-1 High-pass Butterworth component relationships

order	C1	L1	C2	L2	C3	L3	C4	L4
2	1.4142	0.707						
3		2.000	0.750	0.667				
4	2.6130	0.924	0.634	0.653				
5		3.236	1.118	0.724	0.590	0.647		
6	3.8640	1.319	0.832	0.644	0.569	0.652		
7		4.494	1.524	0.949	0.716	0.603	0.556	0.630
8	5.1256	1.731	1.067	0.794	0.654	0.578	0.548	0.641

A.B. Williams, F.J. Taylor, *Electroni Filter Design Handbook 3rd ed*, McGraw Hill, pages 11.5 - 11.11

how big can you dream?™

Example 8-3

a. Design a filter using standard value components, with f_{-3dB} = 3 kHz, and gain = −40 dB at 900 Hz.

b. Confirm the performance by simulation.

Solution

a. A_{max} = −3dB, f_{high} = 3 kHz; A_{min} = −40 dB, f_{low} = 900 Hz

$$n = \frac{\log\left[\left(\dfrac{10^{-0.1\times-40\,dB}-1}{10^{-0.1\times-3\,dB}-1}\right)\right]}{2\log\left(\dfrac{3\,kHz}{900\,Hz}\right)}$$

$$n = \frac{\log\left(\dfrac{9999}{0.995}\right)}{1.0458} = 3.83$$

Select a fourth order filter. The gain at 900 Hz should be a little more negative than the −40 dB specified. By trial and error or with the spreadsheet, pick R = 560 Ω.

$$C_1 = \frac{2.613}{2\pi\times3\,kHz\times560\,\Omega} = 0.247\,\mu F \quad \text{Pick } 0.24\ \mu F$$

$$L_1 = \frac{0.924\times560\,\Omega}{2\pi\times3\,kHz} = 27.45\,mH \qquad \text{Pick } 27\ mH$$

C_2 = 60.1 nF   Pick 56 nF in parallel with 3.9 nF

L_2 = 19.4 mH   Pick 20 mH

b. The simulation is shown in Figure 8-11.

Practice: Design a filter f_{-3dB} = 1 MHz, $A_{500\,kHz}$ = −28 dB, R = 50 Ω

Answer: n = 5, L_1 = 26 μH, C_2 = 3.6 nF, L_2 = 5.8 μH, C_3 = 1.9 nF, L_3 = 5.1 μH

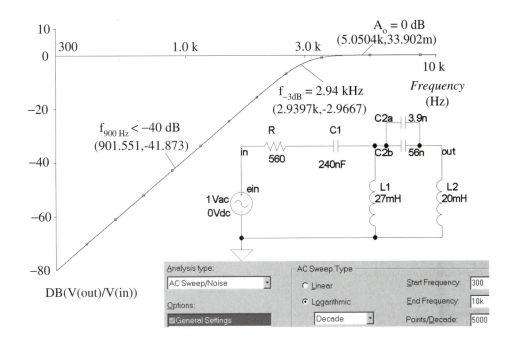

Figure 8-11 Simulation results for Example 8-4

Summary

Begin analyzing a filter with a quick-look. First consider the filter's performance at dc, with all inductors replaced with shorts and all capacitors replaced with opens. Determine the gain for this simplified circuit. Then determine the filter's gain at extremely high frequency. Replace the inductors with opens and the capacitors with shorts. From this brief review you can determine the type of filter, its pass band gain, its order, its roll-off rate, and the phase shift at the critical frequency.

You derive the gain equation by calculating the output voltage in terms of the input signal, the resistors, capacitive reactances, and inductive reactances in the circuit. Remember to *add* phasors in their rectangular form, and to convert them to polar for *division*. The final gain equation can be separated into a magnitude and a phase.

The equation for the half-power point frequency is determined by setting the gain magnitude equation equal to $A_o / \sqrt{2}$. The frequency is encoded in the reactances. The equation for the critical frequency comes from setting the gain phase equation equal to 45° for first order filters or 90° for second order filters, and solving for the frequency.

These steps were applied to several filters.

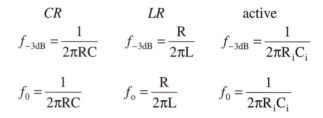

$$CR \qquad\qquad LR \qquad\qquad active$$

$$f_{-3dB} = \frac{1}{2\pi RC} \qquad f_{-3dB} = \frac{R}{2\pi L} \qquad f_{-3dB} = \frac{1}{2\pi R_i C_i}$$

$$f_0 = \frac{1}{2\pi RC} \qquad f_0 = \frac{R}{2\pi L} \qquad f_0 = \frac{1}{2\pi R_i C_i}$$

Higher order filters can be built by configuring a resistor with series capacitors and inductors to common. Use the same configuration as you did for the low-pass higher order filters, replacing each capacitor with an inductor and each inductor with a capacitor. Calculate the order needed from the specified gains at two frequencies. Then scale the coefficients given in Table 8-1 for f_{-3dB} and for the R you want. Repeat this process until you find a set of standard values.

Problems

The *CR* High-pass Filter

8-1   For a simple *CR* high-pass filter with R = 1 kΩ and C = 56 nF, calculate:

A_o
filter order
roll-off rate
f_{-3dB}
f_0
gain magnitude at f_{-3dB}
phase shift at f_0

8-2   Repeat Problem 8-1 if R = 50 Ω, and C = 33 pF.

8-3   Design a *CR* high-pass filter with a minimum input impedance of 10 kΩ, and f_{-3dB} = 280 Hz.

8-4   Design a *CR* high-pass filter with a minimum input impedance of 75 Ω, and f_{-3dB} = 540 kHz.

8-5   For the circuit in Figure 8-12:

    a. Complete a quick-look analysis.

    b. Derive the gain magnitude and gain phase equations.

    c. Derive the equation for f_{-3dB}.

    d. Derive the equation for f_o.

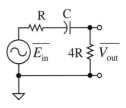

Figure 8-12
Schematic for Problem 8-5

The *RL* High-pass Filter

8-6   For a simple *RL* high-pass filter with R = 1 kΩ and L = 470 mH, calculate:

    A_o
    filter order
    roll-off rate
    f_{-3dB}
    f_o
    gain magnitude at f_{-3dB}
    phase shift at f_o

8-7   Repeat Problem 8-6 if R = 50 Ω, and L = 75 nH.

8-8   Design an *RL* high-pass filter using $R = 4\ \Omega$, and $f_{-3dB} = 5$ kHz.

8-9   Design an *RL* high-pass filter with R = 50 Ω, and f_{-3dB} = 540 kHz

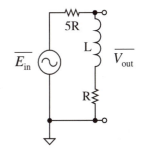

8-10   For the circuit in Figure 8-13:

    a. Complete a quick-look analysis.

    b. Derive the gain magnitude and gain phase equations.

    c. Derive the equation for f_{-3dB}.

    d. Derive the equation for f_o.

Figure 8-13
Schematic for Problem 8-10

The Simple Active High-pass Filter

8-11   For a simple active high-pass filter with R_i = 1 kΩ, R_f = 22 kΩ, and C_f = 1.5 nF, calculate:

    A_o
    filter order
    roll-off rate
    f_{-3dB}
    f_o
    gain magnitude at f_{-3dB}
    phase shift at f_o

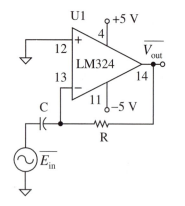

Figure 8-14
Schematic for problem 8-15

8-12   Repeat Problem 8-11, Figure 8-7 if $R_i = 51\ \Omega$, $R_f = 150\ \Omega =$ and $C_i = 100$ pF.

8-13   Design a simple active high-pass filter with a minimum input impedance of 1 kΩ, $A_o = 30.4$ dB, and $f_{-3\text{dB}} = 1.5$ kHz.

8-14   Design a simple active high-pass filter with a minimum input impedance of 75 Ω, $A_o = 45$ dB, and $f_{-3\text{dB}} = 540$ kHz.

8-15   For the circuit in Figure 8-14:
    a.   Complete a quick-look analysis.

    b.   Derive the gain magnitude and gain phase equations

Higher Order High-pass Filters

8-16   Design a Butterworth filter with $A_o = -3$ dB, $f_{-3\text{dB}} = 300$ kHz, and $A_{90\ \text{kHz}} < -39$ dB. Confirm performance by simulation.

8-17   Design a Butterworth filter with $A_o = -3$ dB, $f_{-3\text{dB}} = 2$ MHz, and $A_{500\ \text{kHz}} < -65$ dB. Confirm performance by simulation.

High-pass Filters Lab Exercise

A.   Second Order High-pass Filter.
    1.   Build a second order high-pass filter with

$$R = 56\ \Omega, \quad C = 0.22\ \mu\text{F}, \quad L = 33\ \text{mH}$$

Keep the leads from the signal generator and the leads to each component as short as practical.

2.   Set the signal generator's amplitude to 1 V_{rms} and its frequency to 10 kHz.

3.   Verify the pass band gain (A_o) of your circuit by connecting the oscilloscope and digital multimeter between the output and common. V_{out} should be a maximum at 10 kHz. Record this gain ratio, and compare it to the theoretical pass band gain.

4.   Lower the frequency until the output voltage has fallen to 0.707 of its value in step A3. This frequency is $f_{-3\text{dB}}$. Record it and compare it to the theoretical half-power point frequency.

A_o
measured = _____   theory = _____

$f_{-3\text{dB}}$
measured = _____   theory = _____

f_o
measured= _____   theory = _____

5. Measure, record, and compare the critical frequency (f_o).

6. Lower the frequency to 500 Hz. Record the gain in dB.

7. Lower the frequency to 250 Hz. Record the gain in dB again.

8. Calculate the dB/octave roll-off rate. Compare that to the theoretical roll-off rate.

$A_{@\ 200\ Hz} = $ _____ dB

$A_{@\ 100\ Hz} = $ _____ dB

Roll-off rate
    measured = _____ dB/octave

    theory = _____ dB/octave

9. Construct a spreadsheet with columns for *frequency* (kHz), V_{out} (V_{rms}), *gain* (dB), *phase* (degrees). Provide a row for the column titles and then 19 rows for data. Enter the frequency data, one in each row (0.1, 0.2, 0.3, 0.4, 0.5, 0.6, 0.7, 0.8, 0.9, 1.0, 2.0, 3.0, 4.0, 5.0, 6.0, 7.0, 8.0, 9.0, 10.0).

10. Create a frequency response graph of *frequency* (log scale) on the *x* axis versus *gain* (dB) on the *y* axis. Do this before you begin taking data.

11. Create a frequency response graph of *frequency* (log scale) on the *x* axis versus *phase* on the *y* axis. Do this before you begin taking data.

12. Set the input to 100 Hz, 1 V_{rms}. Measure the output rms voltage magnitude and phase. Allow the spreadsheet to calculate the dB and begin plotting the graph.

13. Change the frequency to that indicated in the next row. Adjust the input voltage magnitude to 1 V_{rms}. Measure the output magnitude and phase. Check the spreadsheet's plots.

14. Compete the table and the plots.

B. Simple Active High-pass Filter
    Repeat all of the steps from Section A for the simple active high-pass filter that you designed in Problem 8-13.
    - Use an LM324 op amp.
    - Set its supplies to ±5 V_{dc}.
    - Set the input signal's amplitude to 50 mV_{rms}.

9

Band-pass Filters and Resonance

Introduction

Band-pass filters combine the function of low- and high-pass filters, blocking both high- and low-frequency signals while passing only those signals whose frequency lies in a narrow band.

Band-pass filters are found in a variety of applications. One of the most common is in the selection of a single communications channel, whether for a garage door opener, a pager, or a television. The mid-range loudspeaker driver in an array can respond only to certain frequency tones. The others must be blocked. Signals from sensors in your car, in your home, or in an industrial manufacturing facility may be carried by an electrical wave of a specific frequency. Lower or higher frequency signals are noise and should be rejected.

Since band-pass filters roll-off on two edges, there are several new parameters to understand. The high-pass and low-pass filters you have already seen may be combined to form broad, flat-topped band-pass filters. However, there are several restrictions. Op amps may also be added to provide gain. Resonance is a phenomenon common to many areas of engineering. Inductors and capacitors in series or in parallel may resonate at a single frequency, which forms a band-pass filter.

Objectives

Upon completion of this chapter, you will be able to do the following for an active high-pass, low-pass filter, series resonant, and parallel resonant circuit:

- Draw the schematic.

- Determine the filter type, pass band gain, order, and roll-off rates.

- Explain when to use the circuit and its limitations.

- Calculate cut-off frequencies, center frequency, bandwidth, and Q.

- Analyze the filter to plot its frequency response.

- Design the filter to meet given specifications.

9.1 Band-pass Terminology

The frequency response plot of a typical band-pass filter is shown in Figure 9-1. The gain begins low, rises to a peak, and then falls off.

The **pass band gain**, A_o is the highest gain.

The **center frequency**, f_{center}, is the frequency at which the gain is a maximum. With a log horizontal axis, it appears to lie in the center. Since the gain drops on each side of f_{center}, there are two half-power point frequencies. The **low frequency cut-off**, f_{low}, is that frequency below f_{center} at which the gain has dropped 3 dB below A_o. On the other side, the **high frequency cut-off**, f_{high}, is that frequency above f_{center} at which the gain has dropped 3 dB below A_o.

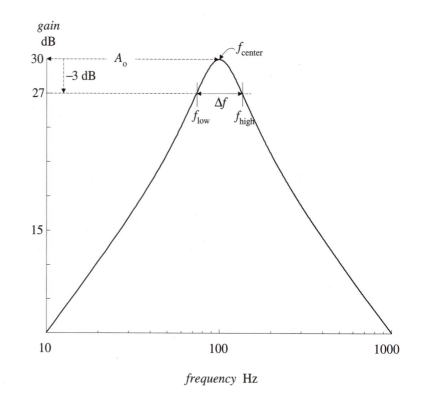

Figure 9-1 Band-pass filter's frequency response

In Figure 9-1 it appears that f_{center} lies precisely half-way between f_{low} and f_{high}. That is because the horizontal axis is scaled logarithmically. Actually the center frequency lies at the **geometric mean** of the two cut-off frequencies (not their algebraic mean).

$$f_{center} = \sqrt{f_{low} \times f_{high}}$$

The distance *between* the cut-off frequencies is the **bandwidth**, Δf.

$$\Delta f = f_{high} - f_{low}$$

Is a filter with a 10 kHz bandwidth considered very broad and open, or is it very narrow and selective? That depends on the center frequency. If the center frequency is 1.7 kHz, and the filter passes everything from 300 Hz to 10.3 kHz, then it appears to be broad, passing almost the entire audio range. However, if the filter is centered around 100 MHz, with cut-offs at 99.995 MHz and 100.005 MHz, it is *very* selective, passing only a narrow band of signals.

A band-pass filter's **selectivity**, Q, is a measure of the filter's sharpness, or narrowness.

$$Q = \frac{f_{center}}{\Delta f}$$

With some algebraic manipulation, these parameters provide

$$f_{low} = f_{center}\left(\sqrt{\frac{1}{4Q^2} + 1} - \frac{1}{2Q} \right)$$

$$f_{high} = f_{center}\left(\sqrt{\frac{1}{4Q^2} + 1} + \frac{1}{2Q} \right)$$

Example 9-1

One form of band-pass filter has the gain magnitude equation

$$|\overline{A}| = \frac{A_o \times R}{\sqrt{R^2 + (X_L - X_C)^2}}$$

- Set $A_o = 31.6$, $R = 1\ k\Omega$, $L = 2.533\ H$, $C = 1\ \mu F$.

- Sweep the frequency from 10 Hz to 400 Hz in steps of 2 Hz.

- At each frequency, calculate the reactances and the dB gain.

- Plot the frequency response.

- Tabulate frequency and dB gain.

- From the table, identify:
    f_{center}, A_o, f_{low}, f_{high}, Δf, and Q.

Solution

The frequency response plot is shown in Figure 9-1. The MATLAB m file is given in Figure 9-2.

```
format short g
f=(10:2:400)';
R=1e3;
C=1e-6;
L=2.533;

Xc=1./(2*pi*f*C);
Xl=2*pi*f*L;

gain=31.62*R./sqrt(R^2+(Xl-Xc).^2);
dB=20*log10(gain);

gain_table=[f dB]
plot(f,dB)
```

Apostrophe places data in a column, rather than a row.

Produces an array with frequency in the first column and dB in the second.

part of gain_table

Hz	dB
70	26.299	← f_{low}
75	27.3
80	28.201
85	28.961
90	29.535
95	29.885
100	29.999	← f_{center} & A_o
105	29.896
110	29.616
115	29.21
120	28.726
125	28.201
130	27.66
135	27.121	← f_{high}
140	26.593

Figure 9-2 MATLAB instructions and partial results for Example 9-1

$$f_{center} = 100 \text{ Hz} \qquad A_o = 30 \text{ dB}$$

The cut-off frequencies occur where the gain falls to 27 dB.

$$f_{low} \approx 73 \text{ Hz} \qquad f_{high} \approx 136 \text{ Hz}$$

The bandwidth is

$$\Delta f = 136 \text{ Hz} - 73 \text{ Hz} = 63 \text{ Hz}$$

Selectivity is

$$Q = \frac{100 \text{ Hz}}{63 \text{ Hz}} = 1.59$$

Practice: Repeat Example 9-1 for $A_o = 178$, $R = 50 \ \Omega$, $L = 600$ nH, $C = 6.6$ pF. Sweep f from 20 MHz to 300 MHz in 1 MHz steps.

Answer: $f_{center} = 80$ MHz, $f_{low} = 73.5$ MHz, $f_{high} = 87$ MHz, $Q = 5.9$

9.2 Active High-pass Low-pass Filter

In Section 7.3 you saw the active low-pass filter. It is built by placing a capacitor in parallel with the feedback resistor, R_f. As frequency increases, the impedance of the parallel combination of R_f and C_f drops, lowering the gain. An active high-pass filter is presented in Section 8.3. In that circuit, a capacitor is placed in series with the input resistor, R_i. At low frequencies, the input capacitor's impedance is much larger than R_i, dropping the input current and therefore the output voltage. In both circuits, the magnitude of the pass band gain is set by R_f/R_i. Each also adds (or subtracts) 180° to the output voltage's phase. Figure 9-3 is the combination of these two filters.

The two resistors set the *magnitude* of the pass band gain.

$$A_o = \frac{R_f}{R_i}$$

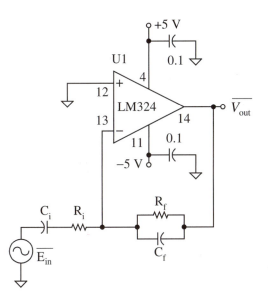

Figure 9-3 Active high-pass low-pass type band-pass filter

$$\overline{Z_{inmin}} = (R_i \angle 0°)$$

$$f_{low} = \frac{1}{2\pi R_i C_i}$$

$$f_{high} = \frac{1}{2\pi R_f C_f}$$

$f_{high} > 9\ f_{low}$
i.e.
$Q < 0.375$

how big can you dream?™

Negative feedback assures that the inverting input pin is held at virtual ground. So, R_i is the minimum input impedance.

The input capacitor and resistor form a first order high-pass filter. Use it to establish the low-frequency edge of the band-pass filter's response. There is a 6 dB/octave roll-off.

The virtual ground also isolates the input from the output. This means that R_f will *not* affect R_i and C_i. So, use

- R_i to set the input impedance.
- C_i to set the low-frequency cut-off, f_{low}.
- R_f to set the pass band gain, A_o.
- C_f to set the high-frequency cut-off, f_{high}.

If you try to bring the cut-off frequencies too close together, one filter begins attenuating the output of the other, lowering the overall gain. You can compensate, to a small degree, by boosting A_o. But changes in the pass band gain also affect the gain at the cut-offs. So, use this only to tweak the circuit's performance a few percent.

Example 9-2

For the circuit in Figure 9-3, given:

$E_{in} = 0.1\ V_p,\ C_i = 36\ nF,\ R_i = 10\ k\Omega,\ R_f = 120\ k\Omega,\ C_f = 270\ pF$

a.   Calculate:

$Z_{in\ min},\ A_o,\ f_{low},\ f_{high},\ f_{center},$ and Q.

b.   Verify the calculations with a simulation.

Solution:

a.

$$Z_{in\ min} = R_i = 10\ k\Omega$$

$$A_o = \frac{R_f}{R_i} = \frac{120\,k\Omega}{10\,k\Omega} = 12$$

$$A_o = 20 \times \log(12) = 21.6\,dB$$

$$f_{low} = \frac{1}{2\pi \times 10\,k\Omega \times 36\,nF} = 442\,Hz$$

$$f_{high} = \frac{1}{2\pi \times 120\,k\Omega \times 270\,pF} = 4.91\,kHz$$

$$f_{center} = \sqrt{442\,Hz \times 4.91\,kHz} = 1.47\,kHz$$

$$Q = \frac{1.47\,\text{kHz}}{4.91\,\text{kHz} - 442\,\text{Hz}} = 0.329 < 0.375$$

b. The simulation frequency response is shown in Figure 9-4. The center frequency occurs at the predicted frequency, and the pass band gain is about 0.8 dB low, as expected. The low frequency cut-off occurs where it should, but the high frequency cut-off is a little low. This is caused by the reduction of the op amp's open loop gain at higher frequencies. This will be discussed in detail in the following chapter.

The data points needed to calculate the roll-off rates on each edge are shown. They are ±6 dB/octave or ±20 dB/decade.

Practice: Repeat Example 9-3 if R_i is changed to 6.8 kΩ, and C_f is changed to 220 pF.

Answer: $Z_{\text{in min}} = 6.8$ kΩ, $A_o = 17.6 = 24.9$ dB, $f_{\text{low}} = 650$ Hz, $f_{\text{high}} = 6.0$ kHz, $f_{\text{center}} = 1.97$ kHz, $Q = 0.369$

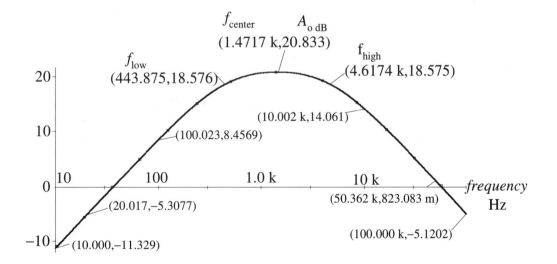

Figure 9-4 Simulation results for Example 9-2

9.3 Resonance

Resonance is a phenomenon in which energy is exchanged back and forth between two forms. This exchange becomes quite protracted when it occurs at a preferred rate, or frequency. Examples are found in mechanics, music, electricity, chemistry, biology, and even in psychology.

A child's swing is a common mechanical resonant system. When the swing is pulled back, *potential* energy is stored in the swing's height. When it is released, it moves downward, exchanging its potential energy (stored energy of position) for *kinetic* energy (the energy of motion). When the swing reaches the bottom of its arc, all of the potential energy is gone. It has become kinetic energy. This momentum carries the swing upward, against gravity. As it rises, it slows, storing potential energy in its rising position. Before long, the swing comes to a stop, at the top of its arc. All of the kinetic energy has been converted into potential energy. Gravity takes over, pulling the swing downward again, converting the potential energy (of position) to kinetic energy (of motion).

The process repeats, over and over; potential energy to kinetic energy to potential energy. Each cycle a little energy is lost to friction and to air resistance. So, each cycle the swing rises just a little less. By pulling back on the rope at just the right time, the swinger can shorten the rope, lifting the swing, and adding back the potential energy lost to friction. The process continues, with very little energy input.

The frequency of the energy exchange is *independent* of the amount of energy. It is set by characteristics of the system. For the swing, the number of cycles per minute does *not* depend on the height from which you start. Starting higher gives you more potential energy, which converts to more kinetic energy (speed) at the bottom, and a higher rise on the other side. You go faster because you began with more potential energy. You go further because you were going faster at the bottom. The *length* of the swing's rope controls its frequency.

To overcome the effects of friction, and to sustain the oscillations, you must pump the swing at just the right time. As it falls, you pull back on the rope, shortening it, and increasing the potential energy. As you cross through the bottom, you stop pumping. Now the swing is on the opposite cycle. In a resonant system, oscillations can be sustained *only* if the energy is added at the proper rate, and phase.

In personal relationships, resonance can occur. When the two people are uniquely attuned to each other, energy, affection, and support flow from one to the other, at just the time, and in just the form that is needed by the recipient. They are stored, and later returned, in the form that the other needs. This resonant system, at its best, is largely immune to external forces, flying higher when more energy is added, moving lower as the energy wanes. But the frequency is constant, and requires only an occasional, properly timed boost, to keep going.

In electrical systems, inductors and capacitors form a resonant system. The inductor stores energy in an electro*magnetic* field, and causes the current through it to *lag* the voltage across it by 90° (ELI). The capacitor stores energy in an electro*static* field. It causes the current through it to *lead* the voltage across it by 90° (ICE).

Combined in a circuit, and driven at just the right frequency, the inductor releases energy at just the rate at which the capacitor stores it. The capacitor discharges at the rate that the inductor charges. Back and forth, charge and discharge, storing energy and releasing energy, they resonate. Resonance occurs at only one frequency. Driven by a signal whose frequency is below or above the circuit's natural resonant frequency, little happens (just like trying to pump the swing at the wrong time). Inductor-capacitor resonant circuits form band-pass filters.

9.4 Series Resonance

Figure 9-5 is the schematic of *LCR* elements forming a band-pass filter. This circuit is resonant. Energy is stored in the electromagnetic field of the inductor. When it discharges, the capacitor charges, storing energy in its electrostatic field. When the capacitor discharges, the inductor is charged. The resistor acts as friction, converting some of the energy every cycle into heat and lowering the amplitude of the oscillations. The input source restores this energy. At low frequencies, the capacitive impedance is large, there is little current, and the voltage across the resistor is low. This is like pumping the swing too slowly. At high frequencies, the inductive impedance is large, there is little current and the voltage across the resistor is low. This is like pumping furiously.

The circuit's total impedance is

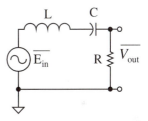

Figure 9-5
Series resonant *LCR* circuit

$$\overline{Z_{total}} = \overline{Z_R} + \overline{Z_L} + \overline{Z_C}$$

$$\overline{Z_{total}} = (R + j0) + (0 + jX_L) + (0 - jX_C)$$

$$\overline{Z_{total}} = [R + j(X_L - X_C)]$$

Figure 9-6 is the plot of the magnitude of this impedance versus frequency for one *LCR* circuit, along with the MATLAB instructions.

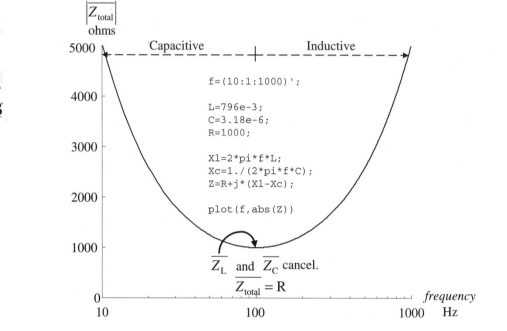

Figure 9-6 Impedance versus frequency for a series resonant circuit

At the resonant frequency, the impedance of the inductor is equal in magnitude and opposite in phase to the impedance of the capacitor. They cancel. Only the opposition of the resistor is left.

$$X_L = X_C \qquad\qquad 2\pi f_r L = \frac{1}{2\pi f_r C}$$

$$f_r^2 = \frac{1}{4\pi^2 LC}$$

Resonant frequency, f_r

$$f_r = \frac{1}{2\pi\sqrt{LC}}$$

At the resonant frequency, the inductive impedance is equal in magnitude and opposite in phase to the capacitive impedance. The current is a maximum. The voltage across the inductor exactly cancels the voltage across the capacitor. This means that the voltage across the resistor, at the output, is equal to the source voltage.

$$\overline{V_{R\,@\,resonance}} = \overline{E_{supply}}$$

Output voltage at resonance

The pass band gain is $A_o = 1 = 0 \text{ dB}$

Gain Derivation

The equation for the output voltage comes from the voltage divider law.

$$\overline{V_{out}} = \overline{E_{in}} \times \frac{\overline{Z_R}}{\overline{Z_R} + \overline{Z_L} + \overline{Z_C}}$$

The gain is needed in order to plot the frequency response.

$$\overline{A} = \frac{\overline{V_{out}}}{\overline{E_{in}}} = \frac{\overline{Z_R}}{\overline{Z_R} + \overline{Z_L} + \overline{Z_C}}$$

Express the impedances in terms of R, X_L, and X_C.

$$\overline{A} = \frac{(R\angle 0°)}{(R\angle 0°) + (X_L\angle 90°) + (X_C\angle -90°)}$$

$$\overline{A} = \frac{(R\angle 0°)}{(R + j0) + (0 + jX_L) + (0 - jX_C)}$$

$$\overline{A} = \frac{(R\angle 0°)}{[R + j(X_L - X_C)]}$$

Place the numerator and denominator in their polar form.

$$\overline{A} = \frac{(R\angle 0°)}{\left(\sqrt{R^2 + (X_L - X_C)^2}\angle \arctan\left(\dfrac{X_L - X_C}{R}\right)\right)}$$

$$\overline{A} = \frac{R}{\sqrt{R^2 + (X_L - X_C)^2}} \angle \left(-\left(\arctan\dfrac{X_L - X_C}{R}\right)\right)$$

The *magnitude* of the gain is

Gain *magnitude*

$$\left|A\right| = \frac{R}{\sqrt{R^2 + (X_L - X_C)^2}}$$

At low frequency, X_C becomes very large. The gain magnitude goes to 0. At very high frequencies, X_L becomes very large, much larger than R or X_C. Again, the gain goes to 0. However, at the frequency where the reactances equal, they cancel. The gain rises to R/R = 1.

The gain's phase shift is

Gain *phase*

$$\Theta = -\arctan\left(\frac{X_L - X_C}{R}\right)$$

At low frequencies, X_C is very large. The arctangent of a very large negative number approaches −90°. When frequency goes up, the inductive reactance becomes dominant. The arctangent becomes a positive angle, eventually 90°. At resonance, the inductive reactance cancels the capacitive reactance. Arctan(0) = 0°. At resonance, the circuit is purely resistive. There is no phase shift between input and output.

Resonant Frequency, $f_r = f_{center}$

At the center frequency, the gain ratio is 1.

$$\left|A_{@\,f\,center}\right| = 1$$

Substitute this into the general gain magnitude equation.

$$1 = \frac{R}{\sqrt{R^2 + (X_L - X_C)^2}}$$

$$R = \sqrt{R^2 + (X_L - X_C)^2}$$

$$R^2 = R^2 + (X_L - X_C)^2$$

$$0 = (X_L - X_C)^2$$

Take the square root of each side.

$$X_L - X_C = 0$$

$$X_L = X_C$$

At f_{center} $X_C = X_L$

Substitute the equation for both reactances.

$$2\pi f L = \frac{1}{2\pi f C}$$

Rearrange the equation to isolate f. This is the center frequency.

$$f_{center}^{\;2} = \frac{1}{2\pi L \times 2\pi C}$$

**Center frequency
(f_{center})**

$$f_{center} = \frac{1}{2\pi\sqrt{LC}}$$

Critical Frequency, f_o

At the critical frequency, the phase shift for a band-pass filter is $0°$. Substitute this into the equation for the gain phase shift.

$$0° = -\arctan\left(\frac{X_L - X_C}{R}\right)$$

$$0° = \arctan\left(\frac{X_L - X_C}{R}\right)$$

Take the tangent of each side.

$$0 = \frac{X_L - X_C}{R}$$

Multiply each side of the equation by R.

$$0 = X_L - X_C$$

$$X_L = X_C$$

$$2\pi f_o L = \frac{1}{2\pi f_o C}$$

$$f_o^{\;2} = \frac{1}{2\pi L \times 2\pi C}$$

$$f_o = \frac{1}{2\pi\sqrt{LC}}$$

Quality and Selectivity, *Q*

The inductor and the capacitor store energy while the resistor converts it into heat. The resistor is the "friction" in the resonant circuit. The **quality** of the resonant circuit indicates how effectively energy is stored in the inductor and capacitor as opposed to being dissipated by the resistor. Since power is the rate of doing work, *Q* can be expressed in terms of the reactive power compared to the resistive power.

$$Q = \frac{reactive\ power}{resistive\ power}$$

For a series circuit, current is the same through all of the elements. At resonance, the inductor and the capacitor have the same reactance.

Quality factor for a series resonant circuit

$$Q_s = \frac{I^2 X}{I^2 R} = \frac{X}{R}$$

This is the same *Q* as the selectivity.

$$Q = \frac{f_{center}}{\Delta f}$$

Adjusting the values of L and C sets f_{center}. The value of the resistor, compared to the reactance at f_{center} determines the location of f_{high} and f_{low}.

Example 9-3

    a. Design a cross-over network for the midrange loudspeaker driver with $f_{low} = 500$ Hz and $f_{high} = 2$ kHz. Use the driver's 8 Ω resistance as the R in the network.

    b. Confirm these two frequencies by simulation.

multiSIM

Solution

    a.

$$f_{center} = \sqrt{500\,\text{Hz} \times 2\,\text{kHz}} = 1\,\text{kHz}$$

$$Q = \frac{1\,\text{kHz}}{2\,\text{kHz} - 500\,\text{Hz}} = 0.67$$

$$Q_s = \frac{X}{R}$$

$$X_L = Q_s \times R = 0.67 \times 8\,\Omega = 5.36\,\Omega$$

This reactance of 5.36 Ω occurs at the center frequency.

$$X_L = 2\pi f_{center} L = 5.36\Omega$$

$$L = \frac{5.36\Omega}{2\pi \times 1\,kHz} = 853\,\mu H$$

At the resonant frequency, $X_C = X_L$.

$$C = \frac{1}{2\pi \times 1\,kHz \times 5.36\Omega} = 29.7\,\mu F$$

b. The simulation results are shown in Figure 9-7. The cursor in the top Bode plotter indicates that the gain has fallen to −3dB at 500 Hz. The lower Bode plotter marks f_{high} at 2 kHz.

Practice: Design a series *LCR* resonant circuit with f_{low} = 1.2 MHz, f_{high} = 1.4 MHz, and R = 75 Ω.

Answer: f_{center} = 1.3 MHz, Q = 6.5, X = 486 Ω, L = 59.7 μH, C = 253 pF

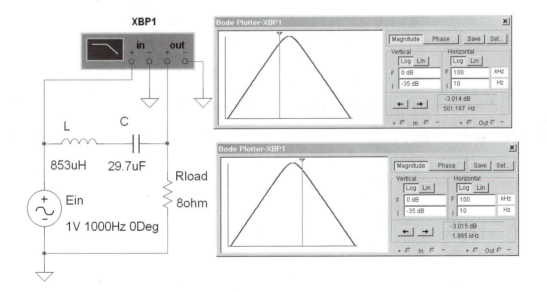

Figure 9-7 Simulation results for Example 9-3

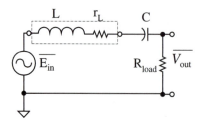

Figure 9-8
LCR circuit with internal r_L

The Effects of the Inductor's Resistance

Inductors are built by looping wire into a coil, often around a core of ferromagnetic material. The more loops of wire in the coil, the higher the inductance. This wire has resistance. The more wire, the more inductance, *and* the more resistance. Inductors with 10 mH or more of inductance may have 5 Ω or more of *internal* resistance. You can get a reasonable measure of the inductor's internal resistance by measuring the inductor with a dc ohmmeter. This effect is shown in Figure 9-8.

The internal resistance alters the pass band gain, A_o. At resonance, the impedance of the inductance cancels the impedance of the capacitor. This leaves r_L between the source and the load, forming a voltage divider.

$$\overline{V}_{out} = \overline{E}_{in} \times \frac{R_{load}}{r_L + R_{load}}$$

Pass band gain with r_L

$$A_o = \frac{R_{load}}{r_L + R_{load}}$$

The other effect of the inductor's internal resistance is to lower the circuit's Q.

$$Q_s = \frac{X}{r_L + R_{load}}$$

Example 9-4

   a. Add a 3 Ω internal resistance to the inductor in Example 9-3, Figure 9-7. Determine the effect on A_o, Q, f_{low}, and f_{high}.

   b. Verify your calculations by a simulation.

Solution

   a.
$$A_o = \frac{8\,\Omega}{3\,\Omega + 8\,\Omega} = 0.727 = -2.77\,\text{dB}$$

$$Q_s = \frac{5.36\,\Omega}{3\,\Omega + 8\,\Omega} = 0.487$$

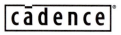

how big can you dream?™

$$f_{low} = f_{center}\left(\sqrt{\frac{1}{4Q^2}+1}-\frac{1}{2Q}\right) = 407\,\text{Hz}$$

$$f_{high} = f_{center}\left(\sqrt{\frac{1}{4Q^2}+1}+\frac{1}{2Q}\right) = 2.46\,\text{kHz}$$

The internal resistance of the inductor creates ~ 30% error in A_o, and over 20% shift in the cut-off frequencies.

b. The simulation results are shown in Figure 9-9.

Practice: Calculate A_o and Q_s for L = 59.7 μH, C = 253 pF, R_{load} = 75 Ω, and r_l = 15 Ω. Compare these to the results in Practice 9-3.

Answer: $A_o = 0.833 = -1.59$ dB, $Q_s = 5.4$. Both have been lowered.

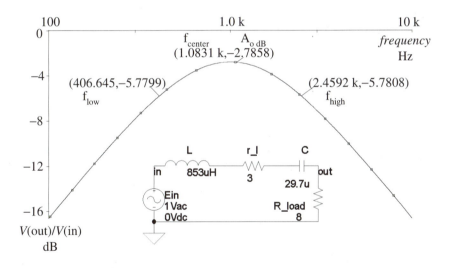

Figure 9-9 Simulation results for Example 9-4

9.5 Parallel Resonance

An inductor placed in *parallel* with a capacitor also forms a resonant circuit. Generally, a resistive load is included, and the parallel elements are driven by a *current* source. A typical schematic is shown in Figure 9-10.

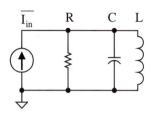

Figure 9-10
Parallel resonant circuit

At low frequencies, the impedance of the inductor is very low. Most of the current from the source flows through that small inductive impedance, producing only a small voltage across the parallel circuit.

At high frequencies, the impedance of the capacitor is very low. Most of the current from the source flows through that small capacitive impedance, producing a small voltage across the parallel circuit.

At the *resonant* frequency, the impedance of the inductor is exactly equal and opposite to the impedance of the capacitor. They cancel each other. Placed in parallel, this means that together they look like an *open*. The impedance, at resonance, is equal to the resistance of the resistor.

$$\overline{Z_{\text{total}}} = \left[\frac{1\angle 0°}{R\angle 0°} + \frac{1\angle 0°}{X_C \angle -90°} + \frac{1\angle 0°}{X_L \angle 90°} \right]^{-1}$$

$$\overline{Z_{\text{total}}} = \left[\frac{1}{R}\angle 0° + \frac{1}{X_C}\angle 90° + \frac{1}{X_L}\angle -90° \right]^{-1}$$

$$\overline{Z_{\text{total}}} = \left[\frac{1}{R} + j\left(\frac{1}{X_C} - \frac{1}{X_L} \right) \right]^{-1}$$

At resonance, the capacitive reactance equals the inductive reactance.

$$\overline{Z_{\text{total}}} = \left[\frac{1}{R} + j\left(\frac{1}{X} - \frac{1}{X} \right) \right]^{-1}$$

$$\overline{Z_{\text{total}}} = \left[\frac{1}{R} + j0 \right]^{-1} = \left[\frac{1}{R}\angle 0° \right]^{-1}$$

$$\overline{Z_{\text{total}}} = \left(R\angle 0° \right)$$

The impedance plot of one such resonant circuit is shown in Figure 9-11. The output voltage is the input current times the impedance.

$$\overline{V_{\text{out}}} = \overline{I_{\text{in}}} \times \overline{Z_{\text{total}}}$$

Gain is the output (voltage) divided by the input (current).

$$\overline{A} = \frac{\overline{V_{\text{out}}}}{\overline{I_{\text{in}}}} = \overline{Z_{\text{total}}}$$

You may hear this "gain" referred to as **transimpedance**. The plot in Figure 9-11, then, is also the graph of the circuit's "gain."

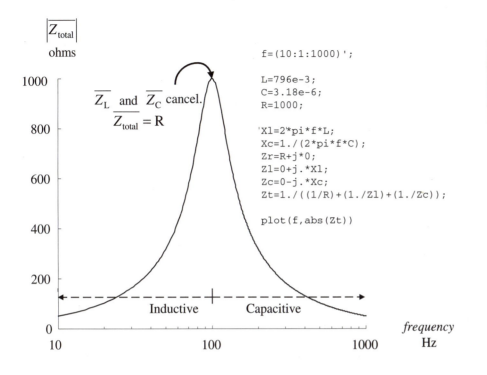

The following MATLAB code is shown in the figure:

```
f=(10:1:1000)';

L=796e-3;
C=3.18e-6;
R=1000;

'Xl=2*pi*f*L;
Xc=1./(2*pi*f*C);
Zr=R+j*0;
Zl=0+j.*Xl;
Zc=0-j.*Xc;
Zt=1./((1/R)+(1./Zl)+(1./Zc));

plot(f,abs(Zt))
```

Labels on the figure: $\overline{Z_L}$ and $\overline{Z_C}$ cancel. $\dfrac{}{Z_{total}} = R$; Inductive; Capacitive; frequency Hz; $|Z_{total}|$ ohms

Figure 9-11 Impedance of a parallel resonant circuit

At resonance, $X_L = X_C$, just like in series resonance.

$$2\pi f_{center}L = \frac{1}{2\pi f_{center}C}$$

$$f_{center} = \frac{1}{2\pi\sqrt{LC}}$$

Center frequency

At that frequency, all of the input current flows through the resistor.

$$\overline{V_{out\,max}} = \overline{I_{in}} \times (R\angle 0°)$$

Maximum output

Quality and Selectivity, Q

The Q of the resonant circuit indicates how effectively energy is stored in the inductor and capacitor as opposed to being dissipated by the resistor. Since power is the rate of doing work, Q can be expressed in terms of the reactive power compared to the resistive power.

$$Q = \frac{reactive\ power}{resistive\ power}$$

For a parallel circuit, *voltage* is the same across all of the elements. At resonance, the inductor and the capacitor have the same reactance.

$$Q_p = \frac{\dfrac{V^2}{X}}{\dfrac{V^2}{R}}$$

$$Q_p = \frac{V^2}{X} \times \frac{R}{V^2}$$

Parallel resonant circuit's quality, Q_p

$$Q_p = \frac{R}{X}$$

Notice that this is the *inverse* of the Q_s for a series resonant circuit. Be careful not to confuse the two. This is the same as the selectivity.

$$Q = \frac{f_{center}}{\Delta f}$$

Adjusting the values of L and C sets f_{center}. The value of the resistor compared to the reactance at f_{center} then determines both the location of f_{high} and f_{low}, and the magnitude of the output voltage.

Example 9-5

    a. For the circuit in Figure 9-12, calculate:

        f_{center}, Q_p, and $v_{load\ max.}$

    b. Confirm your calculations with a simulation.

Solution

    a. The resistor, R_{load} is in parallel with C_{col} and L_{col}, because V+ provides an ac short to common.

$$f_{center} = \frac{1}{2\pi\sqrt{10\,nF \times 2.4\,\mu H}} = 1.027\,MHz$$

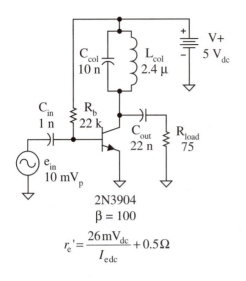

Figure 9-12 Parallel resonant amplifier for Example 9-5

$$X_L = 2\pi \times 1.027\,\text{MHz} \times 2.4\,\mu\text{H} = 15.5\,\Omega$$

$$Q_p = \frac{R_{\text{load}}}{X} = \frac{75\,\Omega}{15.5\,\Omega} = 4.8$$

$$I_{\text{base dc}} = \frac{5\,\text{V}_{\text{dc}} - 0.7\,\text{V}_{\text{dc}}}{22\,\text{k}\Omega} = 195\,\mu\text{A}_{\text{dc}}$$

$$I_{\text{emitter dc}} = (\beta+1)I_{\text{base dc}} = (100+1)\times 195\,\mu\text{A}_{\text{dc}} = 19.7\,\text{mA}_{\text{dc}}$$

$$r_e' = \frac{26\,\text{mV}_{\text{dc}}}{19.7\,\text{mA}_{\text{dc}}} + 0.5\,\Omega = 1.82\,\Omega$$

$$R_{\text{in}} = (\beta+1)r_e' = 101 \times 1.82\,\Omega = 184\,\Omega$$

$$i_{\text{base}} = \frac{e_{\text{in}}}{R_{\text{in}}} = \frac{10\,\text{mV}_p}{184\,\Omega} = 54.3\,\mu\text{A}_p$$

$$i_{\text{col}} = \beta \times i_{\text{base}} = 100 \times 54.3\,\mu\text{A}_p = 5.43\,\text{mA}_p$$

At resonance, all of the collector current flows through R_{load}.

$$v_{\text{load}} = 5.43\,\text{mA}_{\text{p}} \times 75\,\Omega = 408\,\text{mV}_{\text{p}}$$

$$A_{\text{o}} = \frac{408\,\text{mV}_{\text{p}}}{10\,\text{mV}_{\text{p}}} = 40.8 = 32\,\text{dB}$$

b. The simulation results are shown in Figure 9-13. The pass band gain from the simulation matches the manual calculations, as does f_{center}.

$$Q = \frac{f_{\text{center}}}{f_{\text{high}} - f_{\text{low}}} = \frac{1.023\,\text{MHz}}{1.14\,\text{MHz} - 912\,\text{kHz}} = 4.5$$

This is only 6 % below the value predicted.

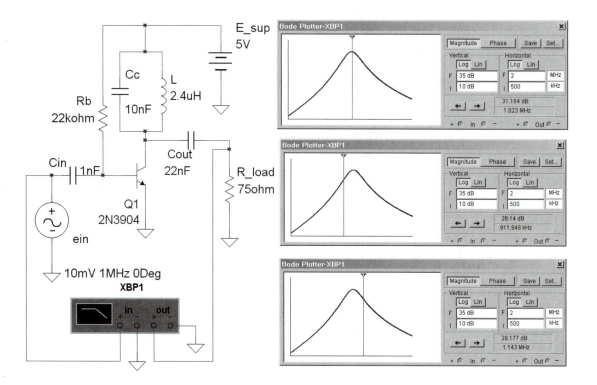

Figure 9-13 Simulation results for Example 9-5

Practice: Work this problem again, with R_b = 47 kΩ, C_{col} = 22 nF and R_{load} = 110 Ω.

Answer: f_{center} = 692 kHz, Q = 10.5, A_o = 33 = 30.4 dB

The Effects of the Inductor's Resistance

The internal resistance of the inductor has a more complicated effect on the parallel resonant circuit than it does on the series circuit. Look at Figure 9-14. The circuit is no longer a simple parallel arrangement.

For the techniques from the previous section to work, you must convert the series *RL* branch into a parallel equivalent.

$$\overline{Z_L} = \left(r_{L\,series} + jX_{L\,series}\right)$$

The admittance of the series branch is

$$\overline{Y_L} = \frac{1}{\left(r_{L\,series} + jX_{L\,series}\right)}$$

When you complete this calculation, express the results in rectangular notation. The real part is the conductance of the parallel equivalent circuit. The imaginary part is the susceptance.

$$\overline{Y_L} = \left(G_{L\,parallel} + jB_{L\,parallel}\right)$$

A *parallel* circuit made of a resistor with this conductance and an inductor of this susceptance has the same admittance and impedance as the original *RL* branch. The resistor of that parallel equivalent circuit is

$$r_{L\,parallel} = \frac{1}{G_{L\,parallel}}$$

The reactance of the inductor in the parallel, equivalent circuit is

$$X_{L\,parallel} = \frac{1}{B_{L\,parallel}}$$

The effect of this conversion on the schematic is shown in Figure 9-15. The final step is to recognize that R_{load} is now in parallel with $r_{L\,parallel}$. These two resistances can be combined.

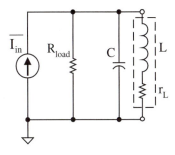

Figure 9-14
Parallel resonant circuit
with a real inductor

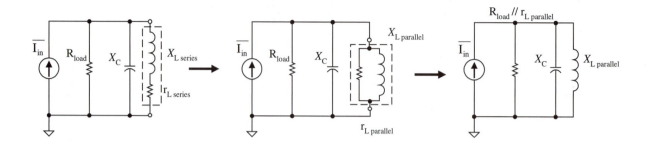

Figure 9-15 Series RL to parallel RL conversion

Example 9-6

Determine the parallel equivalent circuit for $R_{load} = 75\ \Omega$, $X_C = X_{L\,series} = 15\ \Omega$, and $r_{L\,series} = 3\ \Omega$.

Solution

$$\overline{Z_{L\,series}} = (3\,\Omega + j15\,\Omega)$$

$$\overline{Y_L} = \frac{1}{(3\,\Omega + j15\,\Omega)} = (12.8\,\text{mS} - j64.1\,\text{mS})$$

$$r_{L\,parallel} = \frac{1}{12.8\,\text{mS}} = 78.1\,\Omega$$

$$X_{L\,parallel} = \frac{1}{64.1\,\text{mS}} = 15.6\,\Omega$$

$$R_{load}//r_{L\,parallel} = \frac{1}{\dfrac{1}{75\,\Omega} + \dfrac{1}{78.1\,\Omega}} = 38.3\,\Omega$$

Practice: Determine the parallel equivalent circuit for this example if $r_{L\,series} = 8\ \Omega$.

Answer: $r_{L\,parallel} = 36.1\ \Omega$, $X_{L\,parallel} = 19.3\ \Omega$, $R_{load}\,//r_{L\,parallel} = 24.4\ \Omega$

Summary

Band-pass filters pass only a range of signals, rejecting both those of a frequency below and above that band. Maximum gain, A_o, occurs at the center of that band, f_{center}. At f_{low}, the gain is -3dB below A_o. The high frequency cut-off, f_{high}, occurs *above* the center frequency. There too the gain has also fallen to $A_o - 3$ dB. The distance between these two half-power points is the bandwidth, Δf. Selectivity, Q, is determined by how narrow the response is, compared to its center frequency.

The traditional inverting op amp configuration can be modified to make an active band-pass filter. The pass band gain is set by R_f/R_i. An input capacitor in series with R_i forms the high-pass filter and sets f_{low}. Another capacitor in parallel with R_f forms a low-pass filter and sets f_{high}.

Resonance is the phenomena in which energy is transferred from one form to another, and back again. This transfer occurs at a precise rate, dependent on the characteristics of the system. Friction dissipates energy each cycle. A child's swing and an *RLC* circuit are examples.

A resistor, inductor, and capacitor placed in series form a resonant circuit. At resonance, the inductor's impedance cancels the capacitor's impedance. The overall circuit opposition is at a minimum and equals the resistor's resistance. Current and voltage across the resistor are maximum. Resonance occurs at that frequency where the capacitive reactance matches the inductive reactance.

The quality of a resonant circuit is the ratio of the energy stored by the inductor or capacitor, to that dissipated by the resistor. For a *series* resonant circuit, L and C set the center frequency and R the bandwidth.

Practical inductors have internal resistance. This resistance forms a voltage divider with the load resistance, lowering A_o. It must also be added into the R used to calculate Q_s, lowering the selectivity.

Placing an inductor, capacitor, and resistor in *parallel* also forms a resonant circuit and band-pass filter. Usually, this circuit is driven by a current source. The inductor's and the capacitor's impedances cancel at resonance. Again the resistor sets the circuit's total impedance. For this parallel resonant circuit, impedance is a maximum at f_{center}, producing a maximum voltage.

The inductor's internal resistance forms a series circuit for that branch. It must be converted into a parallel equivalent resistance and inductive reactance, and this equivalent resistance must be combined in parallel with the load resistor. As long as the inductor's internal resistance is small compared to the resonant impedance of the inductor, this simple parallel-to-series conversion works well.

Problems

Band-pass Terminology

9-1   One form of band-pass filter has the gain magnitude equation

$$|\overline{A}| = \frac{A_o \times R}{\sqrt{R^2 + (X_L - X_C)^2}}$$

    a.   Set $A_o = 17.8$, $R = 1$ kΩ, $L = 796$ mH, $C = 3.18$ μF.

    b.   Sweep the frequency from 10 Hz to 1000 Hz, in 5 Hz steps.

    c.   At each frequency, calculate the reactances and the dB gain.

    d.   Plot the frequency response.

    e.   Tabulate the frequency and dB gain.

    f.   From the table, identify:
        f_{center}, A_o, f_{low}, f_{high}, Δf, and Q.

9-2   Repeat steps c–f for Problem 9-1 with
    a. $A_o = 2.82$, $R = 50$ Ω, $L = 39.7$ μH, $C = 159$ pF.

    b.   Sweep the frequency from 100 kHz to 8 MHz, in 10 kHz steps.

Active High-pass Low-pass Filter

9-3   For the circuit in Figure 9-16, with
    $C_f = 1$ nF, $R_f = 47$ kΩ, $R_i = 10$ kΩ, $C_i = 0.1$ μF, calculate:
    A_o, f_{low}, f_{high}, Q.

9-4   For the circuit in Figure 9-16, with
    $C_f = 4.7$ nF, $R_f = 330$ kΩ, $R_i = 10$ kΩ, $C_i = 1$ nF, calculate:
    A_o, f_{low}, f_{high}, Q.

9-5   Design a filter using the circuit in Figure 9-16 with
    $f_{low} = 500$ Hz, $f_{high} = 12$ kHz, $R_{in\ min} = 10$ kΩ, $A_o = 33$.

9-6   Design a filter using the circuit in Figure 9-16 with
    $f_{low} = 20$ Hz, $f_{high} = 3$ kHz, $R_{in\ min} = 10$ kΩ, $A_o = 22$.

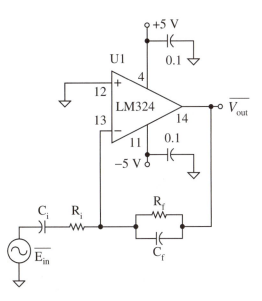

Figure 9-16 Schematic for Problems 9-3 through 9-6

Resonance

9-7 Describe how a note from a flute is an example of resonance. What is vibrating? What determines the resonant frequency?

9-8 Describe another example of resonance not in this chapter. Explain what is vibrating. Between what two elements is energy being exchanged? What determines the resonant frequency?

Series Resonance

9-9 Plot the *magnitude* of the impedance of a series resonant circuit with $L = 33$ mH, $C = 0.1$ μF, $R = 220$ Ω. Identify on the plot: f_{center}, the capacitive region, and the inductive region.

9-10 Plot the *phase* of the impedance of a series resonant circuit with $L = 33$ mH, $C = 0.1$ μF, $R = 220$ Ω. Identify on the plot: f_{center}, the capacitive region, and the inductive region.

9-11 For the circuit in Figure 9-17, with R = 220 Ω, L = 33 mH, C = 0.1 μF, calculate f_{center} and Q.

9-12 For the circuit in Figure 9-17, with R = 50 Ω, L = 3.6 μH, C = 6.8 pF, calculate f_{center} and Q.

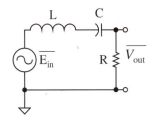

Figure 9-17
Schematic for Problems
9-11 and 9-12

9-13 Design a series resonant band-pass filter with f_{center} = 10 MHz, Q = 20, and R = 75 Ω.

9-14 Design a cross-over network for a midrange loudspeaker driver with f_{low} = 200 Hz, f_{high} = 1 kHz. Use the driver's 4 Ω resistance as the R for the network.

9-15 Repeat the analysis of Problem 9-11. Include the 50 Ω internal resistance of the inductor.

9-16 Repeat the analysis of Problem 9-12. Include the 20 Ω internal resistance of the inductor.

Parallel Resonance

9-17 Plot the *magnitude* of the impedance of a parallel resonant circuit with L = 33 mH, C = 0.1 μF, R = 220 Ω. Identify: f_{center}, the capacitive region, and the inductive region.

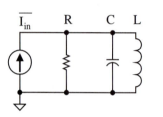

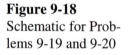

9-18 Plot the *phase* of the impedance of a parallel resonant circuit with L = 33 mH, C = 0.1 μF, R = 220 Ω. Identify on the plot: f_{center}, the capacitive region, and the inductive region.

9-19 For the circuit in Figure 9-18, with R = 220 Ω, L = 33 mH, C = 0.1 μF, calculate f_r and Q.

Figure 9-18
Schematic for Problems 9-19 and 9-20

9-20 For the circuit in Figure 9-18, with R = 50 Ω, L = 3.6 μH, C = 6.8 pF, calculate f_r and Q.

9-21 Design a parallel resonant band-pass filter with f_{center} = 10 MHz, Q = 20, and R = 75 Ω.

9-22 Design a parallel resonant band-pass filter with f_{low} = 200 Hz, f_{high} = 1 kHz and R = 4 Ω.

9-23 Repeat the analysis of Problem 9-19. Include the 50 Ω internal resistance of the inductor.

9-24 Repeat the analysis of Problem 9-20. Include the 20 Ω internal resistance of the inductor.

Band-pass Filters Lab Exercise

A. Series Resonant Circuit Set-up
1. Find the components needed to build the circuit in Figure 9-19.

2. Measure and record the value of each component. Be sure to record both the inductance and the *resistance* of the inductor.

3. Using *measured* values, calculate f_{center}, A_o, and Q. Be sure to include the effect of the inductor's internal resistance.

B. Series Resonant Circuit Quick-look
1. Build the circuit in Figure 9-19.

2. Set the input signal to 1 V_{rms} at 100 Hz, as measured by a digital multimeter.

3. Monitor the input and the output voltages with an oscilloscope. Trigger on the input signal.

4. Monitor the output voltage with the digital multimeter.

5. Raise the input signal's frequency until the output voltage is a maximum. This is the resonant frequency. Measure and record the output voltage's magnitude, phase and frequency.

6. Compare the measured resonant frequency to that calculated in step A3.

7. Calculate the measured pass-band gain. Compare it to the gain calculated in step A3.

8. Lower the input signal's frequency until the output voltage is 0.707 of the value recorded in step 5. This is the low-frequency cut-off, f_{low}. Measure and record the output voltage's magnitude and frequency.

9. Raise the input signal's frequency. The output voltage will rise until you pass f_{center}, then begin to fall. Find the frequency *above* f_{center} where the output voltage is 0.707 of the value recorded in

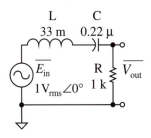

Figure 9-19
Series resonant
circuit schematic

step 5. This is the high-frequency cut-off, f_{high}. Measure and re-cord the output voltage's magnitude and frequency.

10. Calculate the measured bandwidth and the measured Q of your circuit. Compare this Q with that calculated in step A3.

C. Series Resonant Circuit Frequency Response
  1. Construct a spreadsheet with columns for *frequency* (kHz), V_R (V_{rms}), *phase* (degrees), V_C, and V_L. Provide a row for the column titles and then 19 rows for data. Enter the frequency data, one in each row (0.1, 0.2, 0.3, 0.4, 0.5, 0.6, 0.7, 0.8, 0.9, 1.0, 2.0, 3.0, 4.0, 5.0, 6.0, 7.0, 8.0, 9.0, 10.0).

  2. Measure the magnitude of the voltage across each component with the digital multimeter. Also be sure to measure the phase angle of the voltage across the resistor (with respect to the source voltage). Assure that the magnitude of the source remains at 1 V_{rms}.

  3. Produce four plots, one for each voltage magnitude versus frequency, and one for the phase of the voltage across the resistor versus frequency. Scale each horizontal axis logarithmically.

D. Active High-pass Low-pass Filter
  1. Build the circuit in Figure 9-16; E_{in} = 100 mV$_{rms}$, C_i = 33 nF, R_i = 10 kΩ, R_f = 120 kΩ, and C_f = 270 pF.

  2. Set the input amplitude to 100 mV$_{rms}$, at 100 Hz.

  3. Repeat the measurements and comparisons indicated in sets B3 through C2. For the calculated values, use those from Example 9-2.

10

Amplifier Frequency Response

Introduction

Resistor-capacitor networks are placed around op amps to block dc, for single-supply biasing. They are also used to reduce the high-frequency gain, producing a low-pass filter. However, RC networks when incorrectly used cause many op amps to output only their maximum (saturation) voltage, or to enhance rf noise so much that the circuit is useless.

In the previous chapters, it has been assumed that the operational amplifier performs consistently at all frequencies. However, as frequency increases, capacitive reactance decreases. At high frequencies, it becomes likely that a signal *within* the op amp may couple onto an adjacent trace through parasitic capacitance. In the worst case, a large output signal is coupled onto a low-level input line. From there it is amplified, passed to the output, coupled back to the input, amplified, and so on, continuously. The op amp breaks into oscillations.

To prevent this, op amps have a low-pass filter built inside. This filter reduces the gain of high-frequency signals far enough that the op amp is stable. It also affects both the sinusoidal and the transient response of the op amp. Depending on the signal's amplitude and shape, four specifications are provided: gain bandwidth product (small-signal, sinusoidal), rise time (small-signal, step), slew rate (large-signal, step), and full-power bandwidth (large-signal, sinusoidal).

Objectives

Upon completion of this chapter, you will be able to do the following:

- Design and analyze op amp circuits with RC networks in their input or their feedback loop.

- Define and apply the following op amp parameters:
    Gain bandwidth product: $f_{-3dB}, f_{10\%}, f_{5\%}, f_{1\%}$
    Small-signal rise time
    Slew rate
    Full-power bandwidth

10.1 Reactive Networks Around an Op Amp Amplifier

Parallel *RC* in the Feedback Loop

We live in a world filled with high-frequency electromagnetic radiation. From commercial radio and television, to cellular and portable phones, pagers, garage door openers, microwave ovens, personal computers, and even light dimmers and variable speed tools, the circuits you build have to survive a bombardment of rf energy. Figure 10-1 shows a common approach to making your op amps immune to these distractions.

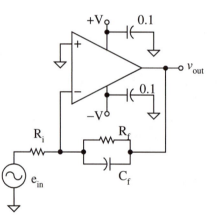

Figure 10-1 Parallel *RC* in the feedback loop

Three capacitors have been added. The two 0.1 μF capacitors are for power supply decoupling. They must be snuggled right up against the power pins of the IC. Leads have inductance, the opposite of capacitance. So trim the leads as short as you can, and connect the capacitors as directly to circuit common as possible. The long line of the power supply bus, as it runs from part to part, acts as an antenna. It picks up noise from these parts and from the environment. The two decoupling capacitors short that noise to circuit common *before* it can enter the IC. Typically, 0.1 μF is appropriate. But this may depend on the signals and the interference present in your circuit. Be prepared to experiment.

The input and the feedback impedances set the gain for this circuit.

$$\overline{A} = -\frac{\overline{Z_f}}{\overline{Z_i}} = \frac{\overline{Z_f}}{\overline{Z_i}} \angle 180°$$

The input impedance is just $(R_i \angle 0°)$. However, the feedback impedance comes from the *parallel* combination of R_f and C_f.

Perform a quick-look analysis. At dc, C_f is an open.

$$\left| \overline{A} \right|_{f \to 0} = A_o = \frac{R_f}{R_i}$$

As the frequency increases, the reactance of C_f *decreases*, shorting out R_f. At a high enough frequency, the capacitor becomes dominant, and the gain rolls off.

$$\left| \overline{A} \right|_{f \to \infty} = \frac{0}{R_i} = 0$$

Placing a capacitor in *parallel* with the feedback resistor converts the amplifier into a low-pass filter. The low-frequency signals receive the gain you intend, but the high-frequency noise is attenuated.

Determining f_{-3dB} of this filter requires a few steps.

$$\overline{Z_f} = \frac{1}{\dfrac{1}{\overline{Z_{Rf}}} + \dfrac{1}{\overline{Z_{Cf}}}}$$

Obtain a common denominator for the denominator.

$$\overline{Z_f} = \frac{1}{\dfrac{\overline{Z_{Cf}}}{\overline{Z_{Rf}} \times \overline{Z_{Cf}}} + \dfrac{\overline{Z_{Rf}}}{\overline{Z_{Rf}} \times \overline{Z_{Cf}}}}$$

$$\overline{Z_f} = \frac{1}{\dfrac{\overline{Z_{Cf}} + \overline{Z_{Rf}}}{\overline{Z_{Rf}} \times \overline{Z_{Cf}}}}$$

Simplify this complex fraction by inverting and multiplying.

$$\overline{Z_f} = \frac{\overline{Z_{Rf}} \times \overline{Z_{Cf}}}{\overline{Z_{Rf}} + \overline{Z_{Cf}}}$$

The gain of the amplifier is

$$\overline{A} = \frac{\overline{Z_f}}{\overline{Z_i}} \angle 180°$$

Substitute for the feedback impedance.

$$\overline{A} = \frac{\dfrac{\overline{Z_{Rf}} \times \overline{Z_{Cf}}}{\overline{Z_{Rf}} + \overline{Z_{Cf}}}}{\overline{Z_{Ri}}} \angle 180°$$

$$\overline{A} = \frac{\overline{Z_{Rf}}}{\overline{Z_{Ri}}} \times \frac{\overline{Z_{Cf}}}{\overline{Z_{Rf}} + \overline{Z_{Cf}}} \angle 180°$$

Substitute appropriate resistances and reactance.

$$\overline{A} = \frac{R_f}{R_i} \times \frac{(X_{Cf} \angle -90°)}{(R_f - jX_C)} \angle 180°$$

The pass-band gain, A_o is R_f/R_i.

$$\overline{A} = A_o \frac{(X_{Cf} \angle 90°)}{\sqrt{R_f^2 + X_{Cf}^2} \angle \arctan\left(\dfrac{-X_{Cf}}{R_f}\right)}$$

$$|\overline{A}| = A_o \frac{X_{Cf}}{\sqrt{R_f^2 + X_{Cf}^2}}$$

The half-power point for this low-pass filter is the frequency at which the gain magnitude has fallen to by $1/\sqrt{2}$ below A_o.

$$\frac{A_o}{\sqrt{2}} = A_o \frac{X_{Cf}}{\sqrt{R_f^2 + X_{Cf}^2}}$$

$$\frac{1}{\sqrt{2}} = \frac{X_{Cf}}{\sqrt{R_f^2 + X_{Cf}^2}}$$

Square both sides.

$$\frac{1}{2} = \frac{X_{Cf}^2}{R_f^2 + X_{Cf}^2}$$

Clear the fractions.

$$R_f^2 + X_{Cf}^2 = 2X_{Cf}^2$$

$$R_f^2 = X_{Cf}^2$$

$$R_f = X_{Cf}$$

Substitute the equation for the reactance of the capacitor at f_{-3dB}.

$$R_f = \frac{1}{2\pi f_{-3dB} C_f}$$

$$f_{-3dB} = \frac{1}{2\pi R_f C_f}$$ **Half-power point frequency**

Example 10-1

    a. Design an inverting amplifier using an LM324 with:
        $Z_{in} = 10$ kΩ, $A_o = 22$, $f_{-3dB} = 1.5$ kHz

    b. Verify your design with a simulation.

multi**SIM**

Solution

    a. Use the schematic in Figure 10-1.

$$R_i = Z_{in} = 10 \text{ k}\Omega$$

$$A_o = \frac{R_f}{R_i}$$

$$R_f = A_o \times R_i$$

$$R_f = 22 \times 10 \text{ k}\Omega = 220 \text{ k}\Omega$$

$$f_{-3dB} = \frac{1}{2\pi R_f C_f}$$

$$C_f = \frac{1}{2\pi \times 220 \text{ k}\Omega \times 1.5 \text{ kHz}} = 482 \text{ pF} \quad \text{Pick 470 pF.}$$

b. The simulation results are shown in Figure 10-2.

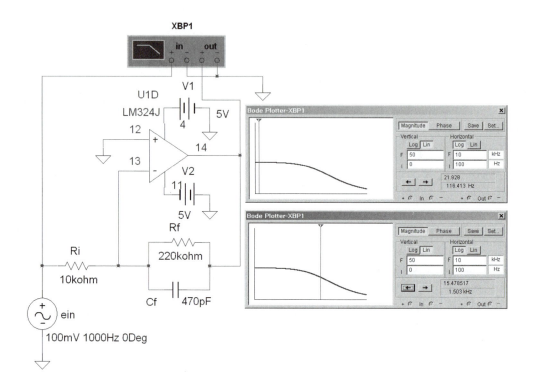

Figure 10-2 Simulation for Example 10-1

The cursor in the upper Bode plotter shows a pass-band gain of 22. The cursor in the lower Bode plotter is set at a frequency of 1.5 kHz. It indicates a gain of 15.5. This is the gain at the half-power point.

$$A_{f-3dB} = \frac{22}{\sqrt{2}} = 15.6$$

Practice: Repeat the design for $A_o = 47$ (inverting), $f_{-3dB} = 300$ Hz

Answer: $R_i = 10$ kΩ, $R_f = 470$ kΩ, $C_f = 1.1$ nF

Series *RC* in the Input Loop

A series *RC* combination can be added to the input branch. You have seen this in several examples already. It is shown again in Figure 10-3.

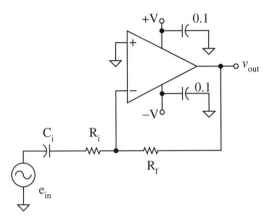

Figure 10-3 Op amp with a series *RC* in the input loop

Both ac and dc negative feedback are provided through R_f. So this amplifier should behave properly. At low frequencies, the capacitor looks like an open. No current flows. So, the output goes to zero.

$$\left.\left|\overline{A}\right|\right|_{f\to dc} = 0$$

At high frequencies, the capacitor looks like a short, leaving R_i and R_f.

$$\left.\left|\overline{A}\right|\right|_{f\to\infty} = A_o = \frac{R_f}{R_i}$$

Placing a series *RC* circuit in the input loop forms a first order *high-pass* filter, with a pass-band gain of R_f/R_i.

To determine its half-power point frequency, the gain is

$$\overline{A} = \frac{\overline{Z_f}}{\overline{Z_i}} \angle 180°$$

Substitute resistance and reactance values.

$$\overline{A} = \frac{(R_f \angle 0°)}{(R_i - jX_C)} \angle 180°$$

Convert the denominator into its polar form.

$$\overline{A} = \frac{(R_f \angle 0°)}{\left(\sqrt{R_i^2 + X_C^2} \angle -\arctan\dfrac{X_C}{R_i}\right)} \angle 180°$$

Separate the equation into its magnitude and phase parts.

$$\overline{A} = \left[\frac{R_f}{\sqrt{R_i^2 + X_C^2}} \angle \left(180° + \arctan\frac{X_C}{R_i}\right)\right]$$

$$|\overline{A}| = \frac{R_f}{\sqrt{R_i^2 + X_C^2}}$$

At the half-power point frequency, the gain magnitude has fallen to

$$|\overline{A}| = \frac{A_o}{\sqrt{2}} = \frac{\dfrac{R_f}{R_i}}{\sqrt{2}} = \frac{R_f}{\sqrt{R_i^2 + X_C^2}}$$

Square both sides.

$$\frac{\dfrac{R_f^2}{R_i^2}}{2} = \frac{R_f^2}{R_i^2 + X_C^2}$$

Multiply to begin to clear the fractions.

$$\frac{R_f^2}{R_i^2}\left(R_i^2 + X_C^2\right) = 2R_f^2$$

Multiply both sides by $\dfrac{R_i^2}{R_f^2}$.

$$R_i^2 + X_C^2 = 2R_f^2 \times \frac{R_i^2}{R_f^2}$$

Simplify the right side of the equation.

$$R_i^2 + X_C^2 = 2R_i^2$$

$$X_C^2 = R_i^2$$

$$X_C = R_i$$

Substitute the equation for the reactance of X_C.

$$R_i = \frac{1}{2\pi f_{-3dB} C_i}$$

$$f_{-3dB} = \frac{1}{2\pi R_i C_i}$$

Half-power point frequency

At f_{-3dB} the gain falls to 0.707 of its high-frequency, pass-band value. That is 30% error. For more precise performance,

$$f_{10\%} = 2 f_{-3dB}$$

Other frequencies

$$f_{5\%} = 3 f_{-3dB}$$

$$f_{1\%} = 7 f_{-3dB}$$

Example 10-2

    a. Calculate A_o, f_{-3dB}, $f_{10\%}$, $f_{5\%}$, and $f_{1\%}$ for the circuit in Figure 10-3, with $C_i = 56$ nF, $R_i = 10$ kΩ, and $R_f = 220$ kΩ.

    b. Verify your calculations with a simulation.

Solution

    a. The pass band gain is

$$A_o = \frac{R_f}{R_i} = \frac{220\,k\Omega}{10\,k\Omega} = 22$$

The half-power point frequency is

$$f_{-3dB} = \frac{1}{2\pi \times 10\,k\Omega \times 56\,nF} = 284\,Hz$$

The other frequencies are

$$f_{10\%} = 2 \times 284\,Hz = 568\,Hz$$

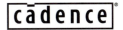
cādence

how big can you dream?™

$$f_{5\%} = 3 \times 284\,\text{Hz} = 852\,\text{Hz}$$

$$f_{1\%} = 7 \times 284\,\text{Hz} = 1.99\,\text{kHz}$$

b. The simulation schematic and the frequency response plot are shown in Figure 10-4. There is good correlation between the manually calculated values and those produced by the simulation. Notice, however, that the gain begins to fall above 10 kHz. This high frequency roll-off is discussed in the second section of this chapter.

Practice: Repeat the problem with $R_i = 3.3\,\text{k}\Omega$

Answer: $A_o = 67$, $f_{-3\text{dB}} = 861$ Hz, $f_{10\%} = 1.72$ kHz, $f_{5\%} = 2.58$ kHz, $f_{1\%} = 6.03$ kHz

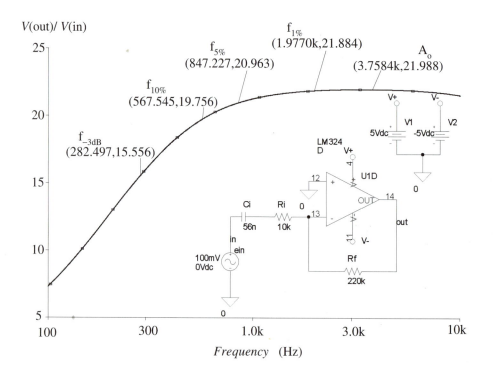

Figure 10-4 Simulation results for Example 10-2

10.2 Gain Bandwidth Product

In the previous chapters, it has been assumed that the operational amplifier performs consistently at all frequencies. However, as frequency increases, capacitive reactance decreases. At high frequencies, it becomes likely that a signal *within* the op amp may couple onto an adjacent trace through parasitic capacitance. In the worst case, a large output signal is coupled onto a low-level input line. From there it is amplified, passed to the output, coupled back to the input, amplified, and so on, continuously. The op amp breaks into oscillations.

Open-loop Frequency Response

To prevent these oscillations, op amps have a low-pass filter built inside. This filter reduces the gain of high-frequency signals far enough that the op amp is stable. The most pronounced effect of this internal filter is to reduce the op amp's **open-loop** gain. Until now, you have assumed that the open-loop gain is fixed at a very large value. In reality, this gain begins to fall at a low frequency, and continues to fall, at a steady roll-off rate, throughout the op amp's useful frequency range. Look at Figure 10-5.

The vertical axis is the op amp's open-loop gain (in ratio, not dB). It is scaled logarithmically, equal distances changing a decade. The horizontal scale is frequency, also scaled logarithmically. At low frequencies, the open-loop gain is indeed very high, 200,000 in this example. However, it begins to fall at 10 Hz, and falls at the same rate that the frequency increases. The slope is −1. So, an increase in frequency produces an identical *decrease* in gain. The result is that at every point on the slope, the product of gain and frequency is a constant. This is the **gain bandwidth product**, and is often specified by the manufacturer.

$$GBW = A \times f$$

At point A:

$$1k \times 1kHz = 1MHz$$

At point B:

$$10 \times 100kHz = 1MHz$$

At point C:

$$1 \times 1MHz = 1MHz$$

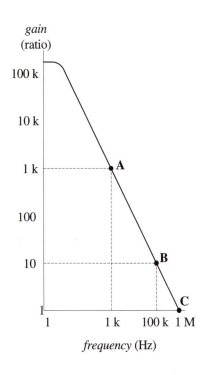

Figure 10-5
Typical op amp open-loop gain frequency response

gain bandwidth

At each of these points, the product of gain and bandwidth is a constant. Eventually, at point C, the open loop gain has fallen to 1. So, this gain bandwidth product specification may also be called the **unity gain bandwidth.**

Closed-loop Frequency Response

Practical op amp based amplifiers have defined, smaller gains than those shown in Figure 10-5. The effect on the frequency response is shown in Figure 10-6. The open-loop gain remains the same, and is shown with a dashed line. The closed-loop gain, $A_{closed\ loop}$, is shown with a solid line. The pass band gain, A_o, has been set to 10 in this illustration. As the closed-loop gain approaches the open-loop gain limit, the actual closed-loop gain begins to bend downward, eventually falling along with $A_{open\ loop}$ as they roll off.

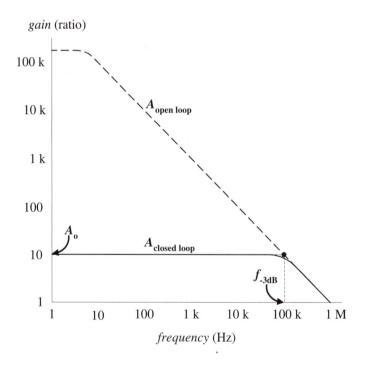

Figure 10-6 Closed-loop frequency response of an op amp

The key parameter is where the *projection* of A_o intersects the open-loop line. The frequency at which that occurs is f_{-3dB} for the closed-loop amplifier. In Figure 10-6, this happens at 100 kHz. Remember, for every point on the open-loop line,

$$GBW = A \times f$$

So, at the intersection of the closed loop line with the open loop line,

$$GBW = A_o \times f_{-3dB}$$

GBW in closed-loop terms

Figure 10-7 shows an enlargement of the frequency response plot, one decade on each side of the half-power point frequency. Remember, at f_{-3dB} the gain has fallen from A_o to 0.707 A_o. Other important frequencies are:

$$f_{10\%} = \frac{f_{-3dB}}{2} \qquad f_{5\%} = \frac{f_{-3dB}}{3} \qquad f_{1\%} = \frac{f_{-3dB}}{7}$$

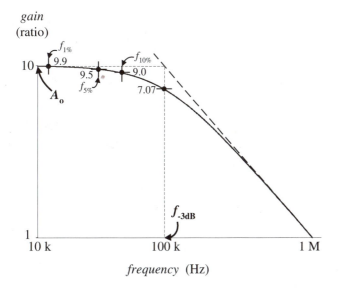

Figure 10-7 Enlargement of the frequency response plot

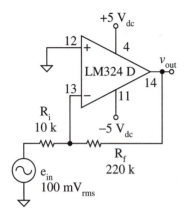

Figure 10-8
Schematic for Example
10-3

Example 10-3

For the circuit in Figure 10-8, given that the op amp's gain bandwidth product is 957 kHz,

Calculate $f_{-3dB}, f_{10\%}, f_{5\%}, f_{1\%}$, and the gain at each frequency

Solution

The pass-band gain is

$$A_o = \frac{R_f}{R_i} = \frac{220\,k\Omega}{10\,k\Omega} = 22$$

The half-power point frequency is

$$f_{-3dB} = \frac{957\,kHz}{22} = 43.5\,kHz$$

At f_{-3dB} the gain has dropped to $A_o/\sqrt{2} = 15.56$.

$$f_{10\%} = \frac{f_{-3dB}}{2} = \frac{43.5\,kHz}{2} = 21.7\,kHz$$

At $f_{10\%}$ the gain has dropped to $0.9\,A_o = 19.80$.

$$f_{5\%} = \frac{f_{-3dB}}{3} = \frac{43.5\,kHz}{3} = 14.5\,kHz$$

At $f_{5\%}$ the gain has dropped to $0.95\,A_o = 20.9$.

$$f_{1\%} = \frac{f_{-3dB}}{7} = \frac{43.5\,kHz}{7} = 6.2\,kHz$$

At $f_{1\%}$ the gain has dropped to $0.99\,A_o = 21.8$.

Practice: Repeat Example 10-3 with $R_f = 470\,k\Omega$.

Answer: $A_o = 47$, $f_{-3dB} = 20.4$ kHz, $f_{10\%} = 10.2$ kHz, $f_{5\%} = 6.8$ kHz, $f_{1\%} = 2.9$ kHz

Example 10-4

An amplifier must have a gain of 33 with no more than 5% error at 2.5 MHz.

a. Calculate the required op amp gain bandwidth product.

b. Locate an op amp that meets that specification.

Solution

a. Since the gain can fall only 5% at 2.5 MHz,

$$f_{5\%} = 2.5 \text{ MHz}$$

This is related to the half-power point frequency by

$$f_{5\%} = \frac{f_{-3dB}}{3}$$

Solve for the half-power point frequency.

$$f_{-3dB} = 3f_{5\%} = 3 \times 2.5\,\text{MHz} = 7.5\,\text{MHz}$$

You can now find the op amp's gain bandwidth.

$$GBW = A_o \times f_{-3dB}$$

$$GBW = 33 \times 7.5\,\text{MHz} = 248 \text{ MHz}$$

b. The LMH6654 has a gain bandwidth of 250 MHz.

Practice: Repeat Example 10-4 for $A_o = 10$ with no more than 10% error at 450 kHz.

Answer: $f_{-3dB} = 900$ kHz, $GBW = 9$ MHz, LM318 has $GBW = 15$ MHz

High gain bandwidth op amps may require special handling procedures, exceptional attention to layout, and are often only available in surface mount packages.

An alternative is to build the amplifier with several stages. The gains of each stage multiply to produce the overall gain.

$$A_{total} = A_{stage\,1} \times A_{stage\,2} \times A_{stage\,3} \times ...$$

So, each stage may have a much lower gain, and its op amp may have a much lower gain bandwidth.

Example 10-5

Select an op amp to be used in an amplifier that covers the entire audio range, with an overall gain of 650.

Solution

The top end of the audio range is 20 kHz. Set

$$f_{-3dB} = 20 \text{ kHz}$$

This requires a *single* op amp with a gain bandwidth product of

$$GBW = A_o \times f_{-3dB} = 650 \times 20 \text{ kHz} = 13 \text{ MHz}$$

The LM318 has a $GBW = 15$ MHz. This could be made to work. However, it costs several dollars and may tend to break into oscillations when hit with the large transients in music.

The LM324 is inexpensive, and quite stable. There are *four* op amps in the IC. So, four stages can be built with a single IC.

$$A_{total} = A_{stage 1} \times A_{stage 2} \times A_{stage 3} \times A_{stage 4} = A_{stage}{}^4$$

$$A_{stage} = \sqrt[4]{A_{total}} = \sqrt[4]{650} = 5.05$$

Each stage must have a *GBW* of

$$GBW = A_o \times f_{-3dB} = 5.1 \times 20 \text{ kHZ} = 102 \text{ kHz}$$

The LM324 has a *GBW* of about 1 MHz. So it can easily provide the needed gain at the required frequency. This is a less expensive and more stable design than using a single LM318.

Practice: What is the highest A_{total} possible from a single LM324 package with $f_{1\%} = 50$ kHz?

Answer: $f_{-3dB} = 350$ kHz, $A_{stage} = 2.85$, $A_{total} = 67$

Small-signal Rise Time

The laboratory measurement of gain bandwidth is difficult and time consuming. Repeated measurements of the output voltage must be made as the frequency is swept across a wide range. And, the circuit is, at best, marginally stable. The circuit in Figure 10-9 offers a better approach.

A *small* step is applied to a voltage follower. The peak-to-peak amplitude must a few hundred millivolts. The first order low-pass filter within the op amp responds just as any *RC* circuit would to a step. The output rises exponentially to its final value, shown in Figure 10-10.

The **rise time** is the time it takes for the output to go from 10% of its final value to 90% of that final value. This is a standard measurement and is automated on many digital oscilloscopes. Figure 10-10 is the response of a LM324. Its rise time is

$$t_{\text{rise}} = 370\,\text{ns} - 50\,\text{ns} = 320\,\text{ns}$$

Determining how this rise time relates to the voltage follower's bandwidth, $f_{-3\text{dB}}$, and therefore to its gain bandwidth product specification requires a little algebra. The equation for the exponential charge of a capacitor is

$$v_{\text{cap}} = V_{\text{p}}\left(1 - e^{-\frac{t}{RC}}\right)$$

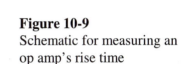

Figure 10-9
Schematic for measuring an op amp's rise time

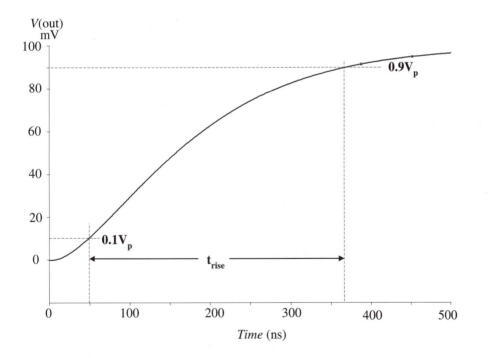

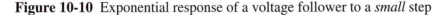
Figure 10-10 Exponential response of a voltage follower to a *small* step

First, determine the *time* it takes to get to 90% of the final value.

$$0.9 \times V_p = V_p \left(1 - e^{-\frac{t}{RC}} \right)$$

$$0.9 = 1 - e^{-\frac{t}{RC}}$$

$$-0.1 = -e^{-\frac{t}{RC}}$$

$$0.1 = e^{-\frac{t}{RC}}$$

Take the natural log (ln) of each side to complement the exponentiation.

$$\ln(0.1) = -\frac{t}{RC}$$

$$t_{90\%} = -RC \times \ln(0.1)$$

$$t_{90\%} = 2.30RC$$

This determines the time to get from $t = 0$ to 90% of the final value. Repeat these steps to determine the time to get to 10% of the final value.

$$t_{10\%} = 0.10RC$$

The time for the output to rise from 10% to 90% of its final value is

$$t_{rise} = t_{90\%} - t_{10\%} = 2.20RC$$

or
$$RC = \frac{t_{rise}}{2.20}$$

The trick is to realize that this simple *RC* circuit is a low-pass filter. When you apply a sine wave to it, the half-power point frequency is

$$f_{-3dB} = \frac{1}{2\pi RC}$$

Substitute the value for *RC* from the rise time calculation.

$$f_{-3dB} = \frac{1}{2\pi \frac{t_{rise}}{2.20}}$$

Clear the complex fraction by combining the constants.

$$f_{-3dB} = \frac{0.350}{t_{rise}}$$

Since the original circuit is a voltage follower,

$$A_o = 1$$

The gain bandwidth product is

$$GBW = A_o \times f_{-3dB}$$

$$GBW = \frac{0.350}{t_{rise}}$$

GBW in terms of t_{rise}

Example 10-6

Compare the *GBW* of the LM324 obtained by closed loop frequency response calculations with the *GBW* obtained from the rise time displayed in Figure 10-10.

Solution

From Example 10-3, *GBW* = 957 kHz. Figure 10-10 shows a rise time of 320 ns.

$$GBW = \frac{0.350}{t_{rise}}$$

$$GBW = \frac{0.350}{320\,ns} = 1.09\,MHz$$

The difference between these two results is

$$\% \ difference = \frac{1.09\,MHz - 957\,kHz}{\left(\dfrac{1.09\,MHz + 957\,kHz}{2}\right)} \times 100\%$$

$$\% \ difference = 13\%$$

Practice: Calculate the small-signal rise time for a LM318 op amp.

Answer: t_{rise} = 23 ns

10.3 Slew Rate

The gain bandwidth product defines how quickly an op amp can respond to *small* signals. Typically, *small* is about 1 V_p. For larger signals, the current available inside the op amp limits how rapidly the output signal can change. This limit is specified by the **slew rate**.

Response to a Step

Slew rate is the maximum rate of change of the op amp's output voltage.

$$SR = \frac{dv_{out}}{dt}\bigg|_{max}$$

The units are V/μs. Be careful. It is easy to handle this number incorrectly since there is a 1E–6 in the denominator.

Slew rate is the maximum slope of the output. When measuring slew rate, use the steepest, *linear* part of the output signal's edge.

Example 10-7

Determine the slew rate of the LM324 by simulation.

Solution

The simulation schematic and the plot of the output voltage are shown in Figure 10-11. The edges rise and fall at

$$SR = \frac{\Delta v}{\Delta t}$$

$$SR = \frac{1.848\,V - 0.397\,V}{3.782\,\mu s - 0.832\,\mu s} = 492\,\frac{kV}{s}$$

The proper units for slew rate are V/μs.

$$SR = 492 \times 10^3\,\frac{V}{s} \times \frac{1s}{1 \times 10^6\,\mu s}$$

$$SR = 492 \times \frac{10^3}{10^6}\,\frac{V}{\mu s} = 0.492\,\frac{V}{\mu s}$$

how big can you dream?™

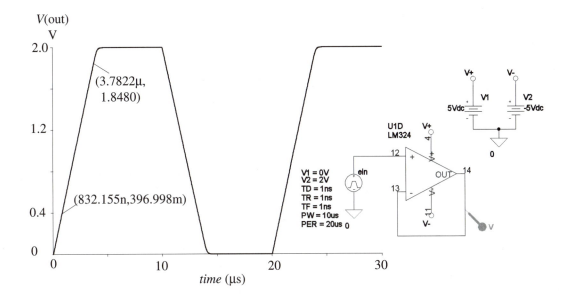

Figure 10-11 Schematic and simulation results for Example 10-7

Practice: Determine the slew rate for the LF411 op amp.

Answer: In the simulation, change **PW = 1us, PER = 2us.** The cursors return points at **(2.0088u,47.403m)** and **(2.1308u, 1.8546)**. This indicates a slew rate of 14.8 V/μs.

Example 10-8

To build an electronic piano tuner, the signal from the microphone must be amplified, and then sent to a comparator where it is converted into a TTL signal for the microprocessor to handle.

It has been decided that the rise time out of the comparator can be no more than 0.1% of the period of the tone for accurate measurements to be made. Since the LM324 is being used as the amplifier, can another one of the op amps in the package be used as the comparator?

Solution

The highest note on the piano is at 6 kHz. Its period is

$$T_{\text{highest}} = \frac{1}{6\,\text{kHz}} = 166.667\,\mu s$$

This gives a minimum time of

$$\Delta t = 0.001 \times 166.667\,\mu s = 167\,\text{ns}$$

The microprocessor recognizes a logic high as soon as the signal passes 2.4 V. The required slew rate is

$$SR = \frac{\Delta V}{\Delta t}$$

$$SR = \frac{2.4\,\text{V}}{167\,\text{ns}} = \frac{2.4\,\text{V}}{0.167\,\mu s} = 14.4\,\frac{\text{V}}{\mu s}$$

The LM324 is too slow.

Practice: Using this criteria of Δt = 0.1% T, what is the highest frequency for which the LM324 should be used as a comparator?

Answer: Δt = 4.8 μs, T_{shortest} = 4.8 ms, f_{highest} = 208 Hz

---

Full-power Bandwidth

The limitation that slew rate imposes on an op amp when processing a *sine* wave is the **full-power bandwidth**, f_{max}. Above this frequency, the sine wave is distorted because the output voltage cannot change fast enough to keep up with the signal. This value can be determined from the basic definition of slew rate.

$$SR = \frac{dv}{dt} \qquad v = V_p \sin(\omega t)$$

$$SR = \frac{d}{dt}\left[V_p \sin(\omega t)\right]$$

When taking the derivative of a constant times a function, the constant can be moved outside of the derivative.

$$SR = V_p \left[\frac{d}{dt} \sin(\omega t) \right]$$

The derivative of the sine is the cosine.

$$SR = V_p \cos(\omega t) \frac{d}{dt} (\omega t)$$

Again, ω is a constant. So, it can be brought out of the derivative.

$$SR = V_p \omega \cos(\omega t) \frac{d}{dt} (t)$$

$$SR = V_p \omega \cos(\omega t)$$

The goal is to determine the *largest* required slew rate to output an undistorted sine wave. The $\cos(\omega t)$ is at its largest value at $t = 0$. That is, a sine wave has its steepest slope at $t = 0$.

$$SR_{max} = V_{p\,max} \omega_{max} = 2\pi f_{max} V_{p\,max}$$

Required slew rate for a sine wave output

This equation can be rearranged to give the maximum frequency at which an op amp can output an undistorted sine wave.

$$f_{max} = \frac{SR_{max}}{2\pi V_{p\,max}}$$

Full-power bandwidth

Notice that the full-power bandwidth depends on *both* the op amp's slew rate and the peak of the output signal. That should make sense. How long it takes you to go somewhere (the period) depends both on the distance (V_p) and how fast you are going (SR).

Example 10-9

    a. What is the full-power bandwidth of a LM324 with ±5 V_{dc}?

    b. Display the output of a voltage follower for a sine wave at this amplitude and frequency, and at twice this frequency.

Solution

    a.
$$f_{max} = \frac{SR_{max}}{2\pi V_{p\,max}}$$

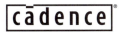

how big can you dream?™

In Example 10-7, the slew rate was determined to be

$$SR = 0.49 \frac{V}{\mu s}$$

The manufacturer's specifications indicate that the positive output may come within 1.5 V of the power supply voltage.

$$V_{p\,max} = 5V - 1.5V = 3.5V_p$$

Combining these parameters gives,

$$f_{max} = \frac{0.49 \frac{V}{\mu s}}{2\pi \times 3.5 V_p} = \frac{490 \frac{kV}{s}}{2\pi \times 3.5 V_p} = 22.3 \text{ kHz}$$

   b. The simulation results are shown in Figure 10-12. At a frequency of 22 kHz, the LM324 outputs a clean sine wave with a peak amplitude of 3.5 V_p. At 44 kHz, the op amp is slewing its output voltage as rapidly as it can, and still cannot keep up. The result is a triangle wave.

Practice: What is the largest V_p, highest frequency sine wave that a LM318 can output?

Answer: $SR = 70 \frac{V}{\mu s}$, $E_{supply\,max} = \pm 20 \text{ V}_{dc}$, $V_{sat} = 3 \text{ V}$, $f_{max} = 655$ kHz

---

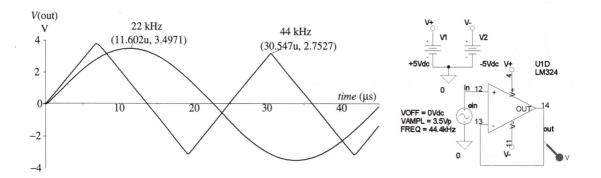

Figure 10-12 Simulation results for Example 10-9

Summary

The frequency response of op amp based amplifiers depends on both the reactive components placed around the op amp and the op amp itself.

Placing a capacitor in *parallel* with the negative feedback resistor is a good practice. As frequency increases, the capacitor's and the feedback network's impedances fall, driving the gain down. This allows you to tailor the amplifier's response to your needs, assuring that the amplifier does *not* respond to high-frequency noise.

Placing a capacitor in *series* with the feedback resistor breaks the dc negative feedback. The circuit has infinite gain at dc. The output is driven into saturation by any small input offset dc voltage. Avoid this connection.

A capacitor in series with the *input* resistor blocks the low frequencies from entering the amplifier. This is an effective way to couple an ac signal while blocking any dc bias that it may contain, and is particularly handy when running the op amp from a single power supply voltage.

A capacitor in *parallel* with the input resistor shorts out that resistor at high frequencies. Since this input network's impedance appears in the denominator of the gain equation, the result is extremely high gain for the high-frequency noise. Avoid this configuration as well.

To assure stability, op amps have a low-pass filter built within the IC. At a low frequency, the open-loop gain begins to fall as the frequency increases. At *any* frequency, the product of that frequency and the open-loop gain is a constant, the gain bandwidth product, *GBW*.

When used in a closed-loop amplifier, this reduction of gain with frequency lowers the closed-loop gain by 0.707 at f_{-3dB}. At that frequency $GBW = A_o \times f_{-3dB}$. Above f_{-3dB}, the gain rolls off with the fall of the open-loop gain. The $f_{10\%}, f_{5\%}$, and $f_{1\%}$ are all defined for the op amp just as they were for the basic low-pass filter.

It is far easier to measure rise time than frequency response. An op amp's *small-signal* rise time is directly related to its gain bandwidth product. So you may see manufacturers provide this specification.

Slew rate defines how rapidly an op amp's output may change in response to a large (>1 V_p) step. It is normally rated in V/μs. This limit imposed on sine waves is the full-power bandwidth. The highest, large-signal sine wave an op amp can output depends on the op amp's slew rate and both the signal's frequency and its peak amplitude.

Problems

Parallel *RC* in the Feedback Loop

10-1   Calculate the feedback resistor and capacitor needed to set $A_o = 22$, and $f_{-3dB} = 300$ Hz. The input resistor, $R_i = 10$ kΩ.

10-2   Calculate the feedback resistor and capacitor needed to set $A_o = 47$, and $f_{-3dB} = 1$ kHz. The input resistor, $R_i = 22$ kΩ.

10-3   An amplifier has $R_i = 3.3$ kΩ, $R_f = 47$ kΩ, and $C_f = 2.2$ nF in parallel with R_f. Calculate A_o, f_{-3dB}, and the gain at f_{-3dB}.

10-4   An amplifier has $R_i = 10$ kΩ, $R_f = 330$ kΩ, and $C_f = 220$ pF in parallel with R_f.. Calculate A_o, f_{-3dB}, and the gain at f_{-3dB}.

Series *RC* in the Input Loop

10-5   Calculate the input resistor and capacitor needed in an inverting amplifier to set $A_0 = 22$, and $f_{-3dB} = 300$ Hz. $R_f = 47$ kΩ.

10-6   Calculate the input resistor and capacitor needed in an inverting amplifier to set $A_0 = 47$, and $f_{-3dB} = 1$ kHz. $R_f = 330$ kΩ.

10-7   An inverting amplifier has $R_i = 3.3$ kΩ, $R_f = 47$ kΩ, and $C_i = 2.2$ nF in series with R_i. Calculate A_o, f_{-3dB}, and the gain at f_{-3dB}.

10-8   An inverting amplifier has $R_i = 10$ kΩ, $R_f = 330$ kΩ, and $C_f = 220$ pF in series with R_i. Calculate A_o, f_{-3dB}, and the gain at f_{-3dB}.

Open-loop Frequency Response

10-9   By simulation, determine the open-loop gain bandwidth product of the LF411. Compare this to manufacturer's specifications.

10-10   By simulation, determine the open-loop gain bandwidth product of the LM741. Compare this to manufacturer's specifications.

Closed-loop Frequency Response

10-11   An inverting amplifier is built with an LM324, $R_i = 3.3$ kΩ, and $R_f = 47$ kΩ. Find $A_o, f_{-3dB}, f_{10\%}, f_{5\%}, f_{1\%}$, and the gain at each.

10-12   An inverting amplifier is built with an LMH 6654, $R_i = 51$ Ω, and $R_f = 11$ kΩ. Calculate $A_o, f_{-3dB}, f_{10\%}, f_{5\%}, f_{1\%}$, and the gains at each of these frequencies.

10-13 Calculate the required *GBW*, and select an op amp for an inverting amplifier, $A_o = 51$, gain error of 30% at 100 kHz.

10-14 Calculate the required *GBW*, and select an op amp for an inverting amplifier, $A_o = 120$, gain error of 10% at 45 kHz.

10-15 Draw the schematic, and explain with calculations how an LM324 can be used to build the amplifier in Problem 10-13.

10-16 Draw the schematic, and explain with calculations how an LF411s can be used to build the amplifier in Problem 10-14.

Small-signal Rise Time

10-17 The manufacturer specifies the small-signal rise time of an op amp to be 70 ns. Calculate the *GBW*.

10-18 What is the small-signal rise time for the LM324?

Response to a Step

10-19 By simulation, determine the slew rate of the LF411. Compare this to the manufacturer's specification.

10-20 By simulation, determine the slew rate of the LM741 (UA741 in Cadence). Compare this to the manufacturer's specification.

10-21 An LM324 is to be used as a comparator. How long does it take its output to rise from 0.8 V to 2.4 V?

10-22 A dedicated comparator IC has a rise time of 250 V/µs. How long does it take its output to rise from 0.8 V to 2.4 V?

10-23 How far can the output of the LM324 travel in 100 ns?

10-24 Repeat Problem 10-23 for the LF411 op amp. Compare the results of the two problems, and provide an application for each.

10-25 It is necessary for an output step to rise from 0.8 V to 2.4 V in 10 µs. Calculate the required slew rate, and select an op amp.

10-26 It is necessary for an output step to rise from 0.8 V to 2.4 V in 12 ns. Calculate the required slew rate, and select an op amp.

Full-power Bandwidth

10-27 What is the highest frequency, f_{max}, at which an LM324 can provide a sine wave of 10 V_p?

10-28 What is the highest frequency, f_{max}, at which an LF411 can provide a sine wave of 10 V_p?

10-29 What is the largest sine wave, $V_{p\ max}$, for which an LM324 can operate at 20 kHz?

10-30 What is the largest sine wave, $V_{p\ max}$, for which an LF411 can operate at 100 kHz?

10-31 Determine the slew rate required to provide a 15 V_p sine wave at 20 kHz. Find an op amp that provides this slew rate.

10-32 Determine the slew rate required to provide a 10 V_p sine wave at 455 kHz. Find an op amp that provides this slew rate.

Op Amp Speed Lab Exercise

A. Closed-loop Frequency Response
   1.   Build the circuit in Figure 10-13.

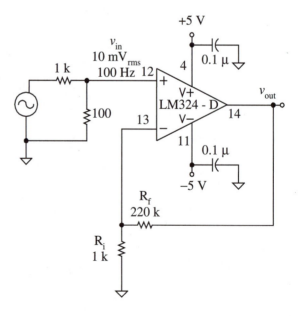

Figure 10-13 Noninverting amplifier

2. Apply dc power, then turn the function generator on. Set the loaded generator to the proper values.

3. Display the amplifier's input on oscilloscope (CH1), and on the digital multimeter, and the amplifier's output on channel 2.

4. Verify the *input's* dc, rms, and frequency with the digital meter.

5. Measure the *output's* dc and rms values with the digital meter. Also assure that the output's phase is correct.

6. Record this data in a table with rows for *theory* and *actual*, and columns for $V_{out\ dc}$, $v_{out\ rms}$, and $\Theta_{out\ degrees}$.

7. Complete the second and third columns of a table similar to Table 10-1.
   a. Change the input frequency to that indicated in each row as measured with the digital meter.

   b. Verify that the input amplitude is $10mV_{rms}$ as measured with the digital meter.

   c. Measure the output amplitude as indicated by the digital meter and then calculate and record the amplifier's gain.

   d. Calculate the gain bandwidth product.

Table 10-1 Closed Loop Frequency Responses

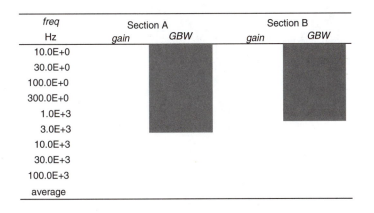

freq	Section A		Section B	
Hz	gain	GBW	gain	GBW
10.0E+0				
30.0E+0				
100.0E+0				
300.0E+0				
1.0E+3				
3.0E+3				
10.0E+3				
30.0E+3				
100.0E+3				
average				

average GBWmeasured = _____

GBWspecifications = _____

8. Calculate the average gain bandwidth products.

B. Frequency Compensation
Place a 1 nF capacitor in parallel with R_f. Repeat all of the steps in Section A. Since the gain bandwidth product has been changed by the feedback capacitor, do your initial check at 10 Hz.

C. Slew Rate
1. Build the circuit in Figure 10-14. Set the input to a 2 V_{pp} square wave.

2. Measure the slope of the output's rising edge. It is handy to use the oscilloscope's paired cursors to identify two points on the *linear* part of the rising edge.

SRmeasured = _____

SRspecifications = _____

3. Calculate the op amp's slew rate. Compare it to specifications

4. Change the input signal to a 5 V_p, 10 kHz sine wave.

5. Monitor the output signal's wave shape on the oscilloscope.

Full power bandwidthmeasured =
_____

Full power bandwidthspecifications =
_____

6. Increase the frequency until the sine appears to distort. Record this frequency and compare it to the expected full-power bandwidth.

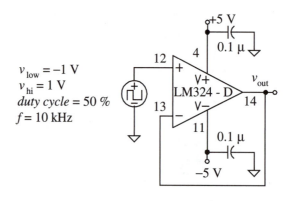

$v_{low} = -1$ V
$v_{hi} = 1$ V
duty cycle = 50 %
$f = 10$ kHz

Figure 10-14 Voltage follower for slew rate measurement

Fourier Series Analysis

The sinusoid is a widely applied signal, the ideal shape of the signal from the power utilities, and a traditional test shape for amplifiers and filters. Calculations of a circuit's response to this shape are straightforward, and were covered in detail in the preceding ten chapters.

However, very few practical signals are sine waves. The pervasive electrical power utility signal may have notable distortion. Digital signals are increasingly replacing analog in communications (digital cell phones, HDTV, pagers, garage door openers), audio (MP3 files and Class D amplifiers), power control (switching power supplies, and H bridge motor drivers), and digital signal processing (DSP). Along the way, these pulses meet amplifiers, transmission paths (resistance, inductance, and capacitance), and are eventually converted into some form of analog signal (more *RCL*). Listening to pure sine waves is *very* boring. A rich *mixture* of tones is what makes a guitar sound different from a saxophone, and *makes* the music. Power supplies use half- and full-wave rectified signals. These are then processed by resistors, inductors, and capacitors. Switching power supplies now dominate the market and handle pulses, sawteeth, and triangles with amplifiers, resistors, capacitors, and inductors. So, in the end, the sine wave is only one of *many* signals that you need to be able to handle.

In your earlier studies, you used the superposition theorem to handle many signals, determining the effect several dc sources had when working together. Chapter 11 reviews these dc calculations and then expands them to include sine waves of different frequencies as well as dc. The result is the beginning of a complex wave shape.

Fourier series allow you to express *any* repetitive signal as a combination of sine waves of different frequencies, different phases and different amplitudes. In Chapter 12 you will learn to calculate these Fourier coefficients (harmonics) for all of the wave shapes common to electronic circuits. You will then use manual calculations and a variety of computer-aided tools to extract these harmonics.

Armed with the superposition theorem and the Fourier series, you can analyze the effect of a wide variety of nonsinusoidal signals on low-pass and high-pass, first and second order passive circuits and op amp based amplifiers. Chapter 13 presents a computational tool that helps keep all of the quantities organized and then compares those solutions to the results from simulation and MATLAB.

11

Superposition

Introduction

All repetitive wave shapes can be constructed from dc and the proper amplitude, frequency, and phase sine waves. The superposition theorem indicates that the composite input wave shape is just the addition of these signals. The effect that each input signal has on the output can be calculated *independently*. Finally, the composite output wave is the sum of these output signals.

Although this may sound trivial and obvious, its implication is enormous. Regardless of the wave shape, all you need to be able to do is to figure out how the circuit responds to dc and handle the ac phasor analysis, several times. Then, add up the results.

This chapter begins with a statement of the superposition theorem, and then applies it to a dc circuit. When time-varying signals are used, such as sine waves, you must begin and end in the time domain. Computers can aid with the plotting of these signals. Both a spreadsheet and MATLAB examples will be illustrated. The circuit analysis consists of simple dc or ac phasor calculations, repeated for each source. The resulting output signals are then added to produce the composite output. Computer-aided analysis and simulations are shown in the examples.

Objectives

Upon completion of this chapter, you will be able to do the following:

- State the superposition theorem.

- Reconstruct a complex wave from its dc and sinusoidal components.

- Apply the superposition theorem to a circuit containing several dc sources to calculate the output voltage or current.

- Apply the superposition theorem to a circuit containing several dc and ac signals of different magnitudes, frequencies, and phases to calculate the output voltage or current.

11.1 Basic Principles

All of the circuits in the preceding chapters have been driven from a single source. But often, several signals are applied to a circuit simultaneously. To handle these circuits, the superposition theorem states that:

> **When multiple *independent* sources drive a linear network, you can determine the effect of all of the sources acting together by calculating the effect of each source individually, and then adding the results.**

To turn off a source, replace it with its characteristic impedance. The impedance of an ideal voltage source is 0 Ω, a short, while the impedance of an ideal current source is an open. The output impedance of an op amp may practically be considered to be 0 Ω. Most function generators have an output impedance of 50 Ω, as does most rf circuitry.

Figure 11-1(a) shows a full schematic with three sources. The sine wave from a function generator is *RC* coupled into the circuit where it is combined with the current from a sensor, i_{sensor}, and delivered to a load. There is also a biasing supply, E_{supply}, to consider.

The effect of the function generator alone is shown in Figure 11-1(b). The current source has been replaced with an open and the bias voltage supply with a short. The circuit is much simpler: R1 is in parallel with R2 and the series combination of R3 and R_{load}.

In Figure 11-1(c) the current source is considered. Both of the voltage sources have been replaced by shorts. However, the function generator's 50 Ω output impedance is still in the circuit. The current source is in parallel with the load. R1 continues to parallel R2. The generator's 50 Ω now is in series with the capacitor.

Figure 11-1 (d) shows the effect of the bias voltage supply. This time, R1 is in series with a branch containing three resistors. Since the capacitor is an open to dc, the branch containing R1 and C_{in} is ignored.

Each of these circuits results in a configuration that is quite different from the original, and from each other. The results from each are added together to produce the overall solution.

When applying superposition, you may turn *off* only the **independent** sources. Independent sources are those whose value do not change as currents or voltages elsewhere in the circuit change. All three of the sources in Figure 11-1 are independent.

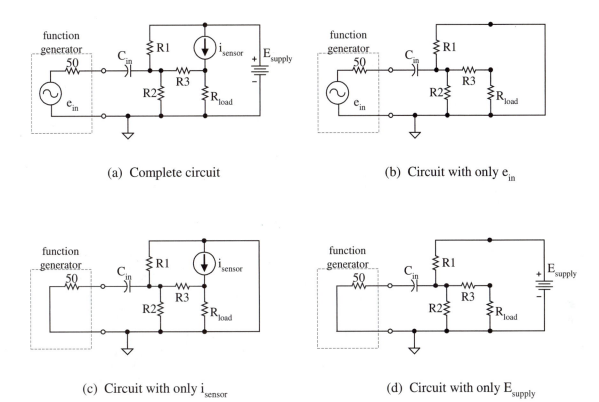

(a) Complete circuit

(b) Circuit with only e_{in}

(c) Circuit with only i_{sensor}

(d) Circuit with only E_{supply}

Figure 11-1 Turning sources *off* when using superposition

Example 11-1

Superposition analysis of the circuit in Figure 11-1 produces:

$$\overline{V_{\text{load from } ein}} = (2.25\,\text{V}_{\text{rms}}\angle 0°) \text{ @ 1kHz}$$

$$\overline{V_{\text{load from } isensor}} = (0.75\,\text{V}_{\text{rms}}\angle 0°) \text{ @ 3kHz}$$

$$V_{\text{load from Esupply}} = 2.5\,\text{V}_{\text{dc}}$$

a. Write the equation for the composite voltage across the load.

b. Plot two cycles of the voltage across the load.

Solution

a. First, each of the phasors must be written in their time domain form.

$$v(t) = V_p \sin(\omega t + \theta)$$

Where $V_p = V_{rms}\sqrt{2}$

$$\omega = 2\pi f$$

Then, they must be added together.

$$\begin{aligned} v_{load} = &\left[2.25\,V_{rms} \times 1.414\sin(2\pi \times 1\,kHz \times t)\right] \\ &+ \left[0.75\,V_{rms} \times 1.414\sin(2\pi \times 3\,kHz \times t)\right] \\ &+ 2.5\,V_{dc} \end{aligned}$$

MATLAB

b. The plot of this composite wave is shown in Figure 11-2. It looks much more like a square wave than a sine. Certainly just adding 2.25 V_{rms} + 0.75 V_{rms} + 2.5 V_{dc} = 5.5 V gives an entirely incorrect result.

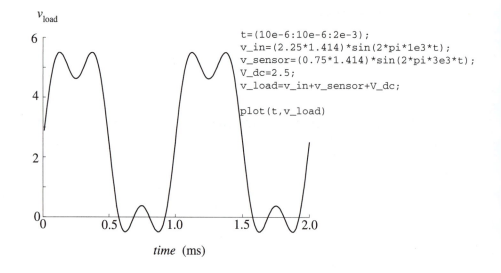

```
t=(10e-6:10e-6:2e-3);
v_in=(2.25*1.414)*sin(2*pi*1e3*t);
v_sensor=(0.75*1.414)*sin(2*pi*3e3*t);
V_dc=2.5;
v_load=v_in+v_sensor+V_dc;

plot(t,v_load)
```

Figure 11-2 Plot and MATLAB commands for Example 11-1

Practice: Plot the result if the magnitude of i_{sensor} drops to 0.25 V_{rms} and its phase lags by 180°.

Answer: The load voltage waveform becomes a triangle.

11.2 AC and DC Circuit Superposition

For circuits that contain both dc and ac signals at different frequencies, several additional steps must be taken. First, express the input signals in their time domain form. For a sinusoidal signal that is

$$e = V_p \sin(2\pi ft + \theta)$$

All of the information you need to solve the problem is contained in this form. However, the analysis of the circuit must be done with the phasor form of the signals.

$$\overline{E} = (V_{rms} \angle \theta)$$

Be careful to convert the peak amplitude from the time domain equation into rms amplitude for the phasor form. Even though the frequency is not included in the quantity, be sure to record it at the beginning of this section of the problem. Remember, the reactance of inductors and capacitors depends on frequency. So, if each source has a different frequency, these components have different reactances.

Once you have completed the dc and phasor analysis for each source, convert each resulting voltage back into the time domain. The superposition theorem then directs you to add these time domain voltages. Finally, this composite equation must be plotted.

Example 11-2

Determine the output of the circuit in Figure 11-3. This is an *attempt* at an audio mixer running from a single supply. Rheostat R1 is the mix control for source e_1, contributing more or less of that signal to the composite. Rheostat R2 is the mix control for source e_2. Rheostat R_f is the overall volume control.

Solution

On first glance, it is tempting to think that with equal inputs and with the mix rheostats (R1 and R2) set equally, the output should contain equal amounts of each signal. And, with R_f at

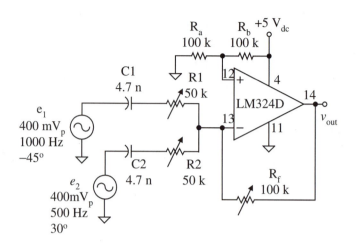

Figure 11-3 Mixer schematic for Example 11-2

twice R1 and R2, the output should be twice these summed inputs. That would be true if the two signals were at the same frequency. However, since frequency affects each of the input capacitors differently, the output signal is more complicated.

a. **V_{dc}**

When considering only the dc source, the capacitors both look like opens. The redrawn circuit is shown in Figure 11-4. This is, indeed, much simpler.

The two 100 kΩ resistors evenly split the +5 V_{dc}, placing 2.5 V_{dc} at the op amp's noninverting input.

There is no difference in potential between the two input pins of an op amp with negative feedback (that is not saturated). The voltage at the op amp's inverting input is 2.5 V_{dc}.

There is no significant current flowing into the input of the op amp. This means that there is no current flowing through R_f. It drops no voltage. The voltage at the left of R_f (2.5 V_{dc}) is the same as the voltage at the right of R_f. This is the output dc voltage.

$$V_{out\,dc} = 2.5\,V_{dc}$$

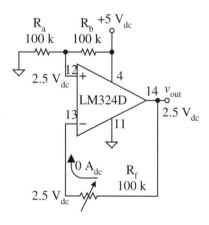

Figure 11-4
Schematic for the dc voltage

b. *V_{out} from e1*
The circuit redrawn with only e_1 *on* is shown in Figure 11-5. The bias dc for the upper two resistors is replaced with a short to common. So, there is no ac voltage at the op amp's noninverting input pin. This means that there is no ac voltage at the inverting pin either. It is at virtual ground.

The time domain equation for e_1 is

$$e_1 = 0.4\,V_p \sin\left(2\pi \times 1000\,\text{Hz} \times t - 45°\right)$$

The phasor version of e_1 is

$$\overline{E1} = \left(\frac{0.4\,V_p}{\sqrt{2}} \angle - 45°\right) = \left(0.2828\,V_{rms} \angle - 45°\right)$$

$$f_1 = 1000\,\text{Hz}$$

The reactance of C1 is

$$X_{C1} = \frac{1}{2\pi \times 1000\,\text{Hz} \times 4.7\,\text{nF}} = 33.86\,\text{k}\Omega$$

Since the op amp's inverting input pin is at virtual ground, all of the input voltage is dropped across the series combination of R1 and C1. The current through these components is

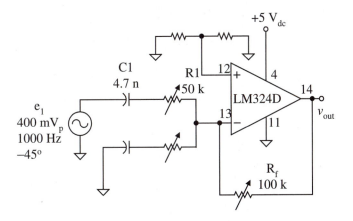

Figure 11-5 Schematic with e_1 *on*

$$\overline{I_1} = \frac{\overline{E_1}}{(R1 - jX_{C1})}$$

$$\overline{I_1} = \frac{(0.2828\,V_{rms}\angle -45°)}{(50\,k\Omega - j33.86\,k\Omega)} = (4.683\,\mu A_{rms}\angle -10.9°)$$

No current flows into the input pin of the op amp. With ground on the left of C2 and R2, and virtual ground on their right, no current flows through them either. So all of I_1 must flow, from left to right, through R_f.

$$\overline{I_{Rf}} = \overline{I_1} = (4.683\,\mu A_{rms}\angle -10.9°)$$

This current produces a voltage drop across R_f.

$$\overline{V_{Rf}} = (4.683\,\mu A_{rms}\angle -10.9°)\times(100\,k\Omega\angle 0°)$$

$$\overline{V_{Rf}} = (468.3\,mV_{rms}\angle -10.9°)$$

The positive input voltage causes current to flow from left to right through R_f. With the left end of R_f at virtual ground, this means that its right end must go negative.

$$\overline{V_{out}} = \overline{V_{op\,amp\,inv}} - \overline{V_{Rf}}$$

$$\overline{V_{out}} = 0 - (468.3\,mV_{rms}\angle -10.9°)$$

$$\overline{V_{out}} = (468.3\,mV_{rms}\angle 169.1°)$$

This must now be written in the time domain.

$$v_{out} = (468.3\,mV_{rms}\times\sqrt{2})\sin(2\pi\times1000\,Hz\times t + 169.1°)$$

$$v_{out} = (662.3\,mV_p)\sin(2\pi\times1000\,Hz\times t + 169.1°)$$

c. *v_{out} from e2*

The circuit redrawn with only e_2 *on* is shown in Figure 11-6. The time domain equation for e_2 is

$$e_2 = 0.4\,V_p\sin(2\pi\times500\,Hz\times t + 30°)$$

The phasor version of e_2 is

$$\overline{E2} = \left(\frac{0.4\,V_p}{\sqrt{2}}\angle 30°\right) = (0.2828\,V_{rms}\angle 30°)$$

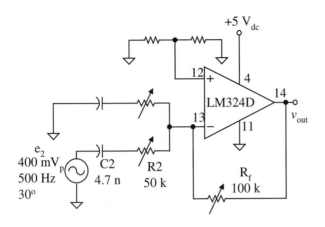

Figure 11-6 Schematic with e_2 *on*

$$f_2 = 500 \text{ Hz}$$

The reactance of C2 is

$$X_{C2} = \frac{1}{2\pi \times 500 \text{ Hz} \times 4.7 \text{ nF}} = 67.73 \text{ k}\Omega$$

Since the op amp's inverting input pin is at virtual ground, all of the input voltage is dropped across the series combination of R2 and C2. The current through these components is

$$\overline{I_2} = \frac{\overline{E_2}}{(R2 - jX_{C2})}$$

$$\overline{I_2} = \frac{(0.2828 \text{ V}_{rms} \angle 30°)}{(50 \text{ k}\Omega - j67.73 \text{ k}\Omega)} = (3.359 \,\mu\text{A}_{rms} \angle 83.6°)$$

No current flows into the input pin of the op amp. With ground on the left of C1 and R1, and virtual ground on their right, no current flows through them either. So all of I_2 must flow, from left to right, through R_f.

$$\overline{I_{Rf}} = \overline{I_2} = (3.359 \,\mu\text{A}_{rms} \angle 83.6°)$$

This current produces a voltage drop across R_f.

$$\overline{V_{Rf}} = (3.359 \,\mu\text{A}_{rms} \angle 83.6°) \times (100 \text{ k}\Omega \angle 0°)$$

$$\overline{V}_{\text{Rf}} = \left(335.9\,\text{mV}_{\text{rms}}\angle 83.6°\right)$$

The positive input voltage causes current to flow from left to right through R_f. With the left end of R_f at virtual ground, this means that its right end must go negative.

$$\overline{V}_{\text{out}} = \overline{V}_{\text{op amp inv}} - \overline{V}_{\text{Rf}}$$

$$\overline{V}_{\text{out}} = 0 - \left(335.9\,\text{mV}_{\text{rms}}\angle 83.6°\right)$$

$$\overline{V}_{\text{out}} = \left(335.9\,\text{mV}_{\text{rms}}\angle -96.4°\right)$$

This must now be written in the time domain.

$$v_{\text{out}} = \left(335.9\,\text{mV}_{\text{rms}} \times \sqrt{2}\right)\sin\left(2\pi \times 500\,\text{Hz} \times t - 96.4°\right)$$

$$v_{\text{out}} = \left(475.0\,\text{mV}_{\text{p}}\right)\sin\left(2\pi \times 500\,\text{Hz} \times t - 96.4°\right)$$

d. Composite wave
The complete signal is the addition of these three solutions, in the *time domain*.

$$v_{\text{out}} = 2.5\,\text{V}_{\text{dc}} + \left(662.3\,\text{mV}_{\text{p}}\right)\sin\left(2\pi \times 1000\,\text{Hz} \times t + 169.1°\right)$$
$$+ \left(475.0\,\text{mV}_{\text{p}}\right)\sin\left(2\pi \times 500\,\text{Hz} \times t - 96.4°\right)$$

e. Plot
The spreadsheet's top section and its graph are shown in Figure 11-7. The first column, column A, is the time. Select time steps that will be small enough to provide about 20 steps over the cycle of the highest frequency signal.

$$T_{\text{fastest}} = \frac{1}{f_{\text{fastest}}} = \frac{1}{1000\,\text{Hz}} = 1\,\text{ms}$$

$$t_{\text{step}} = \frac{1\,\text{ms}}{20\,\text{steps}} = 50\frac{\mu\text{s}}{\text{step}}$$

Plot two full cycles of the lowest frequency signal.

$$T_{\text{longest}} = \frac{1}{f_{\text{slowest}}} = \frac{1}{500\,\text{Hz}} = 2\,\text{ms}$$

$$t_{\text{total}} = 2\frac{\text{ms}}{\text{longest cycle}} \times 2\,\text{cycles} = 4\,\text{ms}$$

time	Vout1	Vout2	Vout
s	v	v	v
0.00E+00	0.13	-0.47	2.15
5.00E-05	-0.08	-0.47	1.95
1.00E-04	-0.28	-0.46	1.76
1.50E-04	-0.45	-0.44	1.61
2.00E-04	-0.58	-0.41	1.51
2.50E-04	-0.65	-0.37	1.48
3.00E-04	-0.66	-0.32	1.53
3.50E-04	-0.60	-0.26	1.64
4.00E-04	-0.48	-0.20	1.82
4.50E-04	-0.32	-0.13	2.06
5.00E-04	-0.13	-0.05	2.32
5.50E-04	0.08	0.02	2.60
6.00E-04	0.28	0.10	2.88
6.50E-04	0.45	0.17	3.12
7.00E-04	0.58	0.23	3.31
7.50E-04	0.65	0.30	3.44
8.00E-04	0.66	0.35	3.51
8.50E-04	0.60	0.40	3.49
9.00E-04	0.48	0.43	3.41
9.50E-04	0.32	0.46	3.28
1.00E-03	0.13	0.47	3.10
1.05E-03	-0.08	0.47	2.89
1.10E-03	-0.28	0.46	2.68
1.15E-03	-0.45	0.44	2.49
1.20E-03	-0.58	0.41	2.33
1.25E-03	-0.65	0.37	2.22
1.30E-03	-0.66	0.32	2.16
1.35E-03	-0.60	0.26	2.16
1.40E-03	-0.48	0.20	2.21
1.45E-03	-0.32	0.13	2.31
1.50E-03	-0.13	0.05	2.43
1.55E-03	0.08	-0.02	2.56
1.60E-03	0.28	-0.10	2.68
1.65E-03	0.45	-0.17	2.78
1.70E-03	0.58	-0.23	2.84
1.75E-03	0.65	-0.30	2.85
1.80E-03	0.66	-0.35	2.81
1.85E-03	0.60	-0.40	2.70
1.90E-03	0.48	-0.43	2.55
1.95E-03	0.32	-0.46	2.36
2.00E-03	0.13	-0.47	2.15

◀ =2.5+B3+C3

=0.4750*SIN(2*PI()*500*A3-96.4/57.3)

=0.6623*SIN(2*PI()*1000*A3+169.1/57.3)

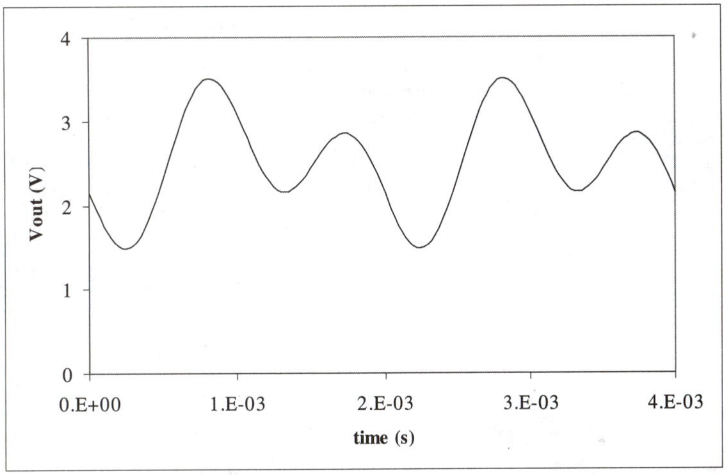

Figure 11-7 Spreadsheet and plot of the output for Example 11-2

f. The simulation results are shown in Figure 11-8. MultiSIM interprets all phase angles as *lag*. So, a phase of −45° is entered as **45Deg**. Leading, positive phase shifts must be converted into their lagging counterpart. Remember that you can subtract 360° from an angle. This just moves to the previous cycle.

$$30° = 30° - 360° = -330°$$

This *lagging* 330° is entered into MultiSIM as **330Deg**.

The output has a dc level of 2.5 V_{dc}. So, when setting up the oscilloscope, be sure to dc couple channel A. At the time $t = 0$, $v_{out} = 2.15$. To make the oscilloscope trace start every sweep when v_{out} crosses 2.15 V, set the **Trigger Edge** to **negative falling**, and the **Level** to **2.15 V**. Trigger mode should be **Nor** (normal).

Practice: Calculate the equation for the composite output waveform from Figure 11-3 for $e_{in1} = 1.92\,V_p \sin(2\pi \times 1000\,Hz \times t + 146°)$ and $e_{in2} = 0.544\,V_p \sin(2\pi \times 3000\,Hz \times t + 167°)$.

Answer: The output equation and the plot are shown in Figure 11-2.

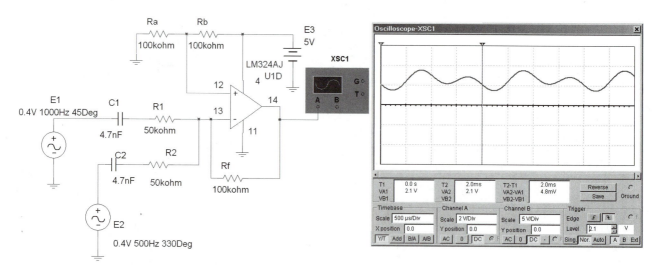

Figure 11-8 Simulation results for Example 11-2

Summary

The superposition theorem states that:

When multiple *independent* sources drive a linear network, you can determine the effect of all of the sources acting together by calculating the effect of each source individually, and then adding the results.

To turn a source *off*, replace it with its internal impedance. For an ideal voltage source, that is a short. Turn an ideal current source *off* by replacing it with an open. Dependent sources, such as those used in modeling transistors and op amps, may *not* be turned *off* because their values are set by voltages or currents elsewhere in the circuit.

Once the analysis is complete for each source, the composite result is calculated by adding the *time domain* expression of the results obtained for each source individually. You may *not* just add the phasors because each phasor may represent a sine wave at a different frequency. The final summation may look rather clumsy. It is important to plot this overall wave versus time for an understanding of the wave shape. You may use a graphing calculator, a spreadsheet, MATLAB, or a simulation program.

An inverted R-2R ladder network was used to illustrate superposition with dc circuits. This network is common in voltage-output-multiplying digital-to-analog converters. Although it may take a bit more effort, and more space, redrawing the schematic at every step of the analysis keeps you from becoming confused as the circuit is repeatedly reduced and reconfigured.

An inverting op amp summer was used to illustrate superposition with dc and ac circuits. The dc was considered first. When each ac source was considered, its value was expressed as a phasor, the phasor output was determined, and the phasor was converted back into the time domain. The total response was the sum of the dc and the time domain equations of the output from each ac source. The composite output waveform was plotted using a spreadsheet. Select the time step size to allow at least 20 steps of the highest frequency. Pull the time steps down until you are calculating two cycles of the lowest frequency. In each column, enter the time domain equation for one of the output sinusoids. In the last column, sum the dc with each of these sinusoid columns. Plot time from the first column (t) against this sum (v_{out}).

Problems

Basic Principles

11-1   Redraw the circuit in Figure 11-9 four times. In each case, simplify the diagram as much as possible, with the single source on the far left of the schematic and a single common bus running across the bottom.

    a. $e_{in\ 1}$ only *on*

    b. $e_{in\ 2}$ only *on*

    c. V_{dc} only *on*

    d. I_{sink} only *on*

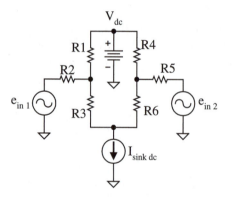

Figure 11-9 Schematic for Problem 11-1

11-2   Redraw the circuit in Figure 11-10 four times. In each case, simplify the diagram as much as possible, with the single source on the far left of the schematic and a single common bus.

    a. $e_{in\ 1}$ only *on*

    b. $E_{in\ 2}$ only *on*

    c. $I_{in\ 1}$ only *on*

    d. $E_{in\ 3}$ only *on*

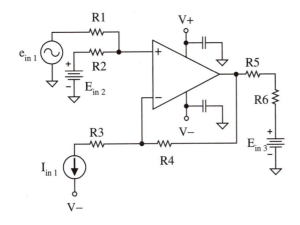

Figure 11-10 Schematic for Problem 11-2

11-3 Superposition analysis of a circuit produces:

$$\overline{V}_{\text{load from } ein\ 1} = \left(3.54V_{\text{rms}} \angle 0°\right) @\ 60\,\text{Hz}$$

$$\overline{V}_{\text{load from } ein\ 2} = \left(1.50V_{\text{rms}} \angle -90°\right) @\ 120\,\text{Hz}$$

$$V_{\text{load from Esupply}} = 3.18\ V_{\text{dc}}$$

   a. Write the equation for the composite voltage across the load.

   b. Plot two cycles of the voltage across the load.

11-4 Superposition analysis of a circuit produces:

$$\overline{V}_{\text{load from } ein\ 2} = \left(5.40V_{\text{rms}} \angle -90°\right) @\ 120\,\text{Hz}$$

$$\overline{V}_{\text{load from } ein\ 2} = \left(1.53V_{\text{rms}} \angle -90°\right) @\ 240\,\text{Hz}$$

$$V_{\text{load from Esupply}} = 11.5\ V_{\text{dc}}$$

   a. Write the equation for the composite voltage across the load.

   b. Plot two cycles of the voltage across the load.

AC and DC Circuit Superposition

11-5 Determine the output from the circuit in Figure 11-11.

   a. Calculate the output voltage produced by the dc supply and each ac generator separately.

b. Plot the output and verify your work by simulation.

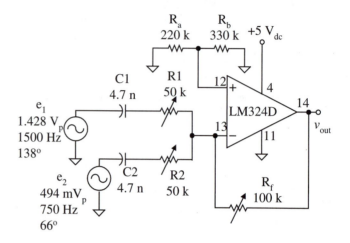

Figure 11-11 Schematic for Problem 11-5

11-6   Determine the output from the circuit in Figure 11-12.
  a. Calculate the output voltage produced by the dc supply and each ac generator separately.

  b. Plot the output and verify your work by simulation.

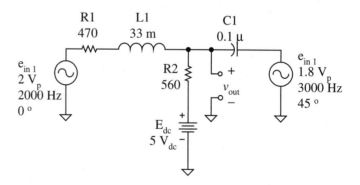

Figure 11-12 Schematic for Problem 11-6

12

Fourier Series of Nonsinusoidal Waveforms

Introduction

All repetitive wave shapes can be constructed from dc and the proper amplitude, frequency, and phase sine waves. This means that you can figure out how a circuit responds to *any* wave shape just by using simple dc and phasor ac calculations. There is no need for special techniques for every different wave. Just break the signal down into its dc and sine wave parts (called **harmonics**), calculate the circuit's response, using phasors, for each of the harmonics, and then superimpose the results. It's *all* just ac phasor analysis.

The *derivation* of the magnitudes, phases, and frequencies of the sine waves that compose each nonsinusoidal signal requires integration. However, you can *use* these series without working through all of these derivations. Combining a sine wave and a cosine wave produces another sine wave with a different amplitude and phase shift, a form more easily used in measurements and in phasor calculations. The rms of any repetitive signal can be calculated from the rms values of its harmonics. The series for square waves, pulses, sawtooth and triangle waves, half- and full-wave rectified sinusoids are all presented, calculated, and reconstructed with a spreadsheet, MATLAB, and two simulation programs.

Objectives

Upon completion of this chapter, you will be able to do the following:

- Write the general Fourier series in its sine and cosine form and in its amplitude and phase form. Convert between forms.

- Calculate the root mean squared value of a signal.

- For square waves, pulses, sawtooth waves, triangles, half- and full-wave rectified sinusoids, do the following:
    Derive the series coefficients' equations.
    Calculate the series coefficient values.
    Draw a frequency spectrum.

12.1 The Fourier Series

In 1826, the French mathematician Baron Jean Baptiste Fourier (1768–1830) was studying heat flow in a metal rod, and discovered that

any repetitive signal may be represented by an infinite series of cosines and sines. The frequencies of these sinusoids are integer multiples of a fundamental frequency.

Today, this sweeping statement forms the core upon which electronics communications is founded. In addition, with the expanding application of digital signals to audio, video, communications, and control systems, as well as the traditional areas of computers, how pulses travel through wires and through the air while interacting with other signals is a major issue in assuring proper, reliable operation of these systems. Fourier's discovery allows you to break complex signals into simple sinusoids, and then determine the effect of a circuit on the signal by using simple phasor algebra. It *all* just becomes ac circuit analysis.

Mathematically, the Fourier series of a repetitive signal with a frequency of ω_1 may be represented as

$$v(t) = A_0 + A_1 \cos(\omega_1 t) + A_2 \cos(2\omega_1 t) + A_3 \cos(3\omega_1 t) + A_4 \cos(4\omega_1 t) + \dots$$
$$+ B_1 \sin(\omega_1 t) + B_2 \sin(2\omega_1 t) + B_3 \sin(3\omega_1 t) + B_4 \sin(4\omega_1 t) + \dots$$

The term in front of the series, A_0, does not vary with time. It is the average value of the nonsinusoidal signal, its dc level.

$$A_0 = V_{dc} = \frac{1}{T} \int_0^T v(t)\,dt$$

Where, T is the period of the nonsinusoidal signal.
             $v(t)$ is the equation of the nonsinusoidal signal.

The angular velocity, ω_1, is the frequency of the nonsinusoidal signal. This is the **fundamental frequency**.

$$\omega_1 = 2\pi f_{nonsinusoid}$$

Each of the coefficients is calculated by an integration.

$$A_n = \frac{2}{T} \int_0^T v(t) \times \cos(n\omega_1 t)\,dt$$

$$B_n = \frac{2}{T} \int_0^T v(t) \times \sin(n\omega_1 t)\,dt$$

Although there are an infinite number of these terms, their magnitudes generally decrease significantly as n increases. Usually calculating the first nine terms provides a good representation of the nonsinusoidal signal.

Expressing the terms as cosines and sines is convenient when calculating the coefficients. But, physically, adding a cosine wave at one frequency to a sine wave at that same frequency produces a single sinusoidal signal, of different amplitude and phase. That composite signal is what you measure in the lab. It is called the **nth harmonic**.

$$A_n \cos(n\omega_1 t) + B_n \sin(n\omega_1 t) = C_n \sin(n\omega_1 t + \theta)$$

Where $\quad C_n = \sqrt{A_n^2 + B_n^2} \qquad \theta = 90° - \arctan\left(\dfrac{B_n}{A_n}\right)$

12.2 Root-mean-squared Value

Two signals that have the same root-mean-squared value will deliver the same power to the same resistor, regardless of the shape of those waves.

$$v_{rms} = \sqrt{\frac{1}{T} \int_0^T [v(t)]^2 dt} \; = \left.\frac{V_p}{\sqrt{2}}\right|_{sine\ wave}$$

But, what about square waves, pulses, triangles, ramps, and rectified sine waves? The rms of the composite signal is the root mean squared of the harmonics.

$$v_{rms} = \sqrt{V_{dc}^2 + V_{rms\,1}^2 + V_{rms\,2}^2 + V_{rms\,3}^2 + V_{rms\,4}^2 + V_{rms\,5}^2 + \dots}$$

Example 12-1

A 0 V to 5 V square wave can be represented by the following Fourier series.

$$v(t) = 2.5V_{dc} + 3.183V_p \sin \omega_1 t + 1.061V_p \sin 3\omega_1 t$$
$$+ 0.637V_p \sin 5\omega_1 t + 0.455V_p \sin 7\omega_1 t$$

a. Calculate the rms value for this square wave.

b. Verify your work by simulation. Calculate the rms value of the original square wave and the Fourier series.

Solution

a. The values needed to calculate the composite's rms are:

$$V_{dc} = 2.5 \, V_{dc}$$

$$V_{rms\,1} = \frac{3.183 \, V_p}{\sqrt{2}} = 2.251 \, V_{rms}$$

$$V_{rms\,3} = \frac{1.061 \, V_p}{\sqrt{2}} = 0.750 \, V_{rms}$$

$$V_{rms\,5} = \frac{0.637 \, V_p}{\sqrt{2}} = 0.450 \, V_{rms}$$

$$V_{rms\,7} = \frac{0.455 \, V_p}{\sqrt{2}} = 0.322 \, V_{rms}$$

The rms value of the composite signal, then, is

$$V_{rms} = \sqrt{(2.5 \, V_{dc})^2 + (2.251 \, V_{rms})^2 + (0.750 \, V_{rms})^2 + (0.450 \, V_{rms})^2 + (0.322 \, V_{rms})^2}$$

$$V_{rms} = 3.491 \, V_{rms}$$

b. The simulation is shown in Figure 12-1.

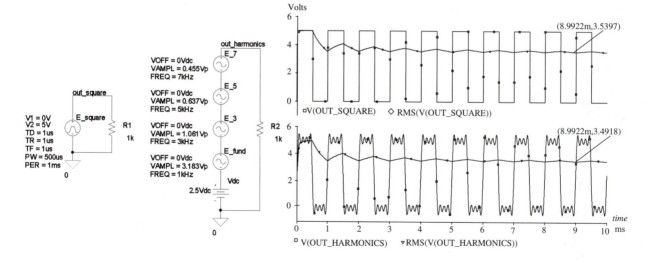

Figure 12-1 Simulation results for Example 12-1

The upper traces are the actual wave shape and its rms value. The lower traces are the composite wave produced by summing the given harmonics, and the resulting rms.

Practice: Repeat the simulation with MultiSIM.

Answer: MultiSIM indicates that the rms value is 2.5 V_{rms}.

12.3 Rectangular Waves

Other than sinusoids, rectangular waves are the most prevalent wave shape. Computers, cell phones, pagers, PDAs, TV remote controls, and even the newer audio systems all convert the information that they are processing into numbers. These data are then sent in streams of pulses. Rectangular waves carry the intelligence of our technological society.

Square Wave

The simplest rectangular wave is a square wave. It is shown in Figure 12-2. At $t = 0$, the voltage steps from 0 V to A V and stays there for half of the period, $T/2$. For the remainder of the period, until T, the signal is 0 V. The equation for this signal is

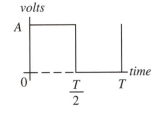

Figure 12-2
Plot of a square wave

$$v(t) = \begin{cases} A & 0 < t < \dfrac{T}{2} \\ 0 & \dfrac{T}{2} < t < T \end{cases}$$

The Fourier series is

$$v(t) = A_0 + \sum_{n=1}^{\infty} \left[A_n \cos(n\omega_1 t) + B_n \sin(n\omega_1 t) \right]$$

The dc level is

$$A_0 = V_{dc} = \frac{1}{T} \int_0^T v(t)\,dt$$

$$A_0 = \frac{1}{T} \left[\int_0^{\frac{T}{2}} A\,dt + \int_{\frac{T}{2}}^{T} 0\,dt \right]$$

$$A_0 = \frac{1}{T}\left(At\Big|_0^{\frac{T}{2}} + 0 \right)$$

$$A_0 = \frac{1}{T}\left(A\frac{T}{2} - 0 \right)$$

**dc level of a
square wave**

$$A_0 = \frac{A}{2}$$

The other A_n coefficients of the $\cos(n\omega_1 t)$ are

$$A_n = \frac{2}{T}\int_0^T v(t) \times \cos(n\omega_1 t)\, dt$$

$$A_n = \frac{2}{T}\left[\int_0^{\frac{T}{2}} A\cos(n\omega_1 t)\, dt + \int_{\frac{T}{2}}^T 0\cos(n\omega_1 t)\, dt \right]$$

$$A_n = \frac{2A}{T}\left[\int_0^{\frac{T}{2}} \cos(n\omega_1 t)\, dt \right]$$

$$A_n = \frac{2A}{n\omega_1 T}\sin(n\omega_1 t)\Big|_0^{\frac{T}{2}}$$

$$A_n = \frac{2A}{n\omega_1 T}\left[\sin\left(n\omega_1 \frac{T}{2} \right) - \sin(0) \right]$$

$$A_n = \frac{2A}{n\omega_1 T}\sin\left(n\omega_1 \frac{T}{2} \right)$$

$$\omega_1 = 2\pi f_1 = \frac{2\pi}{T}$$

$$A_n = \frac{2A}{n\omega_1 T}\sin\left(n\frac{2\pi}{T}\frac{T}{2} \right)$$

$$A_n = \frac{2A}{n\omega_1 T}\sin(n\pi)$$

Remember, n is an integer. Then, $\sin(n\pi) = 0$. So for the square wave, *all* of the A_n coefficients of $\cos(n\omega_1 t)$ are 0.

$$A_n = 0$$

A_n coefficients of a square wave

The B_n coefficients of $\sin(n\omega_1 t)$ are defined by

$$B_n = \frac{2}{T} \int_0^T v(t) \times \sin(n\omega_1 t) dt$$

$$B_n = \frac{2}{T} \left[\int_0^{\frac{T}{2}} A \sin(n\omega_1 t) dt + \int_{\frac{T}{2}}^T 0 \sin(n\omega_1 t) dt \right]$$

$$B_n = \frac{2A}{T} \left[\int_0^{\frac{T}{2}} \sin(n\omega_1 t) dt \right]$$

$$B_n = \frac{-2A}{n\omega_1 T} \cos(n\omega_1 t) \Big|_0^{\frac{T}{2}}$$

$$B_n = \frac{-2A}{n\omega_1 T} \left[\cos\left(n\omega_1 \frac{T}{2}\right) - \cos(0) \right]$$

$$B_n = \frac{-2A}{n\omega_1 T} \left[\cos\left(n\omega_1 \frac{T}{2}\right) - 1 \right]$$

$$\omega_1 = 2\pi f_1 = \frac{2\pi}{T}$$

$$B_n = \frac{2A}{n\frac{2\pi}{T} T} \left[1 - \cos\left(n\frac{2\pi}{T}\frac{T}{2}\right) \right]$$

$$B_n = \frac{A}{n\pi} \left[1 - \cos(n\pi) \right]$$

B_n coefficients of a square wave

These coefficients must be evaluated until a pattern emerges.

At $n = 1$ 　　　　$B_1 = \frac{A}{\pi}[1 - (-1)] = \frac{2A}{\pi}$

For $n = 2$ 　　　　$B_2 = \frac{A}{2\pi}[1 - (1)] = 0$

At $n = 3$

$$B_3 = \frac{A}{3\pi}[1 - (-1)] = \frac{2A}{3\pi}$$

For $n = 4$

$$B_4 = \frac{A}{4\pi}[1 - (1)] = 0$$

At $n = 5$

$$B_5 = \frac{A}{5\pi}[1 - (-1)] = \frac{2A}{5\pi}$$

Look at these coefficients. Not only are all of the $\cos(n\omega_1 t)$ terms zero, but all of the even $\sin(n\omega_1 t)$ terms are zero as well. The series becomes

Fourier series for a square wave

$$v(t) = \frac{A}{2} + \frac{2A}{\pi}\left(\sin \omega_1 t + \frac{1}{3}\sin 3\omega_1 t + \frac{1}{5}\sin 5\omega_1 t + \frac{1}{7}\sin 7\omega_1 t + \frac{1}{9}\sin 9\omega_1 t + ...\right)$$

The **frequency spectrum display** is shown in Figure 12-3.

Example 12-2

a. Calculate the dc and the first nine harmonics of a square wave that goes from 0 V to 5 V, and has a 1 kHz frequency.

b. Draw the composite wave using a spreadsheet.

Solution

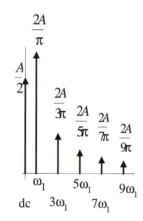

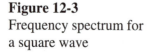

Figure 12-3
Frequency spectrum for a square wave

a. The series is

$$v(t) = \frac{A}{2} + \frac{2A}{\pi}\left(\sin \omega_1 t + \frac{1}{3}\sin 3\omega_1 t + \frac{1}{5}\sin 5\omega_1 t + \frac{1}{7}\sin 7\omega_1 t + \frac{1}{9}\sin 9\omega_1 t + ...\right)$$

$$A = 5\text{ V} \qquad \omega_1 = 2\pi \times 1\,\text{kHz} = 6283\frac{\text{rad}}{\text{s}}$$

$$V_{dc} = \frac{A}{2} = \frac{5\text{ V}}{2} = 2.5\,V_{dc}$$

The fundamental is

$$e_1 = \frac{2 \times 5\text{ V}}{\pi}\sin(6283t) = 3.183\,V_p \sin(6283t)$$

$$e_2 = e_4 = e_6 = e_8 = 0$$

$$e_3 = \frac{2 \times 5\text{ V}}{\pi} \times \frac{1}{3}\sin(3 \times 6283t) = 1.061\,V_p \sin(3 \times 6283t)$$

$$e_5 = \frac{2 \times 5\,\text{V}}{\pi} \times \frac{1}{5} \sin(5 \times 6283t) = 0.637\,\text{V}_p \sin(5 \times 6283t)$$

$$e_7 = \frac{2 \times 5\,\text{V}}{\pi} \times \frac{1}{7} \sin(7 \times 6283t) = 0.455\,\text{V}_p \sin(7 \times 6283t)$$

$$e_9 = \frac{2 \times 5\,\text{V}}{\pi} \times \frac{1}{9} \sin(9 \times 6283t) = 0.354\,\text{V}_p \sin(9 \times 6283t)$$

b. The spreadsheet is shown in Figure 12-4.

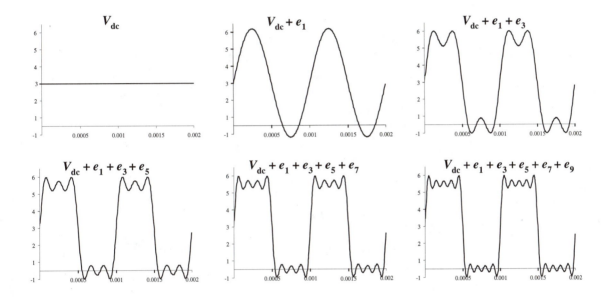

Figure 12-4 Harmonic addition for Example 12-2

Practice: Write the Fourier series for a square wave that goes from 0 V to 15 V with a frequency of 50 kHz.

Answer:

$$v(t) = 7.5\,\text{V}_{\text{dc}} + 9.549\,\text{V}_{\text{p}} \sin\left(314\,\text{k}\frac{\text{rad}}{\text{s}}t\right) + 3.183\,\text{V}_{\text{p}} \sin\left(3\times314\,\text{k}\frac{\text{rad}}{\text{s}}t\right)$$

$$+1.910\,\text{V}_{\text{p}} \sin\left(5\times314\,\text{k}\frac{\text{rad}}{\text{s}}t\right) + 1.364\,\text{V}_{\text{p}} \sin\left(7\times314\,\text{k}\frac{\text{rad}}{\text{s}}t\right)$$

$$+1.061\,\text{V}_{\text{p}} \sin\left(9\times314\,\text{k}\frac{\text{rad}}{\text{s}}t\right)$$

Pulses

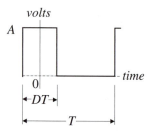
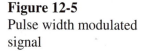

Figure 12-5
Pulse width modulated
signal

The square wave is a simple, standard wave shape, and as such makes a good laboratory test signal. However, the pulse train shown in Figure 12-5 has much wider application. All digital signals are of this form, on every trace of every computer, printer, PDA, elevator control panel, and so on. This is also the shape of pulse width modulated (**pwm**) signals. The amplitude, A, and the period, T, are constant, but the duty cycle, D, changes. These signals are central to all switching power supplies, and the control of small dc motors such as those in printers, copiers, and radio controlled vehicles. Class D audio amplifiers deliver large amounts of power to the loudspeaker by varying the duty cycle at an audio rate.

At $t = -DT/2$, the signal goes from 0 V to A V. It stays at that height during the duty cycle of the signal and then steps back to 0 V for the rest of the period, T. D may be a % (32%) or a decimal (0.32).

The Fourier series is

$$v(t) = AD + \frac{2A}{\pi}\left(\sin(\pi D)\cos\omega_1 t + \frac{1}{2}\sin(2\pi D)\cos 2\omega_1 t + \frac{1}{3}\sin(3\pi D)\cos 3\omega_1 t + \frac{1}{4}\sin(4\pi D)\cos 4\omega_1 t + ...\right)$$

This is a *cosine* series. Typically in lab the magnitude and phase are measured. The shape is assumed to be *sine*. Also, phase calculations use the magnitude and phase of a *sine* wave.

$$\cos(\phi) = \sin(\phi) + 90°$$

$$v(t) = AD + \frac{2A}{\pi}\left[\sin(\pi D)\sin(\omega_1 t + 90°) + \frac{1}{2}\sin(2\pi D)\sin(2\omega_1 t + 90°) + \frac{1}{3}\sin(3\pi D)\sin(3\omega_1 t + 90°) + ...\right]$$

This is the most involved series to be presented. In order to simplify it, the signal has been shifted to the left, starting one-half of a duty cycle early ($-DT/2$). This makes half of the terms fall out. So, the series only consists of $\sin(\phi + 90°)$ terms. However, looking at the full series, it appears that each harmonic has a sine multiplied by $\sin(\phi + 90°)$. In reality, the first sine term is part of the *coefficient*.

The individual coefficients are listed below, and evaluated for a 32% duty cycle, $D = 0.32$, an amplitude of 5 V, $A = 5$.

dc: $AD = 1.600$

fundamental: $\dfrac{2A}{\pi}\sin(\pi D) = 2.688$

$e_1 = 2.688\sin(\omega_1 t + 90°)$

second: $\dfrac{2A}{2\pi}\sin(2\pi D) = 1.440$

$e_2 = 1.440\sin(2\omega_1 t + 90°)$

third: $\dfrac{2A}{3\pi}\sin(3\pi D) = 0.133$

$e_3 = 0.133\sin(3\omega_1 t + 90°)$

fourth: $\dfrac{2A}{4\pi}\sin(4\pi D) = -0.613$

$e_4 = -0.613\sin(4\omega_1 t + 90°)$
$= 0.613\sin(4\omega_1 t + 90° - 180°)$
$= 0.613\sin(4\omega_1 t - 90°)$

fifth: $\dfrac{2A}{5\pi}\sin(5\pi D) = -0.605$

$e_5 = -0.603\sin(5\omega_1 t + 90°)$
$= 0.603\sin(5\omega_1 t + 90° - 180°)$
$= 0.603\sin(5\omega_1 t - 90°)$

There are two additional points. The sine coefficient is evaluated in terms of *radians*. So be sure to properly set the angle mode of your calculator. Secondly, notice that the magnitudes of the fourth and fifth harmonics are *negative*. That is correct. This means that these harmonics are inverted. So, 180° has been subtracted from each to produce a function in a more standard form.

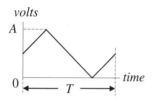

Figure 12-6
Triangle wave

12.4 Ramps

Ramps increase from their minimum value to their maximum *linearly*. Each voltage represents a particular time delay from the start of the wave. These are time-to-voltage converters.

They form the timing core for the pulse width modulators used to create the pulses from the previous section (in switching power supplies, motor speed control, and class D audio amplifiers). Ramps are also at the core of integrating analog-to-digital converters in digital multimeters.

Even astable and monostable multivibrators, found in many digital circuits, rely on the ramp as their key timing signal. So, even though ramps may not appear in the electronics "end-product," they are central to the operation of many of the circuits you build.

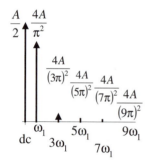

Figure 12-7
Triangle wave spectrum

The Triangle Wave

The triangle wave is shown in Figure 12-6. The Fourier series is

$$v(t) = \frac{A}{2} + \frac{4A}{\pi^2}\left(\sin \omega_1 t - \frac{1}{3^2}\sin 3\omega_1 t + \frac{1}{5^2}\sin 5\omega_1 t - \frac{1}{7^2}\sin 7\omega_1 t + ... \right)$$

Neither laboratory measurements nor phasor calculations use negative magnitude, such as the coefficients for the third and seventh harmonics. Instead, the – sign is included in the phase.

$$-V\sin(\omega t) = V\sin(\omega t + 180°)$$

Since the purpose of calculating the Fourier series for these nonsinusoidal waves is to allow phasor calculations and to better understand the lab measurements, rewrite the series as

$$v(t) = \frac{A}{2} + \frac{4A}{\pi^2}\left[\sin(\omega_1 t) + \frac{1}{3^2}\sin(3\omega_1 t + 180°) + \frac{1}{5^2}\sin(5\omega_1 t) + \frac{1}{7^2}\sin(7\omega_1 t + 180°) + ... \right]$$

The frequency spectrum for the triangle wave is shown in Figure 12-7. The amplitudes, *A*, are equal. The square wave has much larger harmonics. That should seem correct, since the triangle is a sine with its peak sharpened, showing just a little **distortion**. But, a square wave is much steeper on the edges and fatter and flatter in the middle than a sine wave, and is much more distorted.

Example 12-3

    a. Calculate the harmonics of a triangle wave that goes from 0 V to 5 V, with a frequency of 1 kHz.

    b. Verify your answer by reconstructing the triangle wave with MATLAB.

Solution

    a. The series is

$$v(t) = \frac{A}{2} + \frac{4A}{\pi^2}\left[\sin(\omega_1 t) + \frac{1}{3^2}\sin(3\omega_1 t + 180°) + \frac{1}{5^2}\sin(5\omega_1 t) + \frac{1}{7^2}\sin(7\omega_1 t + 180°) + \ldots\right]$$

$$A = 5 \text{ V} \qquad \omega_1 = 2\pi \times 1\text{kHz} = 6283\frac{\text{rad}}{\text{s}}$$

$$V_{dc} = \frac{A}{2} = \frac{5\text{ V}}{2} = 2.5\,V_{dc}$$

$$e_1 = \frac{4\times 5\text{ V}}{\pi^2}\sin(6283t) = 2.026\,V_p\sin(6283t)$$

$$e_2 = e_4 = e_6 = e_8 = 0$$

$$e_3 = \frac{4\times 5\text{ V}}{\pi^2}\times\frac{1}{3^2}\sin(3\times 6283t + 180°)$$
$$= 0.225\,V_p\sin(3\times 6283t + 180°)$$

$$e_5 = \frac{4\times 5\text{ V}}{\pi^2}\times\frac{1}{5^2}\sin(5\times 6283t)$$
$$= 0.082\,V_p\sin(5\times 6283t)$$

$$e_7 = \frac{4\times 5\text{ V}}{\pi^2}\times\frac{1}{7^2}\sin(7\times 6283t + 180°)$$
$$= 0.041\,V_p\sin(7\times 6283t + 180°)$$

$$e_9 = \frac{4\times 5\text{ V}}{\pi^2}\times\frac{1}{9^2}\sin(9\times 6283t)$$
$$= 0.025\,V_p\sin(9\times 6283t)$$

c. The MATLAB file and plot are shown in Figure 12-8. Look at the phase shift in the third and seventh harmonics.

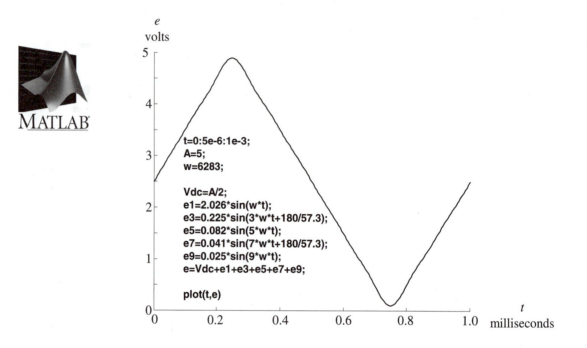

```
t=0:5e-6:1e-3;
A=5;
w=6283;

Vdc=A/2;
e1=2.026*sin(w*t);
e3=0.225*sin(3*w*t+180/57.3);
e5=0.082*sin(5*w*t);
e7=0.041*sin(7*w*t+180/57.3);
e9=0.025*sin(9*w*t);
e=Vdc+e1+e3+e5+e7+e9;

plot(t,e)
```

Figure 12-8 Plot of the sum of the harmonics for Example 12-3

Practice: Repeat Example 12-3 for a 0 V to 8 V, 150 kHz triangle.

Answer: $V_{dc} = 4\ v_{dc}$, $e_1 = 3.242\,V_p \sin(942\,k \times t)$,

$e_3 = 0.361\,V_p \sin(3 \times 942\,k \times t + 180°)$, $e_5 = 0.130\,V_p \sin(5 \times 942\,k \times t)$

Sawtooth

The triangle wave is commonly available from most function generators. Its harmonics are very small, reducing the chance of creating interference in other circuits if any of the high-frequency harmonics accidentally couple onto other traces. However, it is not easy to use this shape as a linear timing signal, or as the core of a pulse width modulator, or for analog-to-digital conversion, or for sweeping a trace across the screen slowly, but retracing it quickly. The sawtooth wave, of Figure 12-9 is preferred.

The sawtooth rises steadily. At the end of its rise, it falls in a step, to its starting voltage, and then begins to rise again. This is ideal for timing, pwm, a-d conversions, and driving a display.

The Fourier series is

$$v(t) = \frac{A}{2} + \frac{A}{\pi}\left(\sin\omega_1 t - \frac{1}{2}\sin 2\omega_1 t + \frac{1}{3}\sin 3\omega_1 t - \frac{1}{4}\sin 4\omega_1 t + ...\right)$$

Every sine coefficient *is* present. These harmonics fall off slowly, like the square wave's harmonics, not nearly as rapidly as the triangle wave. The sawtooth has lots of large harmonics creating considerable interference.

The even harmonics all have a negative magnitude. Neither laboratory measurements nor phasor calculations use the negative magnitude. Instead, the – sign is included in the phase.

$$-V\sin(\omega t) = V\sin(\omega t + 180°)$$

Since the purpose of calculating the Fourier series for these nonsinusoidal waves is to allow phasor calculations and to better understand the lab measurements, rewrite the series as

$$v(t) = \frac{A}{2} + \frac{A}{\pi}\left[\sin(\omega_1 t) + \frac{1}{2}\sin(2\omega_1 t + 180°) + \frac{1}{3}\sin(3\omega_1 t) + \frac{1}{4}\sin(4\omega_1 t + 180°) + ...\right]$$

The frequency spectrum of the ramp is shown in Figure 12-10. Although the ramp's fundamental is smaller than the fundamental for the other signals, there are twice as many harmonics, and their amplitudes do not fall off as rapidly. The sawtooth is very rich in harmonics.

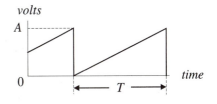

Figure 12-9
Sawtooth

Example 12-4

    a. Calculate the harmonics of a sawtooth wave that goes from 0 V to 5 V, with a frequency of 52 kHz.

    b. Verify your answer by reconstructing the sawtooth wave with a simulator.

Solution

    a. The series is

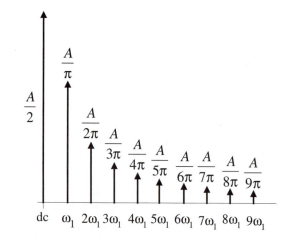

Figure 12-10 Frequency spectrum for a sawtooth wave

$$v(t) = \frac{A}{2} + \frac{A}{\pi}\left[\sin(\omega_1 t) + \frac{1}{2}\sin(2\omega_1 t + 180°) + \frac{1}{3}\sin(3\omega_1 t) + \frac{1}{4}(4\omega_1 t + 180°) + ...\right]$$

$$A = 5\,\text{V} \qquad \omega_1 = 2\pi \times 52\,\text{kHz} = 326.7\,\text{k}\frac{\text{rad}}{\text{s}}$$

$$V_{dc} = \frac{A}{2} = \frac{5\,\text{V}}{2} = 2.5\,\text{V}_{dc}$$

$$e_1 = \frac{5\,\text{V}}{\pi}\sin(326.7\,\text{k} \times t) = 1.592\,\text{V}_p\,\sin(326.7\,\text{k} \times t)$$

$$e_2 = \frac{5\,\text{V}}{\pi} \times \frac{1}{2}\sin(2 \times 326.7\,\text{k} \times t + 180°)$$
$$= 0.796\,\text{V}_p\,\sin(2 \times 326.7\,\text{k} \times t + 180°)$$

$$e_3 = \frac{5\,\text{V}}{\pi} \times \frac{1}{3}\sin(3 \times 326.7\,\text{k} \times t)$$
$$= 0.531\,\text{V}_p\,\sin(3 \times 326.7\,\text{k} \times t)$$

$$e_4 = 0.398\,\text{V}_p\,\sin(4 \times 326.7\,\text{k} \times t + 180°)$$

$$e_5 = 0.318\,\mathrm{V_p}\sin(5\times326.7\,\mathrm{k}\times t)$$

$$e_6 = 0.265\,\mathrm{V_p}\sin(6\times326.7\,\mathrm{k}\times t+180°)$$

$$e_7 = 0.227\,\mathrm{V_p}\sin(7\times326.7\,\mathrm{k}\times t)$$

$$e_8 = 0.199\,\mathrm{V_p}\sin(8\times326.7\,\mathrm{k}\times t+180°)$$

$$e_9 = 0.177\,\mathrm{V_p}\sin(9\times326.7\,\mathrm{k}\times t)$$

b. The schematic and probe simulation results are shown in Figure 12-11. The composite wave is a sawtooth going from 0 V to 5 V at 52 kHz. Even though nine harmonics were calculated, more are required to remove the ripple.

Practice: Calculate the dc and the first four harmonics for a sawtooth that goes from 0 V to 10 V in 167 ms.

Answer: $V_{dc} = 5\,\mathrm{V_{dc}}$,

$e_1 = 3.183\,\mathrm{V_p}\sin(37.7\times t)$, $e_2 = 1.592\,\mathrm{V_p}\sin(2\times37.7\times t+180°)$

$e_3 = 1.061\,\mathrm{V_p}\sin(3\times37.7\times t)$, $e_4 = 0.796\,\mathrm{V_p}\sin(4\times37.7\times t+180°)$

how big can you dream?™

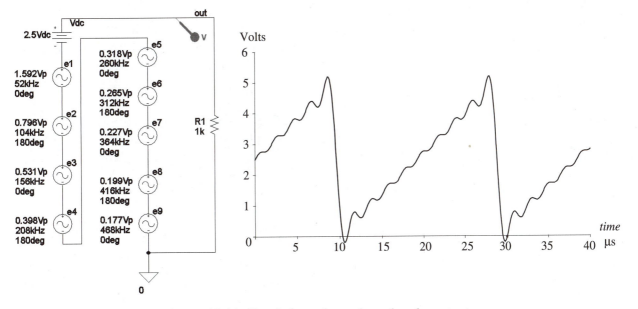

Figure 12-11 Simulation schematic and probe output

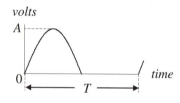

Figure 12-12
Half-wave rectified sinusoid

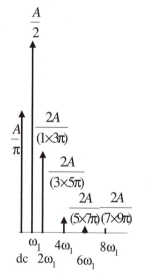

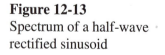

Figure 12-13
Spectrum of a half-wave rectified sinusoid

12.5 Rectified Sine Waves

Commonly, electrical power is distributed as a sine wave. From generator, through step-up and step-down transformers, to the load, the signal has *only* one harmonic, the fundamental. However, once this sinusoid reaches its destination, often it must be converted to dc in order to power electronics and dc machines. The simplest way to convert a sine wave into dc is with a single diode (half-wave rectification) or with four diodes in a bridge (full-wave rectification). As soon as you start chopping up the sinusoid, many upper frequency harmonics are generated. These harmonics affect the subsequent filtering circuit used to extract only V_{dc}. They also have an effect on the line current from the power system.

Half-wave Rectified Sinusoid

A half-wave rectified sinusoid is shown in Figure 12-12. The entire positive half-cycle remains, but all of the negative voltage has been removed. The Fourier series is

$$v(t) = \frac{A}{\pi} + \frac{A}{\pi}\left[\frac{\pi}{2}\sin(\omega_1 t) - \frac{2}{1\times 3}\cos(2\omega_1 t) - \frac{2}{3\times 5}\cos(4\omega_1 t) - \frac{2}{5\times 7}\cos(6\omega_1 t) + ...\right]$$

The first term, A/π, is V_{dc}. In many applications, this is all that you want. Filtering follows the rectifier to remove *all* of the harmonics. All of the odd order harmonics, except the fundamental, are zero.

$$e_3 = e_5 = e_7 = e_9 = ... = 0$$

The fundamental is *sin*. All of the even harmonics are *−cos*. Since both laboratory measurements and phasor calculations require the sine form, convert these cosine functions into their sine form. Delaying the $\sin(\phi)$ by −90° produces an inverted cosine, $-\cos(\phi)$. Making these substitutions into the Fourier series produces

$$v(t) = \frac{A}{\pi} + \frac{A}{\pi}\left[\frac{\pi}{2}\sin(\omega_1 t) + \frac{2}{1\times 3}\sin(2\omega_1 t - 90°) + \frac{2}{3\times 5}\sin(4\omega_1 t - 90°) + \frac{2}{5\times 7}\sin(6\omega_1 t - 90°) + ...\right]$$

The frequency spectrum for the half-wave rectified sine is shown in Figure 12-13. The half-wave rectified sine is not as rich in harmonics (distorted) as is the sawtooth, or even the square wave. However, its harmonics are much larger than those of the triangle wave.

Example 12-5

   a. Calculate the dc and the first nine harmonics of a half-wave rectified sine wave that goes from 0 V to 35 V at 60 Hz.

   b. Draw the composite wave using a spreadsheet.

Solution

   a. The series is

$$v(t) = \frac{A}{\pi} + \frac{A}{\pi}\left[\frac{\pi}{2}\sin(\omega_1 t) + \frac{2}{1\times3}\sin(2\omega_1 t - 90°) + \frac{2}{3\times5}\sin(4\omega_1 t - 90°) + \frac{2}{5\times7}\sin(6\omega_1 t - 90°) + ...\right]$$

$$A = 35\text{ V} \qquad \omega_1 = 2\pi \times 60\text{ Hz} = 377\frac{\text{rad}}{\text{s}}$$

$$V_{dc} = \frac{A}{\pi} = \frac{35\text{ V}}{\pi} = 11.1\text{ V}_{dc}$$

$$e_1 = \frac{35\text{ V}}{2}\sin(377t) = 17.5\text{ V}_p\ \sin(377t)$$

$$e_3 = e_5 = e_7 = e_9 = 0$$

$$e_2 = \frac{35\text{ V}}{\pi} \times \frac{2}{1\times3}\sin(2\times377t - 90°)$$
$$= 7.427\text{ V}_p\ \sin(2\times377t - 90°)$$

$$e_4 = \frac{35\text{ V}}{\pi} \times \frac{2}{3\times5}\sin(4\times377t - 90°)$$
$$= 1.485\text{ V}_p\ \sin(4\times377t - 90°)$$

$$e_6 = \frac{35\text{ V}}{\pi} \times \frac{2}{5\times7}\sin(6\times377t - 90°)$$
$$= 0.637\text{ V}_p\ \sin(6\times377t - 90°)$$

$$e_8 = 0.354\sin\text{V}_p(8\times377t - 90°)$$

   b. The plot is shown in Figure 12-14.

Practice: Calculate the first three harmonics for a half-wave rectified signal that is 6 V_p with a frequency of 450 kHz.

Answer: $V_{dc} = 1.910\,V_{dc}$, $e_1 = 6\,V_p \sin(2.83\,Mt)$

$e_2 = 1.273\,V_p \sin(2 \times 2.83\,Mt - 90°)$, $e_4 = 0.255\,V_p \sin(4 \times 2.83\,Mt - 90°)$

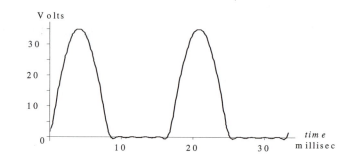

Figure 12-14 Spreadsheet and plot for Example 12-5

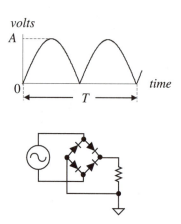

Figure 12-15
Full-wave rectified sine
and bridge rectifier

Full-wave Rectified Sinusoid

The full-wave rectified sinusoid, shown in Figure 12-16, results from the bridge rectifier. This shape is very common at the front-end of many ac-to-dc converters. The output wave has the negative half-cycle inverted and passed to the load, not just blocked. This fills the big blank spot that exists half of the time with the half-wave rectifier. Twice the energy is passed to the load, at twice the frequency. This makes the output easier to smooth, requiring smaller components and delivering lower ripple.

The Fourier series is

$$v(t) = \frac{2A}{\pi} + \frac{2A}{\pi}\left[-\frac{2}{1 \times 3}\cos(2\omega_1 t) - \frac{2}{3 \times 5}\cos(4\omega_1 t) - \frac{2}{5 \times 7}\cos(6\omega_1 t) - ... \right]$$

Convert this to the magnitude and angle form used in measurements and phasor calculations by substituting

$$-\cos(\phi) = \sin(\phi) - 90°)$$

$$v(t) = \frac{2A}{\pi} + \frac{2A}{\pi}\left[\frac{2}{1\times3}\sin(2\omega_1 t - 90°) + \frac{2}{3\times5}\sin(4\omega_1 t - 90°) + \frac{2}{5\times7}\sin(6\omega_1 t - 90°) + ...\right]$$

Compare this to that of a *half*-wave rectified sine.

$$v(t) = \frac{A}{\pi} + \frac{A}{\pi}\left[\frac{\pi}{2}\sin(\omega_1 t) + \frac{2}{1\times3}\sin(2\omega_1 t - 90°) + \frac{2}{3\times5}\sin(4\omega_1 t - 90°) + \frac{2}{5\times7}\sin(6\omega_1 t - 90°) + ...\right]$$

The two series have the same upper frequency harmonics. The only differences are that the full-wave rectified signal has twice the amplitude, and the harmonic at ω_1 is missing.

Example 12-6

   a.   Calculate the dc and the first nine harmonics of a full-wave rectified sine wave that goes from 0 V to 35 V at 60 Hz.

   b.   Draw the composite wave using simulation.

Solution

   a.   The series is

$$v(t) = \frac{2A}{\pi} + \frac{2A}{\pi}\left[\frac{2}{1\times3}\sin(2\omega_1 t - 90°) + \frac{2}{3\times5}\sin(4\omega_1 t - 90°) + \frac{2}{5\times7}\sin(6\omega_1 t - 90°) + ...\right]$$

$$A = 35 \text{ V} \qquad \omega_1 = 2\pi \times 60\,\text{Hz} = 377\,\frac{\text{rad}}{\text{s}}$$

$$V_{dc} = \frac{2A}{\pi} = \frac{2\times35\,\text{V}}{\pi} = 22.2\,V_{dc}$$

$$e_1 = e_3 = e_5 = e_7 = e_9 = 0$$

$$e_2 = \frac{2\times35\,\text{V}}{\pi} \times \frac{2}{1\times3}\sin(2\times377t - 90°)$$

$$= 14.854\,V_p\,\sin(2\times377t - 90°)$$

$$e_4 = \frac{2\times35\,\text{V}}{\pi} \times \frac{2}{3\times5}\sin(4\times377t - 90°)$$

$$= 2.971\,V_p\,\sin(4\times377t - 90°)$$

$$e_6 = \frac{2 \times 35\,\text{V}}{\pi} \times \frac{2}{5 \times 7} \sin(6 \times 377t - 90°)$$

$$= 1.273\,\text{V}_p \sin(6 \times 377t - 90°)$$

$$e_8 = \frac{2 \times 35\,\text{V}}{\pi} \times \frac{2}{7 \times 9} \sin(8 \times 377t - 90°)$$

$$= 0.707 \sin \text{V}_p (8 \times 377t - 90°)$$

b. The MultiSIM simulation results are shown in Figure 12-16. There are several points to notice. Remember that MultiSIM interprets angles as lagging. A negative sign is ignored. Both cursors indicate a peak of 34.6 V$_p$, close to the 35 V$_p$ expected. The period between the cursors is 8.3 ms. This is a frequency of 120 Hz, correct for a full-wave rectified 60 Hz sine wave.

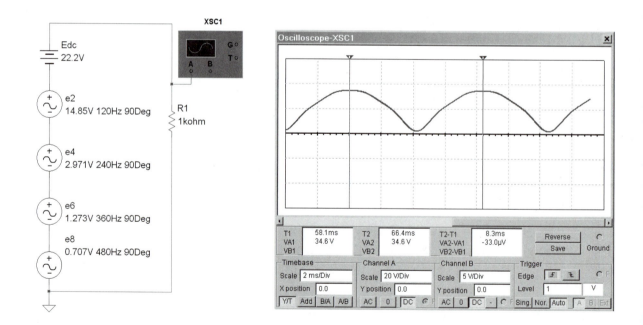

Figure 12-16 Simulation results for Example 12-6

Practice: Calculate the harmonics for a full-wave rectified signal that is 6 V_p, derived from an unrectifed sinusoid of 450 kHz.

Answer: $V_{dc} = 3.82$ V_{dc},
$e_2 = 2.55\,V_p\sin(2\times2.83\,Mt - 90°)$, $e_4 = 0.510\,V_p\sin(4\times2.83\,Mt - 90°)$,
$e_6 = 0.218\,V_p\sin(6\times2.83\,Mt - 90°)$, $e_8 = 0.122\,V_p\sin(8\times2.83\,Mt - 90°)$

12.6 Summary Table

The harmonics of the common wave shapes presented in the preceding sections of this chapter are tabulated in Table 12-1. The magnitudes are all normalized. $A = 1$. To determine the coefficients for any other amplitude, just multiply each term from Table 12-1 by the amplitude of the wave shape you are given. To make this more convenient, Table 12-1 is included as an Excel file on the compact disk provided with this text. All of the values are linked. Just change the value in the **Amplitude (V)** cell, and all of the coefficients are recalculated.

Table 12-1 Harmonics of common wave shapes

harmonic	square		triangle		Amplitude (V) 1 sawtooth		half-wave rectified		full-wave rectified	
	V	deg	V	deg	V	deg	V	deg	V	deg
V_{dc}	0.5000		0.5000		0.5000		0.3183		0.6366	
1st	0.6366	0	0.4053	0	0.3183	0	0.5000	0		
2nd					0.1592	180	0.2122	−90	0.4244	−90
3rd	0.2122	0	0.0450	180	0.1061	0				
4th					0.0796	180	0.0424	−90	0.0849	−90
5th	0.1273	0	0.0162	0	0.0637	0				
6th					0.0531	180	0.0182	−90	0.0364	−90
7th	0.0909	0	0.0083	180	0.0455	0				
8th					0.0398	180	0.0101	−90	0.0202	−90
9th	0.0707	0	0.0050	0	0.0354	0				

Summary

Any repetitive signal may be represented by an infinite series of cosines and sines. The frequencies of these sinusoids are integer multiples of a fundamental frequency.

$$v(t) = A_0 + \sum_{n=1}^{\infty} \left[A_n \cos(n\omega_1 t) + B_n \sin(n\omega_1 t) \right]$$

The coefficients A_n and B_n are calculated through integration.

These harmonics may also be represented as an infinite series of sine waves, each shifted in phase.

$$A_n \cos(n\omega_1 t) + B_n \sin(n\omega_1 t) = C_n \sin(n\omega_1 t + \theta)$$

Where
$$C_n = \sqrt{A_n^2 + B_n^2}$$

$$\theta = 90° - \arctan\left(\frac{B_n}{A_n} \right)$$

The root mean squared value of a signal may be calculated from

$$v_{rms} = \sqrt{V_{dc}^2 + V_{rms\,1}^2 + V_{rms\,2}^2 + V_{rms\,3}^2 + V_{rms\,4}^2 + V_{rms\,5}^2 + \ldots}$$

The Fourier series for a square wave, rectangular pulse, triangle, sawtooth, half-wave, and full-wave rectified sine wave were presented. The derivation of the coefficients for the square wave was completed.

The frequency spectrum is a plot of each harmonic's magnitude versus frequency. The spectrum for each of these waves was presented.

Once the harmonics have been calculated, they can be summed graphically to confirm that together they add up to the original wave shape. A spreadsheet, MATLAB, and either simulation program may be used to perform this addition and plot.

To determine the response of a circuit to a nonsinusoidal signal, first calculate the harmonics of the signal. Then, one at a time, calculate the output from the circuit for each of the harmonics. Finally, applying the superposition theorem, add all of these output harmonics. In the next chapter, you will apply these techniques.

Problems

The Fourier Series

12-1   **a.**   Write the following sinusoid in magnitude and angle format.

$$v(t) = 7\,V_p \cos(1000t) + 5\,V_p \sin(1000t)$$

         b.   Draw the resulting wave with a spreadsheet.

12-2   **a.**   Write the following sinusoid in magnitude and angle format.

$$i(t) = 25\,A_p \cos(377t) + 30\,A_p \sin(377t)$$

         b.   Draw the resulting wave with MATLAB.

Root-mean-squared Value

12-3   **a.**   Calculate the rms value of a square wave with an amplitude $A = 5$ V and the following harmonics.

$$V_{dc} = 2.5\,V_{dc}$$

$$e_1 = 3.183\,V_p \sin(6283t)$$

$$e_3 = 1.061\,V_p \sin(3 \times 6283t)$$

$$e_5 = 0.637\,V_p \sin(5 \times 6283t)$$

$$e_7 = 0.455\,V_p \sin(7 \times 6283t)$$

$$e_9 = 0.354\,V_p \sin(9 \times 6283t)$$

         b.   Compare this to the equation for a square wave's rms value.

$$V_{rms\,square} = \sqrt{0.5} \times A$$

12-4   **a.**   Calculate the rms value of a triangle wave with an amplitude $A = 5$ V and the following harmonics.

$$V_{dc} = 2.5\,V_{dc}$$

$$e_1 = 2.026\,V_p \sin(6283t)$$

$$e_3 = 0.225\,V_p \sin(3 \times 6283t + 180°)$$

$$e_5 = 0.082 \, V_p \sin(5 \times 6283t)$$

$$e_7 = 0.041 \, V_p \sin(7 \times 6283t + 180°)$$

    b. Compare this to the equation for a triangle's rms value.

$$v_{\text{rms triangle}} = \frac{A}{\sqrt{3}}$$

Rectangular Waves

12-5   **a.** Calculate the dc and the first nine harmonics of a square wave that goes from 0 V to 12 V, at 150 kHz.

        b. Draw the composite wave using a simulation program.

12-6   **a.** Calculate the dc and the first nine harmonics of a square wave that goes from 0 V to 8 V, at 60 Hz.

        b. Draw the composite wave using a simulation program.

12-7   **a.** Calculate the dc and the first nine harmonics of a rectangular pulse that goes from 0 V to 12 V, at 150 kHz, $D = 25\%$.

        b. Draw the composite wave using a spreadsheet.

12-8   **a.** Calculate the dc and the first nine harmonics of a rectangular pulse that goes from 0 V to 8 V, at 60 Hz, $D = 65\%$.

        b. Draw the composite wave using MATLAB.

Ramps

12-9   **a.** Calculate the dc and the first nine harmonics of a triangle wave that goes from 0 V to 12 V, at 150 kHz.

        b. Draw the composite wave using a simulation program.

12-10   **a.** Calculate the dc and the first nine harmonics of a triangle wave that goes from 0 V to 8 V, at 60 Hz.

        b. Draw the composite wave using a simulation program.

12-11   **a.** Calculate the dc and the first nine harmonics of a sawtooth wave that goes from 0 V to 12 V, at 150 kHz.

        b. Draw the composite wave using a spreadsheet.

12-12   **a.**   Calculate the dc and the first nine harmonics of a sawtooth wave that goes from 0 V to 8 V, at 60 Hz.

   b.   Draw the composite wave using MATLAB.

Rectified Sine Waves

12-13   **a.**   Calculate the dc and the first nine harmonics of a half-wave rectified sine wave that goes from 0 V to 12 V, at 150 kHz.

   b.   Draw the composite wave using a simulation program.

12-14   **a.**   Calculate the dc and the first nine harmonics of a half-wave rectified sine wave that goes from 0 V to 8 V, at 60 Hz.

   b.   Draw the composite wave using a simulation program.

12-15   **a.**   Calculate the dc and the first nine harmonics of a full-wave rectified sine wave that goes from 0 V to 12 V, at 150 kHz.

   b.   Draw the composite wave using a spreadsheet.

12-16   **a.**   Calculate the dc and the first nine harmonics of a full-wave rectified sine wave that goes from 0 V to 8 V, at 60 Hz.

   b.   Draw the composite wave using MATLAB.

Harmonics Lab Exercise

Theory of Operation

A band-pass filter with a narrow bandwidth and a steep roll-off can be used to isolate each harmonic of a nonsinusoidal signal. Figure 12-17(a) is the spectrum of a square wave. Superimposed over it is the frequency response of a high Q band-pass filter. When the center frequency of the filter matches the fundamental frequency of the square wave, only that one sinusoid is passed by the filter. The filter's output is a *sine wave*, the amplitude of which is set by the harmonic's amplitude and the filter's gain.

As you raise the filter's center frequency, most of the fundamental is rejected, but a little of the next harmonic is allowed to pass. The output is a small combination of first and next harmonics. This is shown in Figure 12-17 (b).

When the filter's center frequency is adjusted to lie directly over the next harmonic, as in Figure 12-17(c), all other harmonics are rejected. The output is a pure sine wave at the frequency and amplitude of that harmonic.

This process can be continued until the harmonics are so small that they are hidden in the circuit's noise.

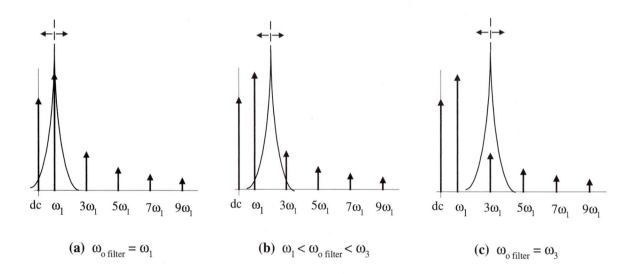

(a) $\omega_{\text{o filter}} = \omega_1$ (b) $\omega_1 < \omega_{\text{o filter}} < \omega_3$ (c) $\omega_{\text{o filter}} = \omega_3$

Figure 12-17 Harmonic separation with a band-pass filter

The schematic in Figure 12-18 illustrates a tunable two stage band-pass filter. Two stages are used in order to produce a narrow pass band with steep roll-off. These filters also have considerable pass-band gain. The input attenuator block reduces the *overall* pass-band gain to 1.

The center frequency is set by the voltage controlled clock.

$$f_{\text{o}} = \frac{f_{\text{clock}}}{50}$$

Adjusting potentiometer R12 changes the clock's frequency from below 40 kHz to above 500 kHz. This, in turn, allows you to sweep the band-pass filter's center frequency from below 800 Hz to above 10 kHz.

The upper half of the MF10, U1, forms a second order band-pass filter. Its output is on pin 2. This is sent to the lower half of the MF10, also a second order band-pass filter. Its output is at pin 19.

Since both low level analog and TTL digital signals are being used, layout, decoupling, and grounding are *critical*. Decouple both ICs precisely as shown, placing the capacitors *directly* to the ICs power pins. Trim all leads as short as possible. Keep the layout compact and neat.

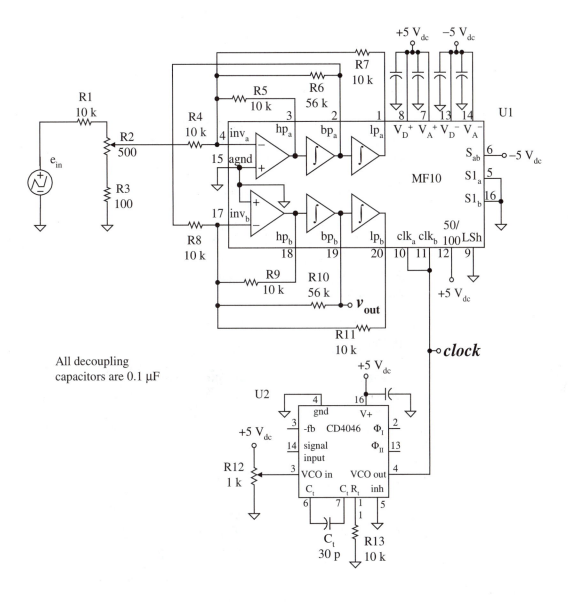

Figure 12-18 Tunable band-pass filter schematic

A. Band-pass Filter Operation

1. Build the circuit in Figure 12-18.

2. Apply only the dc power.

3. Adjust R12 to produce about 2 V_{dc} at pin 3 of U2.

4. Display the clock signal (U2 pin 4) on the oscilloscope. Also measure its frequency.

5. Confirm that the clock signal is TTL.

6. Adjust R12 across its full range. Record the minimum and maximum clock frequencies. This range should go from below 40 kHz to above 500 kHz.

7. Adjust R12 to produce a 50 kHz clock frequency.

8. Apply a 100 mV$_{rms}$, 0 V_{dc}, 1 kHz sine wave as e_{in} while monitoring the filter's output (U1 pin 19) with the oscilloscope

9. Adjust the clock frequency control, R12, *slightly*, to produce a maximum output from the filter.

10. Adjust the input attenuator, R2, until the output, U1 pin 19, is 100 mV$_{rms}$

11. Change the input to 1 V$_{rms}$, and confirm that the output of the filter is also 1 V$_{rms}$. Adjust the frequency control, R12, *slightly*, if necessary to produce a maximum size output. Adjust the input attenuator, R2, *slightly* to set the output equal to the input.

You have just calibrated the input attenuator. Be careful not to change its setting throughout the rest of this exercise.

12. Adjust the frequency of the *input* sine wave and measure the filter's output as indicated in Table 12-2. At each frequency, assure that the output is an undistorted sine wave.

13. Calculate the selectivity, Q, and compare it to the design parameters, also indicated in Table 12-2.

14. Complete a semilog frequency response plot of the filter. Continue to the next section *only* when the filter works correctly.

Table 12-2 Filter performance

measurement	frequency Hz	output V_{rms}	gain dB
0.25 x f_{low}			
0.5 x f_{low}			
f_{low}		0.71	−3
f_{center}		1.00	0
f_{high}		0.71	−3
2 x f_{high}			
4 x f_{high}			

$Q = f_{center}/(f_{high} - f_{low})$

theory	8.7
actual	

B. Harmonics of a Square Wave

1. Change the input signal to a 50% duty cycle square wave that goes from 0 V to 3 V, at 1 kHz.

2. Measure the input signal's dc level with a voltmeter. Record this and compare it to the theoretical value.

3. Monitor the filter's output with both an oscilloscope and an ac voltmeter.

4. Lower the filter's center frequency by adjusting R12 to its minimum.

5. Increase the filter's center frequency from the bottom, by adjusting R12 until the output is the largest 1 kHz *sine* wave possible.

6. Record this frequency and magnitude. Compare them to the theoretical values.

7. Raise the filter's center frequency gradually while watching its output on the oscilloscope. When the signal is the largest possible 3 kHz sine wave, record the frequency and the magnitude. Compare them to the theoretical values.

8. Continue increasing the filter's center frequency to capture the fifth, seventh, and ninth harmonics.

9. Present your measurements, theoretical values, and comparisons in a table.

C. Harmonics of a Triangle Wave
1. Change the input signal to a triangle wave that goes from 0 V to 3 V, at 1 kHz.

2. Repeat the measurements of Section B to obtain the dc level and the harmonics of this triangle wave. Since the harmonics of a triangle wave decrease *very* rapidly, you may not be able to obtain good measurements at the higher frequencies.

3. Present your measurements, theoretical values, and comparisons in a table.

D. Harmonics of a Half-wave Rectified Sine Wave
1. Replace the triangle wave with the half-wave rectifier shown in Figure 12-19.

2. Adjust the signal generator, while monitoring the waveform across the 1 kΩ load, until the signal goes from 0 V to 3 V.

3. Repeat the measurements of Section B to obtain the dc level and the harmonics of this half-wave rectified wave. Remember that this wave shape has *even* rather than odd harmonics.

4. Present your measurements, theoretical values, and comparisons in a table.

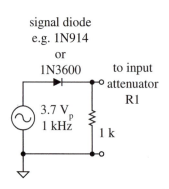

signal diode
e.g. 1N914
or
1N3600

Figure 12-19
Half-wave rectifier

13

Circuit Analysis with Nonsinusoidal Waveforms

Introduction

All repetitive wave shapes can be constructed from dc and the proper amplitude, frequency, and phase sine waves, their harmonics. So, you can calculate how a circuit responds to *any* wave shape just by breaking it down into its dc and sine wave parts, calculating the circuit's response, using phasors for each of the harmonics, and then superimposing the results. It's *all* just ac phasor analysis.

The Fourier series, for square waves, pulses, sawtooth, and triangle waves, half-, and full-wave rectified sinusoids were all presented in Chapter 12. In this chapter, you will apply those wave shapes to a variety of low-pass and high-pass filters and determine the output. This technique requires *many* calculations. The Fourier analysis worksheet is a single page that allows you to organize all of your calculations and results on a single page.

The worksheet is applied to both *RC* and *RLC* low-pass filters used to remove digital noise from dc, create a triangle wave or a sine wave from a square wave, and to filter the output in a power supply. High-pass filters (*CR* and *RL*) are used with nonsinusoidal signals in oscilloscope input couplers, with op amps to create a square wave from a triangle, and in driving motors from power switches.

Finally, the response of both low-pass and high-pass filters to nonsinusoidal signals is related to the signal and the filter's cut-off frequency.

Objectives

Upon completion of this chapter, you will be able to do the following:

*   Use Fourier series and superposition to calculate the output produced by a nonsinusoidal signal applied to a variety of low-pass and high-pass filters.

*   Describe, in general, the effect that a low-pass and a high-pass filter has on a given wave shape, and relate that effect to the harmonics' frequencies and the filter's cut-off frequency.

13.1 The Fourier Analysis Worksheet

Analyzing even a simple circuit driven with a nonsinusoidal signal can quickly become a very involved process. None of the steps, *individually*, are difficult. However, most of the calculations are repeated for each of the harmonics. Along the way, you must keep straight whether you are working with the input or the output, whether the voltage is being expressed in V_p or V_{rms}, and which frequency you are currently using. It is also quite easy to misplace a key intermediate result in the host of calculations.

To help you keep all of this work organized, and avoid clerical errors, use the Fourier analysis worksheet shown in Figure 13-1. Make a fresh, blank copy each time you begin to manually work a problem. This worksheet will also be used to document the work for each of the examples in this chapter.

In general, eight steps are used. It is important to have *all* of the information together. So, begin by *neatly* redrawing the schematic with all values, at the top of the worksheet.

Next, carefully draw the input wave shape. Be sure to clearly label the amplitude, *A*, the dc level, V_{dc}, and the period, *T*. On the following line, record the numeric value for these three items, and then calculate the input's angular velocity, ω. Since all of your subsequent calculations are based on these values, be sure to get them right. Record at least four digits to reduce accumulated rounding error.

Break the input into its harmonics and record this equation in step 4. In step 5, solve the problem in *equations*. Write every equation, and get the order correct. This serves are your plan for the rest of the problem.

The bulk of the calculator work is documented in step 6. You may either work horizontally, calculating all of the appropriate quantities for one harmonic, or you may work vertically, repeating the same calculation for each harmonic before moving to the next column. Be *sure* to convert the V_p from step 4 into V_{rms} for the phasor calculations.

Once the table is complete, you have the output voltage in phasor form in the last two columns. All that remains is to convert these phasors into a single, time domain equation in step 7. Remember to record V_p in the time domain equation.

Finally, plot the resulting equation using a graphics calculator, spreadsheet, or MATLAB. Then carefully look at the result. Does it make sense?

1. Schematic: Draw the schematic *neatly* with component values.

2. Input signal: Draw the input signal *neatly* with A, V_{dc}, and T defined.

3. Parameters:

V_{dc} = _____ A = _____ T = _____ ω = _____
Calculation needed.

4. $e(t)_{input}$ = ____ V_{dc} + ____ sin (___$\omega t \pm$ ____°) + ___ sin (___$\omega t \pm$ ____°) +
___ sin (___$\omega t \pm$ ____°) + ___ sin (___$\omega t \pm$ ____°) + ___ sin (___$\omega t \pm$ ___°)

5. Calculate the output using *equations.*

6. Calculate each harmonic's output voltage.

ω (r/s)	X (Ω)	V_{rms}	E_{in}	°	V_{rms}	V_{out}	°
0							

7. Write the time domain equation for the output voltage.
$v(t)_{out}$ = ____ V_{dc} + ____ sin (___$\omega t \pm$ ____°) + ___ sin (___$\omega t \pm$ ____°) +
___ sin (___$\omega t \pm$ ____°) + ___ sin (___$\omega t \pm$ ____°) + ___ sin (___$\omega t \pm$ ___°)

8. Plot it.

Figure 13-1 Fourier analysis worksheet

13.2 Low-pass Filters

Low-pass filters pass dc and the low-frequency harmonics, while attenuating and shifting the phase of the high-frequency harmonics. This means that the output from a low-pass filter driven by a nonsinusoidal signal will *not* have the same shape as its input. The dc level should be the same. The slow, rounded parts of the input (i.e. the fundamental) will pass through the filter without much change. But the sharp, high-speed edges (the upper harmonics) will all be attenuated. So, the overall effect of a low-pass filter is to smooth out and round a wave's shape.

RC Low-pass Filters

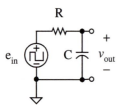

Figure 13-2
RC low-pass filter

To confirm the validity of using Fourier series and superposition to determine the effects of a circuit on a nonsinusoidal waveform, let's first analyze a simple *RC* low-pass filter driven by a square wave. The schematic is shown in Figure 13-2.

When you first studied capacitors, you learned that capacitors respond to a step by charging exponentially through the resistor.

$$v_{out} = E_{in\,p}\left(1 - e^{-\frac{t}{RC}}\right)$$

When the input voltage falls to common, at T/2, the discharge is

$$v_{out} = V_{out\,p}\left(e^{-\frac{\frac{T}{2}-t}{RC}}\right)$$

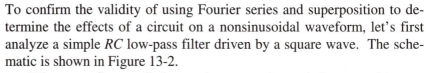

Example 13-1

Plot the output for the circuit in Figure 13-2, given that:

$$E_{in\,p} = 5\ V_p,\ f = 1\ kHz,\ R = 1.5\ k\Omega,\ C = 0.1\ \mu F$$

Solution

a. The spreadsheet and its plot are shown in Figure 13-3. From the time $t = 0$ until $t = 500\ \mu s$, the charge equation is used. The remainder of the time uses the second, discharge equation. $V_{out\,p}$ is the charge on the capacitor at $t = 500\ \mu s$,

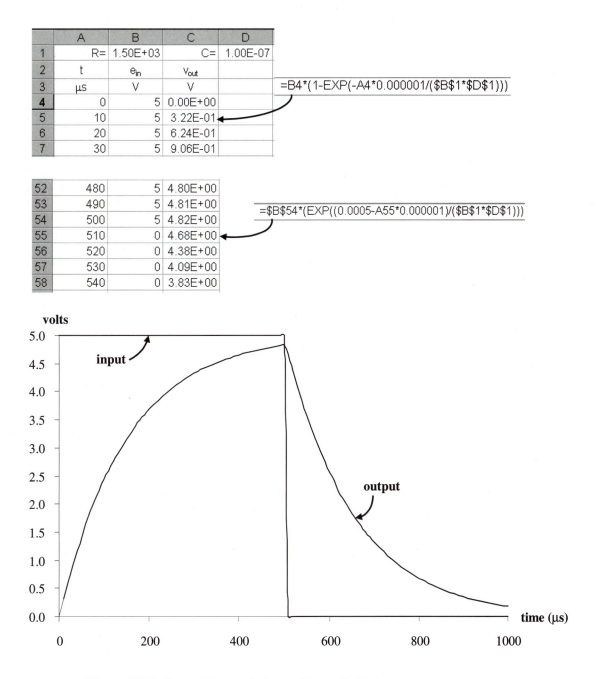

	A	B	C	D
1	R=	1.50E+03	C=	1.00E-07
2	t	e_{in}	v_{out}	
3	μs	V	V	
4	0	5	0.00E+00	
5	10	5	3.22E-01	
6	20	5	6.24E-01	
7	30	5	9.06E-01	

=B4*(1-EXP(-A4*0.000001/($B$1*$D$1)))

52	480	5	4.80E+00
53	490	5	4.81E+00
54	500	5	4.82E+00
55	510	0	4.68E+00
56	520	0	4.38E+00
57	530	0	4.09E+00
58	540	0	3.83E+00

=$B$54*(EXP((0.0005-A55*0.000001)/($B$1*$D$1)))

Figure 13-3 Spreadsheet solution to Example 13-1

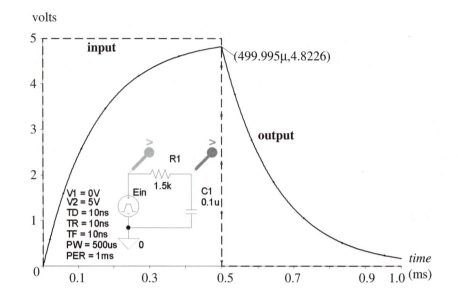

Figure 13-4 Simulation of Example 13-1

b. The simulation schematic and resulting plot are shown in Figure 13-4. Carefully note the parameters used to create the input. The voltage across the capacitor, the output, matches the waveform calculated and plotted with the spreadsheet, including the $V_{out\,p}$.

c. Finally, look at the completed Fourier worksheet shown in Figure 13-5, and the resulting MATLAB plot in Figure 13-6. There are several points to notice.

- $V_{in\,dc}$ shows up at the output unaffected by the filter.

- Four digits of magnitude and 0.1° angle are used.

- Convert the *phasor* inputs in step 6 to rms.

- Remember that the input magnitude and reactance changes for each harmonic (row in the table).

- Convert the *phasor* magnitudes back to V_p for the composite equation in step 7.

- The phase angles in MATLAB are in radians.

1. **Schematic:** Draw the schematic *neatly* with component values.

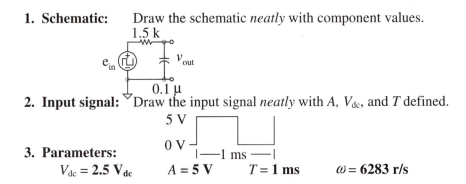

2. **Input signal:** Draw the input signal *neatly* with A, V_{dc}, and T defined.

3. **Parameters:**

$$V_{dc} = 2.5\ V_{dc} \qquad A = 5\ V \qquad T = 1\ ms \qquad \omega = 6283\ r/s$$

4. $e(t)_{input} = 2.5\ V_{dc} + 3.183\ V_p \sin(1\ \omega_1 t) + 1.061\ V_p \sin(3\ \omega_1 t)$
   $+ 0.6365\ V_p \sin(5\ \omega_1 t) + 0.4545\ V_p \sin(7\ \omega_1 t) + 0.3535\ V_p \sin(9\ \omega_1 t)$

5. Calculate the output using *equations.*

$$X_C = \frac{1}{\omega C}$$

$$\overline{V_{out}} = \frac{(X_C \angle -90°)}{(1.5\,k\Omega - X_C)} \times \overline{E_{in}}$$

6. Calculate each harmonic's output voltage.

ω (r/s)	X (Ω)	E_{in} (V_{rms})	E_{in} (°)	V_{out} (V_{rms})	V_{out} (°)
0		2.5 V_{dc}		2.5 V_{dc}	
1 x 6283	1592	2.251	0	1.638	−43.3
3 x 6283	531	0.750	0	0.250	−70.5
5 x 6283	318	0.450	0	0.093	−78.0
7 x 6283	227	0.321	0	0.048	−81.4
9 x 6283	177	0.250	0	0.029	−83.3

7. Write the time domain equation for the output voltage.
   $v(t)_{out} = 2.5\ V_{dc} + 2.316\ V_p \sin(1\ \omega_1 t - 43.3°)$
   $+ 0.3536\ V_p \sin(3\ \omega_1 t - 70.5°) + 0.1315\ V_p \sin(5\ \omega_1 t - 78.0°)$
   $+ 0.0679\ V_p \sin(5\ \omega_1 t - 81.4°) + 0.0410\ V_p \sin(9\ \omega_1 t - 83.3°)$

8. Plot it.

Figure 13-5 Fourier analysis worksheet for Example 13-1

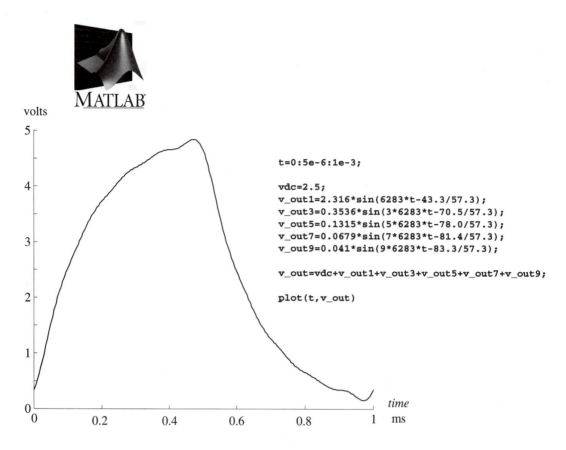

volts

```
t=0:5e-6:1e-3;

vdc=2.5;
v_out1=2.316*sin(6283*t-43.3/57.3);
v_out3=0.3536*sin(3*6283*t-70.5/57.3);
v_out5=0.1315*sin(5*6283*t-78.0/57.3);
v_out7=0.0679*sin(7*6283*t-81.4/57.3);
v_out9=0.041*sin(9*6283*t-83.3/57.3);

v_out=vdc+v_out1+v_out3+v_out5+v_out7+v_out9;

plot(t,v_out)
```

time
ms

Figure 13-6 Plot of the Fourier series solution for Example 13-1

Except for a little wiggle at each peak, the results from applying the square wave's harmonics each individually to the filter then adding them up are the same as using the exponential charge and discharge equations, or using circuit simulation.

Practice: Repeat the Fourier analysis part of Example 13-1 if the square wave goes from 0 V to 8 V, its period is 750 μs, and R changes to 1 kΩ, to determine the effect at the output of the seventh harmonic.

Answer: $v_{out\,7} = 0.123\ V_p \sin(7 \times 8378\ t - 80.3°)$, $V_{out\,p} = 7.812\ V_p$

Op Amp Integrator

From Examples 13-1 it appears that applying a square wave to an *RC* low-pass filter produces a triangle wave. This is also true for the op amp based *RC* **integrator** shown in Figure 13-7. Negative feedback is provided through C_f. So, the op amp's inverting input pin is held at virtual ground. The constant high level from e_{in} produces a constant current,

$$I = \frac{E_{in\,high}}{R_i}$$

This constant current flows through C_f, charging the capacitor, + to −. Since the capacitor's left end is held at virtual ground, its right end (which is v_{out}) goes negative. For a capacitor,

$$I = -C_f \frac{dv}{dt}$$

The capacitor and the output voltage ramp down at a constant rate.

$$\frac{dv_{out}}{dt} = -\frac{I}{C_f} = -\frac{E_{in\,high}}{R_i C_f}$$

This is the slope of the triangle output wave. Its amplitude depends on how long it is allowed to ramp, before the input reverses polarity and the capacitor begins to charge in the opposite direction.

$$distance = rate \times time$$

For a 50% duty cycle wave, $V_{pp} = \frac{E_{in\,high}}{R_i C_f} \times \frac{T}{2}$

The feedback resistor, R_f, is needed to compensate for the nonideal dc performance of *real* op amps. Without it, the dc gain is infinite, since the dc resistance of a good capacitor is infinite. So, any small offsets at the op amp's inputs would drive the output into saturation.

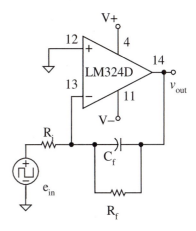

Figure 13-7
Op amp integrator

In general:

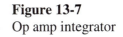

$$v_{out} = -\frac{1}{R_i C_f} \int e_{in} dt$$

$R_f > 10\ R_i$

Example 13-2

    a. Calculate the V_{pp} of the output from the integrator in Figure 13-7, with $R_i = 1.5\ k\Omega$, and $C_f = 0.1\ \mu F$. The input is a square wave that goes from −2.5 V to 2.5 V, at 1 kHz.

    b. Verify your results by simulation. Set $R_f = 150\ k\Omega$.

c. Calculate the output using Fourier analysis and superposition. Plot the result, and compare it to the simulation.

Solution

a.
$$V_{pp} = \frac{E_{in\ high}}{R_i C_f} \times \frac{T}{2} = \frac{2.5\,V}{1.5\,k\Omega \times 0.1\,\mu V} \times \frac{1\,ms}{2} = 8.33\,V_{pp}$$

b. The simulation results are shown in Figure 13-8. It must run for some time as the capacitor settles. The two cursors indicate a V_{pp} of 8.5 V_{pp}, as compared to the 8.33 V_{pp} predicted.

c. The Fourier analysis worksheet is shown in Figure 13-9. The dc level has been set to 0 V_{dc}. The feedback resistor, R_f, has been omitted. The input is exactly the same as that used in Example 13-1, so the first half of the tables are the same. Since the output load is purely capacitive, each of the output signals is shifted by 90°.

multi**SIM**

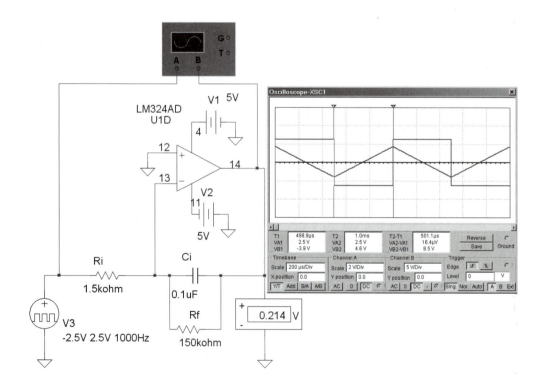

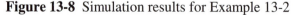

Figure 13-8 Simulation results for Example 13-2

1. **Schematic:**   Draw the schematic *neatly* with component values.

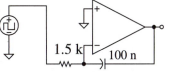

2. **Input signal:** Draw the input signal *neatly* with A, V_{dc}, and T defined.

3. **Parameters:**
   $V_{dc} = 0$ **Vdc**   A = **5 V**   T = **1 ms** ω = **6283 r/s**

4. $e(t)_{input}$ = **2.5** V_{dc} + **3.183** V_p sin (**1** ω_1 t) + **1.061** V_p sin (**3** ω_1 t)
   + **0.6365** V_p sin (**5** ω_1 t) + **0.4545** V_p sin (**7** ω_1 t) + **0.3535** V_p sin (**9** ω_1 t)

5.   Calculate the output using *equations*.

   $$X_C = \frac{1}{\omega C}$$

   $$\overline{I_{in}} = \frac{\overline{E_{in}}}{(1.5\,k\Omega\angle 0°)} \qquad \overline{I_C} = \overline{I_{in}}$$

   $$\overline{V_{out}} = -\overline{I_C} \times (X_C\angle -90°)$$

6.   Calculate each harmonic's output voltage.

ω (r/s)	X (Ω)	E_{in} (V_{rms})	E_{in} (°)	V_{out} (V_{rms})	V_{out} (°)
0				**0**	
1 x 6283	**1592**	**2.251**	**0**	**2.389**	**90**
3 x 6283	**531**	**0.750**	**0**	**0.266**	**90**
5 x 6283	**318**	**0.450**	**0**	**0.095**	**90**
7 x 6283	**227**	**0.321**	**0**	**0.049**	**90**
9 x 6283	**177**	**0.250**	**0**	**0.030**	**90**

7.   Write the time domain equation for the output voltage.

$v(t)_{out}$ = **0** V_{dc} + **3.379** V_p sin (**1** ω_1 t + **90°**) + **0.377** V_p sin (**3** ω_1 t + **90°**)
+**0.134** V_p sin (**5** ω_1 t + **90°**) + **0.069** V_p sin (**7** ω_1 t + **90°**)

+ **0.042** V_p sin (**9** ω_1 t ± **90°**)

Figure 13-9 Fourier analysis worksheet for Example 13-2

The MATLAB plot is given in Figure 13-10. Even though only five harmonics are used, the composite is an 8 V_{pp}, 1 kHz triangle, a good match to the manual calculations and the simulation.

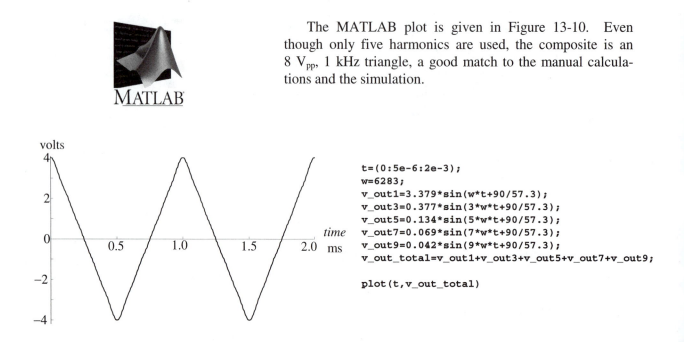

```
t=(0:5e-6:2e-3);
w=6283;
v_out1=3.379*sin(w*t+90/57.3);
v_out3=0.377*sin(3*w*t+90/57.3);
v_out5=0.134*sin(5*w*t+90/57.3);
v_out7=0.069*sin(7*w*t+90/57.3);
v_out9=0.042*sin(9*w*t+90/57.3);
v_out_total=v_out1+v_out3+v_out5+v_out7+v_out9;

plot(t,v_out_total)
```

Figure 13-10 MATLAB plot of the Fourier analysis worksheet results for Example 13-2

Practice: Repeat Example 13-2 if the input is an 8 V_{pp}, 0 V_{dc}, 500 Hz triangle wave.

Answer: The wave shape is almost sinusoidal, with a 6.7 V_{pp} amplitude, and the fifth harmonic is $v_{out\,5} = 0.0275\,V_p \sin(5\omega_1 t + 90°)$.

Effects of a Low-pass Filter

Throughout this section you have seen a variety of wave shapes applied to several low-pass filters. In general, the effect is to round off the sharp edges, smoothing and slowing the transitions. The degree of the rounding depends on the relationship of the signal's frequency and the filter's half-power point frequency.

In Example 13-1, a square wave was turned into a sawtooth. The filter's f_{-3dB} equaled the fundamental's frequency. So, it passed to the output with little attenuation. Many of the upper frequency harmonics also passed through the filter since its roll-off rate was only −6 dB/octave.

Moving the filter's half-power point frequency far below the fundamental allows only the dc component and a little of the fundamental to sneak through. Example 13-2 extracts dc from a composite waveform.

Increasing the order of the filter allows the fundamental to pass, but the higher frequency harmonics to be severely rejected. The result is a sine wave from a square wave.

Setting the filter's f_{-3dB} well *above* the input signal's frequency means that most of the harmonics can pass without attenuation. The output is very similar to the input. As you lower f_{-3dB}, more and more of the upper frequency harmonics are removed. The output begins to lose its sharp edges, rounding more and more as f_{-3dB} falls. Eventually, only the fundamental sine wave is left. Further reduction in f_{-3dB} means that the filter passes only the input's dc with a little output ripple.

13.3 High-pass Filters

The low-pass filter passes the dc and perhaps the fundamental harmonic of its input. The upper frequency harmonics are attenuated, more and more severely as the frequency goes up. The result is to *deemphasize* the signal's sharpness, rounding it off.

The high-pass filter does just the opposite. DC and perhaps some of the fundamental are blocked, while the high-frequency harmonics are usually passed. The result is to *emphasize* the signal's edges and sharp peaks while seriously distorting the basic shape that comes from the fundamental. The DC at the output is almost always zero.

CR High-pass Filters

One of the most common filters is the *CR* coupler. Its schematic is shown in Figure 13-11. Its purpose is to block the $V_{\text{in dc}}$ that the input signal may have, and to pass the time varying part of the signal, e_{in}, completely. This can be done by setting the filter's f_{-3dB} significantly below the fundamental's frequency. For 1% error, set

$$f_{-3dB} = \frac{1}{2\pi RC} < \frac{f_{\text{fundamental}}}{7}$$

Figure 13-12 shows the spectrum of a square wave, and the frequency response of a properly designed *CR* coupler (high-pass filter). Notice that the filter passes *all* of the harmonics and only begins to roll-off *below* the fundamental.

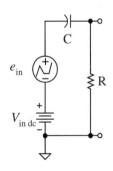

Figure 13-11
CR high-pass filter

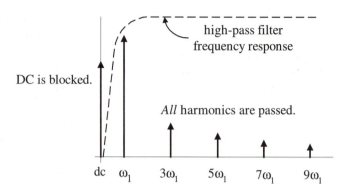

Figure 13-12 *CR* coupler's effect on harmonics

Example 13-3

The *CR* coupler at the input of an oscilloscope using ac input coupling consists of a 22 nF capacitor and a 1 MΩ resistance.

a. Calculate the lowest frequency signal that can be passed without noticeable distortion.

b. Use Fourier analysis to verify that a triangle wave that goes from 0 V to 5 V at that frequency is passed.

Solution

a.
$$f_{-3dB} = \frac{1}{2\pi RC} < \frac{f_{fundamental}}{7}$$

$$f_{fundamental} > \frac{7}{2\pi \times 1\,M\Omega \times 22nF} = 50\,Hz$$

b. The Fourier analysis worksheet is shown in Figure 13-13. Look carefully at the Fourier series for the input in step 4, and the Fourier series for the output in step 7. The dc has been removed from the input, and the fundamental has been attenuated 1%. The output is identical to the input

---

Practice: What is the effect if the input's frequency drops to 25 Hz?

Answer: $v_{out\,\omega 3}(t) = 0.215\,V_p \sin(3\omega_1 t - 163°)$

1. **Schematic** : Draw the schematic *neatly* with component values.

2. **Input signal:** Draw the input signal *neatly* with A, V_{dc}, and T defined.

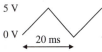

3. **Parameters:**

$V_{dc} = 2.5 \text{ V}_{dc}$ $A = 5 \text{ V}$ $T = 20 \text{ ms}$ $\omega = 314.2 \text{ r/s}$

4. $e(t)_{input} = 2.5 \text{ V}_{dc} + 2.027 \text{ V}_p \sin(\omega_1 t) + 0.225 \text{ V}_p \sin(3\,\omega_1 t + 180°)$

$+ 0.081 \text{ V}_p \sin(5\,\omega_1 t + 180°) + 0.042 \text{ V}_p \sin(7\,\omega_1 t + 180°)$
$+ 0.025 \text{ V}_p \sin(9\,\omega_1 t + 180°)$

5. Calculate the output using *equations*.

$$X_C = \frac{1}{\omega C}$$

$$\overline{V_{out}} = \overline{E_{in}} \times \frac{(1\,M\Omega\angle 0°)}{(1\,M\Omega - jX_C)}$$

6. Calculate each harmonic's output voltage.

ω (r/s)	X (kΩ)	E_{in} (V$_{rms}$)	E_{in} (°)	V_{out} (V$_{rms}$)	V_{out} (°)
0		2.5 V		0	
1 x 314.2	144.7	1.422	0	1.418	8.2
3 x 314.2	48.2	0.159	180	0.159	−177.2
5 x 314.2	34.6	0.057	180	0.057	−178.0
7 x 314.2	20.7	0.030	180	0.030	−179.0
9 x 314.2	16.1	0.018	180	0.018	−179.1

7.   Write the time domain equation for the output voltage.

$v(t)_{output} = 0 \text{ V}_{dc} + 2.005 \text{ V}_p \sin(\omega_1 t + 8°) + 0.225 \text{ V}_p \sin(3\,\omega_1 t - 177°)$
$+ 0.081 \text{ V}_p \sin(5\,\omega_1 t - 178°) + 0.042 \text{ V}_p \sin(7\,\omega_1 t - 179°)$
$+ 0.025 \text{ V}_p \sin(9\,\omega_1 t - 179°)$

Figure 13-13 Fourier analysis worksheet for Example 13-3

Example 13-3 and the Fourier analysis worksheet in Figure 13-13 show that properly setting the relationship between the high-pass filter's half-power point frequency and the signal's frequency allows coupling with no loss of fidelity. However, when the signal's frequency is too low, the filter begins to seriously attenuate the fundamental.

Example 13-4

Calculate and plot the output from the circuit in Example 13-3 with a 0 V to 5 V, 10 Hz square wave.

Solution

Figure 13-14 shows plots of the output voltage. Figure 13-14(a) uses MATLAB to recombine and plot the sinusoids. Figure 13-14 (b) is the probe result from the simulation. They match well.

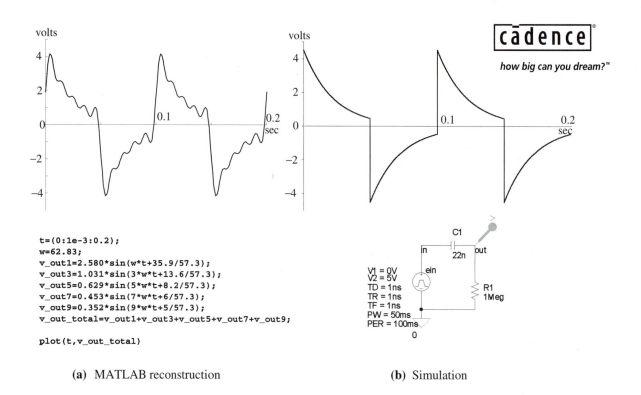

```
t=(0:1e-3:0.2);
w=62.83;
v_out1=2.580*sin(w*t+35.9/57.3);
v_out3=1.031*sin(3*w*t+13.6/57.3);
v_out5=0.629*sin(5*w*t+8.2/57.3);
v_out7=0.453*sin(7*w*t+6/57.3);
v_out9=0.352*sin(9*w*t+5/57.3);
v_out_total=v_out1+v_out3+v_out5+v_out7+v_out9;

plot(t,v_out_total)
```

(a) MATLAB reconstruction   **(b)** Simulation

Figure 13-14 Plots of the results of Example 13-4

The Fourier analysis worksheet is shown in Figure 13-15. Again, compare the input series, in step 4 to the output in step 7. The dc level has been blocked. The fundamental has been attenuated almost 20%, and the third harmonic dropped 30 mV. All of the upper harmonics are unaffected.

3. Parameters:

$V_{dc} = $ **2.5 V**$_{dc}$ $A = $ **5 V** $T = $ **100 ms** $\omega = $ **62.83 r/s**

4. $e(t)_{input}$ = **2.5 V**$_{dc}$ + **3.183 V$_p$** sin $(\omega_1 t)$ + **1.061 V$_p$** sin $(3 \omega_1 t)$
 + **0.637 V$_p$** sin $(5 \omega_1 t)$ + **0.455 V$_p$** sin $(7 \omega_1 t)$ + **0.354 V$_p$** sin $(9 \omega_1 t)$

5. Calculate the output using *equations*.

$$X_C = \frac{1}{\omega C}$$

$$\overline{V_{out}} = \overline{E_{in}} \times \frac{(1\,M\Omega \angle 0°)}{(1\,M\Omega - jX_C)}$$

6.   Calculate each harmonic's output voltage.

ω (r/s)	X (kΩ)	E_{in} (V$_{rms}$)	E_{in} (°)	V_{out} (V$_{rms}$)	V_{out} (°)
0		2.5 V$_{dc}$		0	
1 x 62.83	723.5	2.251	0	1.824	35.9
3 x 62.83	241.2	0.750	0	0.729	13.6
5 x 62.83	144.6	0.450	0	0.445	8.2
7 x 62.83	103.4	0.322	0	0.320	5.9
9 x 62.83	80.4	0.250	0	0.249	4.6

7.   Write the time domain equation for the output voltage.

$v(t)_{output}$ = **0 V**$_{dc}$ + **2.580 V$_p$** sin $(\omega_1 t + 36°)$ + **1.031 V$_p$** sin $(3 \omega_1 t + 14°)$
        + **0.629 V$_p$** sin $(5 \omega_1 t + 8°)$ + **0.453 V$_p$** sin $(7 \omega_1 t + 6°)$
        + **0.352 V$_p$** sin $(9 \omega_1 t + 5°)$

Figure 13-15 Fourier analysis for Example 13-4

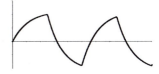

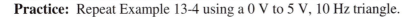

Practice: Repeat Example 13-4 using a 0 V to 5 V, 10 Hz triangle.

Answer: The fundamental is 2.32 V$_p$ sin($\omega_1 t + 35.9°$). A sketch of the waveform is shown in Figure 13-16.

Figure 13-16
Output with a triangle input

Op Amp Differentiator

Placing a capacitor in the input loop of an inverting op amp produces a high-pass filter. This configuration was discussed in the first section of Chapter 10. It is shown again in Figure 13-17. This circuit is also called an op amp differentiator. In its simplest form, R_i is omitted. Because of negative feedback, the inverting input pin is held at virtual ground. All of e_{in} is dropped across C_i.

$$v_C = e_{in}$$

The current through a capacitor depends on the *rate of change* of the voltage across it.

$$i_C = C \frac{dv_C}{dt}$$

For the circuit in Figure 13-17 (still ignoring R_i) this becomes

$$i_{in} = C_i \frac{de_{in}}{dt}$$

All of this current must flow through R_f, left to right. This produces a voltage drop + to −, driving v_{out} negative. The output voltage is

$$v_{out} = -i \times R_f$$

$$v_{out} = -R_f C_i \frac{de_{in}}{dt}$$

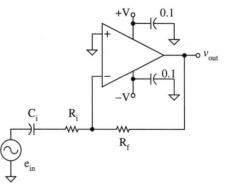

Figure 13-17 Schematic of an op amp based differentiator

The output voltage of a differentiator depends on how rapidly the input is *changing*. A constant voltage, regardless of its value, produces an output of 0 V. Only when that input *changes* does the output become a value other than 0 V.

At very high frequencies, such as those produced by radiated interference and transients, the capacitor's impedance falls drastically. This allows large noise currents into the amplifier. These currents flow through R_f, producing large noise voltages at the output. The op amp tends to break into oscillations. The input resistor, R_i, limits this current. It should be in the range of

$$R_i \approx \frac{R_f}{250} \text{ to } \frac{R_f}{10}$$

At this value, it is small enough to not interfere with the differentiation produced by C_i and R_f, but limits the high-frequency gain.

It is theoretically possible to calculate the output voltage of this circuit from the derivative,

$$v_{out} = -R_f C_i \frac{d e_{in}}{dt}$$

However, often, even standard test signals for e_{in} are not easily defined in terms that can be readily differentiated. Instead, use the Fourier series analysis technique as in the previous examples.

Example 13-5

Determine the performance of the circuit in Figure 13-17, with:

$$C_i = 1\ \mu F, \quad R_i = 220\ \Omega, \quad R_f = 47\ k\Omega$$

The input is a 60 Hz triangle wave that goes from 0 V to 1 V.

a. Calculate the result using a derivative.

b. Apply Fourier analysis to determine the harmonics.

c. Plot the resulting output and compare that to simulation.

Solution

a. The input has a constant slope, changing from 0 V to 1 V in half of a period.

$$v_{out} = -47\ k\Omega \times 1\mu F \times \frac{1\,V}{8.3\,ms} = -5.66\,V$$

When the input ramps down, the output steps up to +5.66 V.

b. The Fourier analysis worksheet is shown in Figure 13-18.

3. Parameters:

$$V_{dc} = 0.5\ V_p \qquad A = \quad 1\ V \qquad T = 16.7\ ms \qquad \omega = 377\ r/s$$

4. $e(t)_{input} = 0.5\ V_{dc} + 0.4053\ V_p \sin(\omega_1 t) + 0.0450\ V_p \sin(3\ \omega_1 t - 180°)$
$+ 0.0162\ V_p \sin(5\ \omega_1 t) + 0.0083\ V_p \sin(7\ \omega_1 t - 180°)$
$+ 0.0050\ V_p \sin(9\ \omega_1 t)$

5. Calculate the output using *equations*.

$$X_C = \frac{1}{\omega C}$$

$$\overline{I} = \frac{\overline{E_{in}}}{(220\Omega - jX_C)}$$

$$\overline{V_{out}} = -\overline{I} \times (47k\Omega\angle0°) = -\overline{E_{in}}\frac{(47k\Omega\angle0°)}{(220\Omega - jX_C)}$$

6. Calculate each harmonic's output voltage.

ω (r/s)	X (Ω)	E_{in} (mV$_{rms}$)	E_{in} (°)	V_{out} (V$_{rms}$)	V_{out} (°)
0		500 mV$_{dc}$		0	
1 x 377.0	2653	286.6	0	5.060	−94.7
3 x 377.0	884	31.8	180	1.641	76.0
5 x 377.0	531	11.5	0	0.940	−112.5
7 x 377.0	379	5.9	180	0.633	59.9
9 x 377.0	295	3.5	0	0.447	−126.7

7. Write the time domain equation for the output voltage.

$v(t)_{out} = 0\ V_{dc} + 7.156\ V_p \sin(\omega_1 t - 95°) + 2.321\ V_p \sin(3\ \omega_1 t + 76°)$
$+ 1.329\ V_p \sin(5\ \omega_1 t - 113°) + 0.895\ V_p \sin(7\ \omega t + 60°)$
$+ 0.624\ V_p \sin(9\ \omega_1 t - 127°)$

Figure 13-18 Fourier analysis worksheet for Example 13-5

c. The plots of the reconstructed Fourier analysis wave and the simulation are shown in Figure 13-19.   Differentiating a ramp produces a square wave.

In the simulation, to produce a triangle wave, the rise and fall times of the pulse source were set almost equal to a half-period, and the pulse width was set very short.

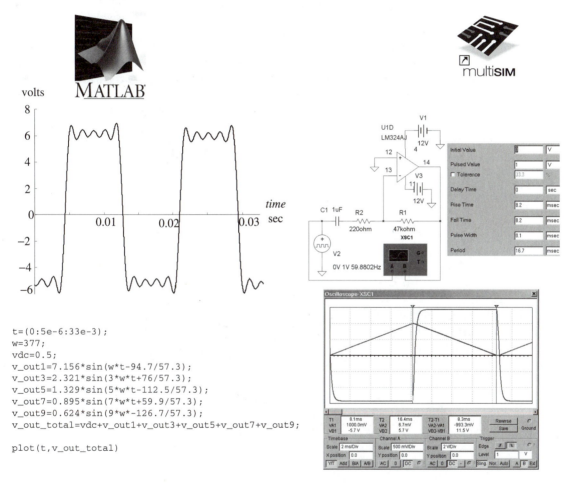

```
t=(0:5e-6:33e-3);
w=377;
vdc=0.5;
v_out1=7.156*sin(w*t-94.7/57.3);
v_out3=2.321*sin(3*w*t+76/57.3);
v_out5=1.329*sin(5*w*t-112.5/57.3);
v_out7=0.895*sin(7*w*t+59.9/57.3);
v_out9=0.624*sin(9*w*t-126.7/57.3);
v_out_total=vdc+v_out1+v_out3+v_out5+v_out7+v_out9;

plot(t,v_out_total)
```

(a) MATLAB reconstruction

(b) Simulation

Figure 13-19 Plots of the results of Example 13-5

RL High-pass Filters

Taking the voltage across the inductor of a series *RL* circuit forms a high-pass filter. At dc, the inductor looks like a short, providing no output. At very high frequencies, the inductor is an open, outputting all of the voltage. A motor is a common source of inductance. Figure 13-20 (a) shows the schematic of a transistor driving a motor, typical in many automotive applications. The circuit analysis model is a simplification that allows you to determine the effect of the inductance on the signal applied to the motor.

Example 13-6

For the circuit analysis model in Figure 13-20 (b), with

$$R_{switch} = 10 \ \Omega, \ R_{motor} = 5 \ \Omega, L_{motor} = 13 \ mH$$
$$e_{in} = 0 \ V \ to \ 12 \ V, \ 100 \ Hz, \ 50\% \ duty \ cycle$$

a. Determine the wave shape across the motor using Fourier analysis.

b. Verify your answer by simulation.

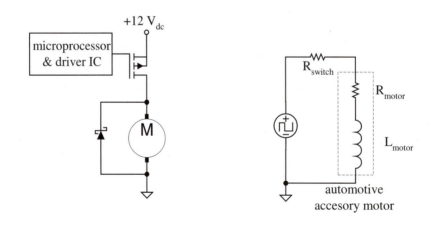

(a) Transistor driving a motor   **(b)** Circuit analysis model

Figure 13-20 A motor driver is an *RL* high-pass filter

Solution

a. The Fourier analysis worksheet is shown in Figure 13-21. The first two items have been omitted. As with the other high-pass filter examples, compare the input harmonics in step 4 with the output harmonics in step 7. There *is* a dc output, which is unusual for a high-pass filter. The fundamental is strongly attenuated. However, the upper frequency harmonics pass intact.

3. Parameters:

$$V_{dc} = 6 \ V_{dc} \quad A = 12 \ V \quad T = 10 \ ms \quad \omega = 628.3 \ r/s$$

4. $e(t)_{input} = 6 \ V_{dc} + 7.639 \ V_p \sin (\omega_1 t) + 2.546 \ V_p \sin (3 \ \omega_1 t)$
$+ 1.528 \ V_p \sin (5 \ \omega_1 t) + 1.091 \ V_p \sin (7 \ \omega_1 t) + 0.848 \ V_p \sin (9 \ \omega_1 t)$

5. Calculate the output using *equations.*

$$V_{dc} = 6 \ V_{dc} \frac{5\Omega}{(10\Omega + 5\Omega)}$$

$$X_L = \omega L$$

$$\overline{V_{out}} = \overline{E_{in}} \frac{(5\Omega + jX_L)}{(15\Omega + jX_L)}$$

6. Calculate each harmonic's output voltage.

ω (r/s)	X (Ω)	E_{in} (V$_{rms}$)	E_{in} (°)	V_{out} (V$_{rms}$)	V_{out} (°)
0		6 V$_{dc}$		2 V$_{dc}$	
1 x 628.3	8.16	5.402	0	3.028	30.0
3 x 628.3	24.51	1.800	0	1.567	19.9
5 x 628.3	40.84	1.080	0	1.021	13.2
7 x 628.3	57.18	0.771	0	0.749	9.7
9 x 628.3	73.51	0.600	0	0.589	7.6

7. Write the time domain equation for the output voltage.
$v(t)_{out} = 2 \ V_{dc} + 4.282 \ V_p \sin (\omega_1 t + 30°) + 2.216 \ V_p \sin (3 \ \omega_1 t + 20°)$
$+ 1.444 \ V_p \sin (5 \ \omega_1 t + 13°) + 1.059 \ V_p \sin (7 \ \omega_1 t + 10°)$
$+ 0.833 \ V_p \sin (9 \ \omega_1 t + 8°)$

Figure 13-21 Fourier analysis worksheet for Example 13-6

b. The plot of the composite wave, and the simulation are shown in Figure 13-22. There is good agreement in the general shape and the levels. Including more upper frequency harmonics would produce a taller, steeper composite wave.

In both examples of square waves into a high-pass filter, the output has drooped because the fundamental was attenuated. In this application, the droop causes the motor to run more slowly than predicted. The large negative spike on the falling edge may damage parts operating only from a positive supply voltage.

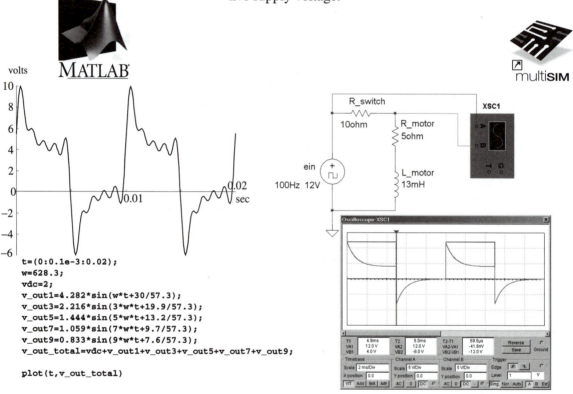

```
t=(0:0.1e-3:0.02);
w=628.3;
vdc=2;
v_out1=4.282*sin(w*t+30/57.3);
v_out3=2.216*sin(3*w*t+19.9/57.3);
v_out5=1.444*sin(5*w*t+13.2/57.3);
v_out7=1.059*sin(7*w*t+9.7/57.3);
v_out9=0.833*sin(9*w*t+7.6/57.3);
v_out_total=vdc+v_out1+v_out3+v_out5+v_out7+v_out9;

plot(t,v_out_total)
```

(a) MATLAB reconstruction   **(b)** Simulation

Figure 13-22 Plots of the results of Example 13-6

Practice: A solution to the droop and the negative spike is to lower the *on* resistance of the transistor and raise the switching frequency. Rework the example with $R_{switch} = 2 \ \Omega$, and $f = 1.5$ kHz.

Answer: The output only droops from 12 V to 11.4 V. The negative spike is -0.6 V. The fundamental is 7.63 V_p sin(9.4 kr/s $t + 1°$). There is virtually *no* attenuation.

Effects of a High-pass Filter

In general, the effect of a high-pass filter on a nonsinusoidal waveform is to attenuate the fundamental and perhaps second and third harmonics while passing the upper frequency harmonics intact. The sharp edges remain, while the flat or slightly rounded sections droop. Usually, but not always, the dc component of the input is blocked.

To pass the entire wave while blocking *only* the dc part, set the filter's half-power point frequency well *below* the fundamental's frequency.

$$f_{-3dB} < \frac{f_{fundamental}}{7}$$

To create spikes, move the filter's half-power point frequency far *above* the fundamental. Only the sharp edges survive. These spikes are handy for timing and triggering pulses.

A differentiator is also a type of high-pass filter. The output depends on how rapidly the input changes. So the high-frequency harmonics produce large results, while the slower low-frequency harmonics are attenuated. The output of the op amp differentiator is inverted and proportional to the rate of change of the input voltage. This allows the circuit to convert a ramp into a square wave. Be sure, however, to include a small series input resistor. This sets the gain for very high frequency harmonics and noise. Otherwise the op amp may oscillate.

A motor is inherently an *RL* high-pass filter. If you drive it with a step, be sure to lower any series resistance and choose a driving frequency whose fundamental is above the filter's cut-off.

Summary

The response of a circuit to a nonsinusoidal signal can be computed using phasor circuit analysis. First determine the Fourier series of the input wave. Then calculate the effect each of these harmonics has on the output. Finally, add all of the output harmonics together, and plot the result. Although these steps are straightforward, it is easy to become lost in the many equations and numbers needed. With the Fourier analysis worksheet you can compile the equations and results on one page.

Low-pass filters pass (more or less) the dc and fundamental harmonics while attenuating the upper frequency elements. This rounds the output, removing the sharp edges. The lower the filter's f_{-3dB} is set with respect to the signal's frequency, the more severely the upper frequency harmonics are attenuated, and the more rounded the output becomes. A square wave may be converted to a sawtooth, triangle, or even sine wave as the low-pass filter removes more of the upper frequency harmonics.

Increasing the order of the low-pass filter enhances the effect. The fundamental is passed with little attenuation while the upper frequencies are attenuated even further than with a first order filter. The result may be an output with only dc and a sine wave at the fundamental frequency.

The high-pass filter passes the upper frequency harmonics with little or no attenuation, but attenuates the dc and the fundamental (and perhaps second and third harmonics). The sharp edges remain, but the dc and flat or rounded middles of the input signal may be removed.

Coupler:

$$f_{-3dB} = \frac{1}{2\pi RC} < \frac{f_{fundamental}}{7}$$

An *CR* coupler removes only the dc and passes the rest of the signal. To assure that even the fundamental is not attenuated, set the filter's half-power point to 1/7 of the fundamental's frequency. Moving the filter's f_{-3dB} well beyond the fundamental frequency leaves only the sharp edges, an appropriate signal for triggering and timing circuits.

Differentiator:

$$5RC < t_{min\,pulse\,width}$$

The output from a differentiator is proportional to how *rapidly* the input is changing. The high-frequency harmonics are passed while the fundamental and lower frequency harmonics are blocked. To set up a differentiator, set its f_{-3dB} well above the input signal's frequency. This is opposite to the *CR* coupler.

A differentiator can be made with an op amp. This gives you more control over the output amplitude. However, the signal is inverted and precautions must be taken to assure that the circuit does not oscillate.

Problems

RC Low-pass Filters

13-1 Use Fourier analysis to calculate the output from the circuit in Figure 13-23, with R = 390 Ω, C = 0.1 μF. The input source, e_{in}, is a triangle that goes from 0 V to 12 V at 2 kHz.

13-2 Repeat Problem 13-1 with e_{in} going from 0 V to 3 V at 5 kHz.

13-3 Use Fourier analysis to calculate the output from the circuit in Figure 13-23, with R = 390 Ω, C = 22 μF. The input source, e_{in}, is a half-wave rectified sinusoid with a 16 V_p at 60 Hz.

13-4 Repeat Problem 13-3 with e_{in} a full-wave rectified sinusoid with a 16 V_p at 60 Hz.

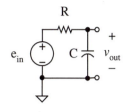

Figure 13-23
Schematic for Problems 13-1 through 13-4

Op Amp Integrator

13-5 Use Fourier analysis to calculate the output from the circuit in Figure 13-24, with R_i = 390 Ω, C_f = 0.1 μF. The input source, e_{in}, is a triangle that goes from −6 V to 6 V at 2 kHz. Ignore R_f.

13-6 Repeat Problem 13-5 with e_{in} going from −1.5 V to 1.5 V at 5 kHz.

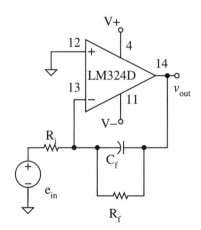

Figure 13-24 Schematic for Problems 13-5 through 13-8

13-7   Repeat Problem 13-5 with R = 390 Ω, and C = 4.7 μF.   The source, e_{in}, is a square wave going from −10 V to 10 V at 60 Hz.

13-8   Repeat Problem 13-5 with R = 390 Ω, and C = 1 μF.   The source, e_{in}, is a square wave going from −10 V to 10 V, at 60 Hz.

CR High-pass Filters / Couplers

13-9   Use Fourier analysis to calculate the output from the circuit in Figure 13-25, with R = 4.7 kΩ, C = 0.1 μF.   The input source, e_{in}, is a triangle that goes from 0 V to 12 V at 2 kHz.

13-10   Repeat Problem 13-9 with R = 1 kΩ, C = 0.1 μF, and e_{in} as a triangle that goes from 0 V to 12 V at 2 kHz.

Op Amp Differentiator

13-11   Use Fourier analysis to calculate the output from the circuit in Figure 13-26, with R_i = 22 Ω, C = 33 nF, and R_f = 390 Ω.   The input source, e_{in}, is a triangle going from −6 V to 6 V at 2 kHz.

13-12   Use Fourier analysis to calculate the output from the circuit in Figure 13-26, with R_i = 100 Ω, C = 33 nF, and R_f = 2.2 kΩ.   The input source, e_{in}, is a triangle going from −6 V to 6 V at 2 kHz.

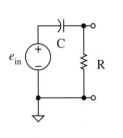

Figure 13-25
Schematic for Problems 13-9 and 13-10

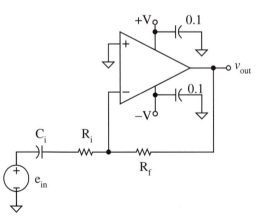

Figure 13-26 Schematic for Problems 13-11 and 13-12

RL High-pass Filters

13-13   Use Fourier analysis to calculate the output from the circuit in Figure 13-27, with $R_{switch} = 3\ \Omega$, $R_{motor} = 6\ \Omega$, and $L_{motor} = 20$ mH. The input source, e_{in}, is a square wave from 0 V to 12 V, 500 Hz.

13-14   Repeat Problem 13-13 with $R_{switch} = 12\ \Omega$, and the source at 60 Hz.

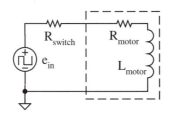

Figure 13-27 Schematic for Problems 13-13 and 13-14

Transient Analysis Lab Exercise

A. *RC* Low-pass Filter

1.   Using Fourier analysis, calculate and plot the waveform at the output of the circuit in Figure 13-28.

2.   From the plot in step 1, determine the theoretical output voltage at 75 μs, 150 μs, 225 μs, ... , 975 μs. Place these values in a spreadsheet.

3.   Build the circuit in Figure 13-28. Assure that the function generator is properly configured.

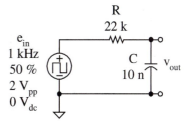

Figure 13-28
RC Low-pass Filter

4.   Monitor the input signal on channel 1 of the oscilloscope, and the output channel on channel 2. Be sure to:
   Set 0 V_{dc} across the center of the screen.
   DC couple the oscilloscope's inputs.
   Adjust the V/div to give the largest display possible.
   Adjust the time/div to provide one full cycle on the screen.

5. Save this oscilloscope display.

6. From the oscilloscope display of the output voltage, determine the actual output voltage at 75 μs, 150 μs, 225 μs, ... , 975 μs. Place these values beside the theoretical values from step 2.

7. In another column in the spreadsheet, calculate the % full scale error of each output voltage.

$$\%_{full\ scale} = \frac{theory - actual}{1\,V} \times 100\%$$

B. Op Amp Differentiator
1. Using Fourier analysis, calculate and plot the waveform at the output of the circuit in Figure 13-29. The input is a *triangle*.

2. From the plot in step 1, determine the theoretical output voltage at 75 μs, 150 μs, 225 μs, ... , 975 μs. Place these values in a spreadsheet.

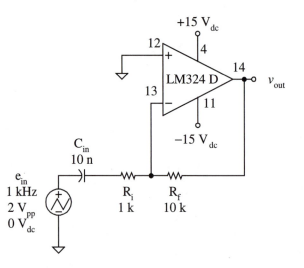

Figure 13-29 Op amp integrator

3. Build the circuit in Figure 13-29. Assure that the function generator is properly configured. The input is a *triangle*.

4. Monitor the input signal on channel 1 of the oscilloscope, and the output channel on channel 2. Be sure to:

   Set 0 V_{dc} across the center of the screen.
   DC couple the oscilloscope's inputs.
   Adjust the V/div to give the largest display possible.
   Adjust the time/div to provide one full cycle on the screen.

5. Save this oscilloscope display.

6. From the oscilloscope display of the output voltage, determine the actual output voltage at 75 µs, 150 µs, 225 µs, ... , 975 µs. Place these values beside the theoretical values from step 2.

7. In another column in the spreadsheet, calculate the % full scale error of each output voltage.

$$\%_{\text{full scale}} = \frac{theory - actual}{1\,V} \times 100\%$$

Advanced Analysis Techniques

All of the circuits you have seen so far have contained simple series or parallel elements. Even the op amp based circuits have a single input and a single feedback loop. Their performance was analyzed with Ohm's law, the voltage divider law, and Kirchhoff's voltage and current laws. You learned the principles of phasor algebra, filter performance, and Fourier analysis without being overwhelmed by circuit complexity.

While you have seen many useful single-loop circuits, often practical, real-world systems fill an entire page or more and contain *many* loops. Understanding how these circuits behave requires the use of phasor algebra, the application of filter principles, superposition (for multiple sources), and Fourier analysis (if the sources are nonsinusoidal). Now it is time to apply these tools to multi-loop circuits.

Impedance combination extends Ohm's law, the voltage divider law, and Kirchhoff's laws to single source, multiple loop circuits. Multiple source circuits can be analyzed by applying these techniques over and over, then adding the results. The core of impedance combination is consistent nomenclature and rigorous documentation. The circuit is folded up as impedances are combined, then expanded as the effect of the source is determined on each layer of increasing complexity.

The rigorous application of Kirchhoff's laws allows you to write a series of simultaneous equations, one for each loop. The analysis of the circuit, then, is the solution of these simultaneous equations. Many calculators and math programs automate this mathematical operation. With a little practice, you can write the equations by inspection. Mesh analysis uses voltage sources and Kirchhoff's voltage law. This is typical for commercial power production and distribution systems. Nodal analysis uses current sources and Kirchhoff's current law. It is used for transistor-based circuits.

14

Series-Parallel Analysis by Impedance Combination

Introduction

When faced with a *huge* network, it is very tempting to panic. "Oh no! I have *no* idea where to start, or what to do! I can't do this. Shoot!" In reality, you *do* know how to solve the problem. You are just being overwhelmed by its magnitude.

The solution is to combine and reduce, combine and reduce, combine and reduce the circuit until it becomes a simple series or parallel circuit. Then, work your way back out, with Ohm's law, Kirchhoff's laws, and the voltage divider law. Determine the effect the source has on each layer as you reintroduce the complexity by undoing the combinations you created to simplify the problem.

The trick in all of this combining is the documentation. Only the most gifted can keep all of the simplifications and combinations in their heads. It is critical to develop a consistent way of naming the intermediate combinations of components, and to redraw the circuit *every* time you combine elements, changing its topology.

Once you have developed a clear, workable plan, it is then just a matter of plugging in the numbers, step-by-step, combination-by-combination, working your way through the simplified schematics out to the full schematic.

Objectives

Upon completion of this chapter, you will be able to do the following:

- Describe how to use impedance combination to solve a multi-loop series-parallel circuit.

- Properly develop nomenclature to define intermediate elements.

- Draw each of the intermediate schematics correctly.

- Apply basic circuit laws to calculate the component voltages and currents in the intermediate and final circuits.

14.1 High-frequency Cabling

When interconnecting pieces of electronic equipment, it is critical that the signals be passed accurately. As the signal's frequency exceeds 1 MHz, you must consider the effects of the signal generator's output resistance, the cable's *RLC* model, and the load's resistance, capacitance, and inductance. A circuit that appears as a simple generator driving a load resistor becomes much more complex as you consider the high-frequency effects. Figure 14-1 shows both the simple and the first approximation of the high-frequency model for the interconnections.

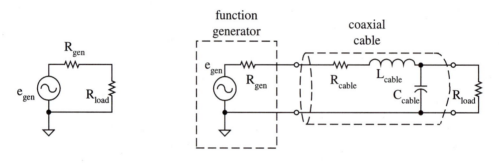

(a) Simple circuit   **(b)** Circuit with cable characteristics

Figure 14-1 First approximation of a high-frequency connection

Specifications for high-frequency cables include:

- Resistance per unit length.
- Capacitance per unit length.
- Characteristic impedance, Z_o.

The characteristic impedance of the cable plays a significant role in accurately transmitting high-frequency signals. Most of the details are best left to a complete text on rf electronics. For our purposes, the characteristic impedance can be used to determine the cable's inductance.

$$Z_o = \sqrt{\frac{L_{cable}}{C_{cable}}}$$

$$L_{cable} = C_{cable} \times Z_o^2$$

Example 14-1

The following specifications are given for a coaxial cable:

$$R = 275\ \Omega/1000\ \text{ft}, \quad Z_o = 50\ \Omega, \quad C = 24.2\ \text{pF/ft}$$

a. Calculate the components in the model from Figure 14-1(b) for a 6-foot-long cable.

b. Calculate the output voltage for
$$e_{gen} = 1\ V_{rms},\ 15\ \text{MHz}, \quad R_{gen} = 50\ \Omega, \quad \text{and } R_{load} = 50\ \Omega.$$

c. Verify your calculations by a simulation.

Solution

a.
$$R_{cable} = 275\frac{\Omega}{1000\,\text{ft}} \times 6\,\text{ft} = 1.65\,\Omega$$

$$C_{cable} = 24.2\frac{\text{pF}}{\text{ft}} \times 6\,\text{ft} = 145\,\text{pF}$$

$$L_{cable} = C_{cable} \times Z_o^2 = 145\,\text{pF} \times (50\,\Omega)^2 = 363\,\text{nH}$$

b. Begin by redrawing the diagram in terms of the impedances. This is shown in Figure 14-2.

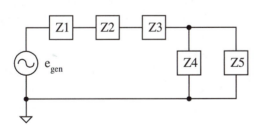

$$\overline{Z1} = (50\,\Omega\angle 0°)$$
$$\overline{Z2} = (1.7\,\Omega\angle 0°)$$
$$\overline{Z3} = (X_{L\,cable}\angle 90°)$$
$$\overline{Z4} = (X_{C\,cable}\angle -90°)$$
$$\overline{Z5} = (50\,\Omega\angle 0°)$$

Figure 14-2 Block diagram for Example 14-1

Without becoming bogged down in the numbers, yet, Figure 14-2 shows that Z4 and Z5 are in parallel. So, the next step is to combine these two impedances into a single impedance, Z_{45}. The circuit is shown in Figure 14-3.

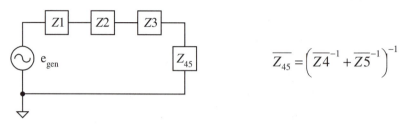

$$\overline{Z_{45}} = \left(\overline{Z4}^{-1} + \overline{Z5}^{-1}\right)^{-1}$$

Figure 14-3 Simplification of Figure 14-2

The problem is to find the voltage across Z5, which is also the voltage across Z_{45}. No further simplification is needed. The circuit in Figure 14-3 is a simple series circuit. Apply the voltage divider law to calculate the voltage across Z_{45}.

$$\overline{V_{Z5}} = \overline{V_{Z45}} = e_{gen} \times \frac{\overline{Z_{45}}}{\overline{Z1} + \overline{Z2} + \overline{Z3} + \overline{Z_{45}}}$$

That's it! You have developed a *plan* for simplifying and solving this problem. It is documented in clear, unambiguous steps that anyone using your results can easily follow. All that remains is to plug in values. But, do *not* be tempted to calculate as-you-go. Develop the plan *first*. Once you know you have a good map out of the maze, then you can grind out the numbers. Otherwise, you may spend all your time running out numbers that end up leading you nowhere.

When you perform these calculations manually, it is a good idea to place your work beside each of the drawings (Figure 14-2 and Figure 14-3) in a third column.

$$X_{L\,cable} = 2\pi \times 15\,MHz \times 363\,nF = 34.21\,\Omega$$

$$X_{C\,cable} = \frac{1}{2\pi \times 15\,MHz \times 145\,pF} = 73.17\,\Omega$$

$$\overline{Z1} = (50\,\Omega\angle0°)$$

$$\overline{Z2} = (1.7\,\Omega\angle0°)$$

$$\overline{Z3} = (34.21\,\Omega\angle90°)$$

$$\overline{Z4} = (73.17\,\Omega\angle-90°)$$

$$\overline{Z5} = (50\,\Omega\angle0°)$$

$$\overline{Z_{45}} = \left[(73.17\,\Omega\angle-90°)^{-1} + (50\,\Omega\angle0°)^{-1}\right]^{-1} = (41.28\,\Omega\angle-34.3°)$$

$$\overline{V_{Z5}} = \overline{V_{Z45}} = \frac{(1\,V_{rms}\angle0°)\times(41.28\,\Omega\angle-34.3°)}{(50\,\Omega\angle0°)+(1.7\,\Omega\angle0°)+(34.21\,\Omega\angle90°)+(41.29\,\Omega\angle-34.3°)}$$

$$\overline{V_{Z5}} = (477.2\,mV_{rms}\angle-41.6°)$$

b. The simulation results are shown in Figure 14-4 and agree with the calculations above.

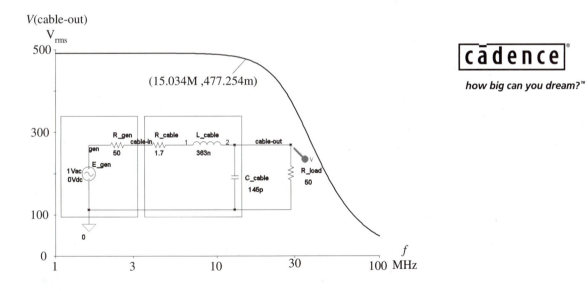

Figure 14-4 Simulation frequency response for Example 14-1

The cable forms a second order low-pass filter, as described in Section 7.4. It has a pass band gain of 0.5, just as you would expect from the simple circuit in Figure 14-1. But the signal out of the cable rolls off at −40 dB/decade, even though you may have thought that the cable is only an expensive piece of wire.

Practice: Change the load to 500 Ω, and the frequency to 12 MHz. Calculate the voltage at the load. Also repeat the simulation.

Answer: $\overline{V_{R\,load}} = (984.8\,\text{mV}\angle -37.6°)$ The frequency response plot begins at 900 mV and *rises* to 1 V before falling. This boost is *very* undesirable.

Terminated in its characteristic impedance (50 Ω in Example 14-1) a cable provides a well behaved, critically damped, low-pass performance. However, as soon as the termination no longer matches the cable's characteristic impedance, the resulting *RLC* filter may become underdamped. The voltage delivered to the load can even be *greater* than that provided by the function generator. Look back at Figure 7-16.

The circuit becomes even more complicated when you begin to use hook-up wire to connect from the cable's BNC connector to the input of your circuit. In Section 3.5 the inductance of wire is given as

$$L \approx 2\times10^{-7}\frac{\text{H}}{\text{m}}\,l\left[2.303\log_{10}\left(\frac{4l}{d}\right)-1\right]$$

where l = length of the wire.
  d = diameter of the wire. See Table 3-1.

At high frequencies, this inductance adds another rung to the circuit diagram ladder, and another set of calculations

Example 14-2

Using the same cable from Example 14-1, a 50 Ω terminator has been added to the end of the cable, followed by a 12-inch piece of wire leading to a 100 Ω load. See Figure 14-5.

a. Calculate the voltage across the 100 Ω load at 15 MHz.

b. Verify your answer with a simulation.

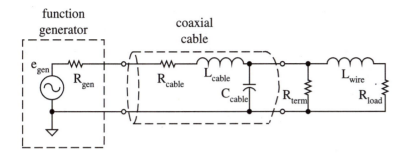

Figure 14-5 Schematic for Example 14-2

Solution

a. Begin by converting the wire's length from inches to meters.

$$l = 12\,\text{inches} \times \frac{2.54\,\text{cm}}{\text{inch}} \times \frac{1\,\text{m}}{100\,\text{cm}} = 0.3045\,\text{m}$$

The inductance of the 12-inch piece of wire is

$$L_{\text{wire}} \approx 2 \times 10^{-7}\,\frac{\text{H}}{\text{m}} \times 0.305\,\text{m} \left[2.303\log_{10}\left(\frac{4 \times 0.305\,\text{m}}{0.643 \times 10^{-3}\,\text{m}} \right) - 1 \right]$$

$$L_{\text{wire}} = 399.5\,\text{nH}$$

Redraw the schematic in terms of impedances, Figure 14-6.

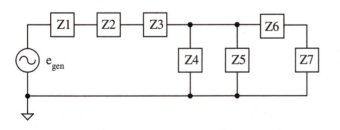

Figure 14-6 Block diagram for Example 14-2

$$\overline{Z1} = (50\,\Omega\angle 0°)$$
$$\overline{Z2} = (1.7\,\Omega\angle 0°)$$
$$\overline{Z3} = (34.21\angle 90°)$$
$$\overline{Z4} = (73.17\angle -90°)$$
$$\overline{Z5} = (50\,\Omega\angle 0°)$$
$$\overline{Z6} = (X_{\text{L wire}}\angle 90°)$$
$$\overline{Z7} = (100\,\Omega\angle 0°)$$

Begin the impedance combination as far from the source as possible. Combine Z6 and Z7 to form Z_{67}. Draw the diagram to keep track of the circuit's configuration, shown in Figure 14-7.

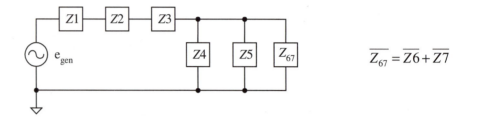

$$\overline{Z_{67}} = \overline{Z6} + \overline{Z7}$$

Figure 14-7 Impedance combination of Z6 and Z7

Impedances Z4, Z5, and Z_{67} are in parallel. They can be combined into a single impedance, Z_{4-7}. The new configuration is shown in Figure 14-8.

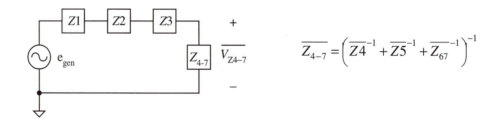

$$\overline{Z_{4-7}} = \left(\overline{Z4}^{-1} + \overline{Z5}^{-1} + \overline{Z_{67}}^{-1} \right)^{-1}$$

Figure 14-8 Impedance combination of Z4, Z5, and Z_{67}

This is a simple series circuit. No further impedance combinations are needed. Apply the voltage divider law to calculate V_{Z4-7}, the voltage across Z_{4-7}.

This is the voltage across Z4, Z5, and Z_{67}. Figure 14-9 is a repeat of Figure 14-7, with this voltage added.

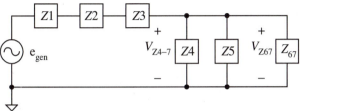

$$\overline{V_{Z4\text{-}7}} = e_{gen} \times \frac{\overline{Z_{4\text{-}7}}}{\overline{Z1} + \overline{Z2} + \overline{Z3} + \overline{Z_{4\text{-}7}}}$$

Figure 14-9 Voltage across $Z_{4\text{-}7}$

The voltage V_{Z4-7} is across the impedance Z_{67}.

$$\overline{V_{Z67}} = \overline{V_{Z4-7}}$$

This voltage produces the drop across Z7, and can be calculated by another application of the voltage divider law. This is shown in Figure 14-10.

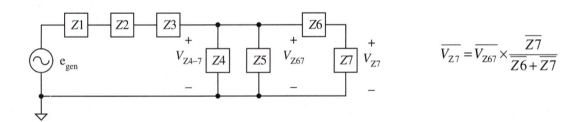

$$\overline{V_{Z7}} = \overline{V_{Z67}} \times \frac{\overline{Z7}}{\overline{Z6} + \overline{Z7}}$$

Figure 14-10 Voltage across Z7

You have a plan to calculate the voltage across Z7, the output. Only after the plan is completed and reviewed should you begin to insert numbers and calculate results.

$$X_{L\,wire} = 2\pi \times 15\,\text{MHz} \times 399.5\,\text{nH} = 37.65\,\Omega$$

$$\overline{Z1} = (50\,\Omega\angle 0°)$$

$$\overline{Z2} = (1.7\,\Omega\angle 0°)$$

$$\overline{Z3} = (34.21\angle 90°)$$

$$\overline{Z4} = (73.17\angle -90°)$$

$$\overline{Z5} = (50\,\Omega\angle 0°)$$

$$\overline{Z6} = (37.65\angle 90°)$$

$$\overline{Z7} = (100\,\Omega\angle 0°)$$

$$\overline{Z_{67}} = (37.65\,\Omega\angle 90°) + (100\,\Omega\angle 0°)$$

$$\overline{Z_{67}} = (106.9\,\Omega\angle 20.6°)$$

$$\overline{Z_{4-7}} = \left[(73.17\,\Omega\angle -90°)^{-1} + (50\,\Omega\angle 0°)^{-1} + (106.9\,\Omega\angle 20.6°)^{-1}\right]^{-1}$$

$$\overline{Z_{4-7}} = (32.71\,\Omega\angle -19.8°)$$

$$\overline{V_{Z4-7}} = \frac{(1\,\text{V}_{\text{rms}}\angle 0°) \times (32.71\,\Omega\angle -19.8°)}{(50\,\Omega\angle 0°) + (1.7\,\Omega\angle 0°) + (34.21\,\Omega\angle 90°) + (32.71\,\Omega\angle -19.8°)}$$

$$\overline{V_{Z4-7}} = (381.9\,\text{mV}_{\text{rms}}\angle -35.5°)$$

$$\overline{V_{Z7}} = \frac{(381.9\,\text{mV}_{\text{rms}}\angle -35.5°) \times (100\,\Omega\angle 0°)}{(37.65\,\Omega\angle 90°) + (100\,\Omega\angle 0°)}$$

$$\overline{V_{Z7}} = (357.4\,\text{mV}_{\text{rms}}\angle -56.1°)$$

b. The simulation result is shown in Figure 14-11 and agrees with the calculation above. When using MultiSIM, remember that the generator voltage is displayed in V_p. The panel meters normally display dc. You must adjust them to indicate V_{rms}.

Practice: Repeat Example 14-2, with R_{term} removed, $R_{\text{load}} = 50\ \Omega$, and $f = 20$ MHz.

Answer: $\overline{V_{Z7}} = (490.2\,\text{mV}_{\text{rms}}\angle -82.8°)$

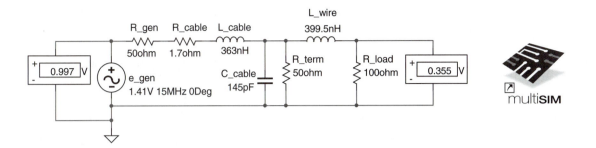

Figure 14-11 Simulation of Example 14-2

14.2 The x10 Oscilloscope Probe

A BNC cable presents considerable capacitance between the signal, on its center conductor, and common, on its shield. Combined with the other parasitic capacitances around an amplifier, the cable often makes the amplifier break into oscillations. That is, the amplifier is working fine until you connect a cable to its output to measure it. Then it breaks into oscillations. Remove the cable and the oscillations stop. The very act of trying to measure the amplifier's performance destroys that performance.

Instead of connecting to the amplifier's output directly with a BNC cable, use a ×10 oscilloscope probe. A 9 MΩ resistor is added at the tip of the probe, and considerable effort is made to lower the cable's inductance and capacitance. A simplified schematic is shown in Figure 14-12.

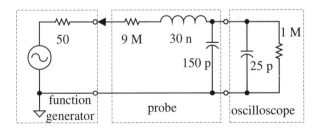

Figure 14-12 Simple ×10 oscilloscope probe

At first glance, this seems reasonable. The 9 MΩ resistor separates the cable's capacitance from the circuit, eliminating oscillations. At dc, the inductor is a short and the capacitors are opens. This just leaves a voltage divider of 9 MΩ and 1 MΩ, producing a ×10 attenuation.

However, the probe's resistance along with the cable's inductance and capacitance form a second order *RLC* low-pass filter. The probe's frequency response may be seriously degraded.

cādence®

how big can you dream?™

Example 14-3

Plot the frequency response of the simple probe in Figure 14-12.

Solution

The frequency response plot using Cadence PSpice is shown in Figure 14-13. Its half-power point frequency is 1 kHz. This makes the probe worthless at high frequencies.

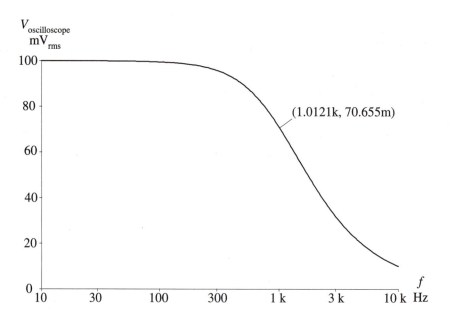

Figure 14-13 Frequency response of simple ×10 probe

The solution to this severe low-pass filter performance is the *addition* of two more capacitors. Capacitor C1 is placed in parallel with the 9 MΩ input resistor. Combined with the oscilloscope's 1 MΩ input resistor, C1 forms a high-pass filter. It boosts the high frequencies. The 9 MΩ input resistor, the cable's (150 pF) capacitance, the oscilloscope's input capacitance (25 pF), and C2 form a low-pass filter. As this low-pass filter begins to attenuate the high-frequency signals, the high-pass filter (C1 and the 1 MΩ resistor) begins to boost. When the values are properly set, the attenuation from the low-pass filter is precisely offset by the boost from the high-pass filter. The frequency response is flat out to 100 MHz or more.

For the capacitors to provide a ×10 attenuation, the impedance of the three parallel capacitors must be ⅑ the impedance of C1. This allows the capacitive impedance to match the attenuation provided by the resistors. Since impedance and capacitance are reciprocals, the entire parallel capacitance must be nine times larger than C1.

$$X_{C\,\text{paralllel}} = \frac{1}{2\pi f\,C_{\text{parallel}}} = \frac{1}{9}X_{C1} = \frac{1}{9(2\pi f\,C1)}$$

$$C_{\text{parallel}} = 9 \times C1$$

The exact value of the cable's capacitance and the oscilloscope's input are rarely known. So C2 is a variable capacitor built into the connector of the probe. To adjust C2 to the correct value, apply a square wave

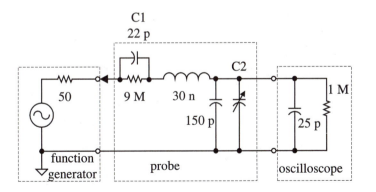

Figure 14-14 Frequency compensated ×10 oscilloscope probe

(a) C2 is too large, the low-pass filter is dominant

(b) C2 is correctly adjusted

(c) C2 is too small, the high-pass filter is dominant

Figure 14-15 Oscilloscope probe adjustment

to the oscilloscope and probe. Adjust the capacitor in the probe's connector until the display is a clean square wave, with no rounding (too much low-pass) or overshoot (too much high-pass). Look at Figure 14-15.

The impedance from the parallel capacitance is now $\frac{1}{9}$ that from C1. This precisely matches the resistor divider. Since frequency affects both C1 and the parallel capacitance the same, regardless of the frequency, the attenuation is $\frac{1}{10}$.

Example 14-4

    a. Calculate the correct value for C2 in the probe schematic of Figure 14-14.

    b. Use the impedance combination technique to calculate the voltage across the oscilloscope's 1 MΩ input resistor at 2 kHz.

    c. Confirm your calculation with a simulation. Also use the simulation to display the response of the oscilloscope's input voltage as the frequency sweeps from 1 Hz to 100 MHz.

Solution

    a.
$$C_{\text{parallel}} = 9 \times 22\,\text{pF} = 198\,\text{pF}$$
$$C2 = 198\,\text{pF} - 150\,\text{pF} - 25\,\text{pF} = 23\,\text{pF}$$

    b. The block diagram is given in Figure 14-16.

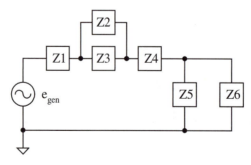

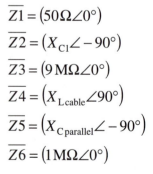

$$\overline{Z1} = (50\,\Omega\angle 0°)$$
$$\overline{Z2} = (X_{C1}\angle -90°)$$
$$\overline{Z3} = (9\,\text{M}\Omega\angle 0°)$$
$$\overline{Z4} = (X_{L\,\text{cable}}\angle 90°)$$
$$\overline{Z5} = (X_{C\,\text{parallel}}\angle -90°)$$
$$\overline{Z6} = (1\,\text{M}\Omega\angle 0°)$$

Figure 14-16 Block diagram for Example 14-4

Begin combining impedances at the far right. Since there are two sets of parallel components, simplify them both in the same step. The resulting diagram is shown in Figure 14-17.

Again, the circuit has simplified into a series circuit. Apply the voltage divider law.

$$\overline{V_{\text{oscilloscope}}} = e_{\text{gen}} \times \frac{\overline{Z_{56}}}{\overline{Z1} + \overline{Z_{23}} + \overline{Z4} + \overline{Z_{56}}}$$

Now, follow your plan, inserting the numbers and completing the calculations. The three reactances are:

$$X_{C1} = \frac{1}{2\pi \times 2\,\text{kHz} \times 22\,\text{pF}} = 3.617\,\text{M}\Omega$$

$$X_{L\,\text{cable}} = 2\pi \times 2\text{kHz} \times 30\,\text{nH} = 377\,\mu\Omega \approx 0\,\Omega$$

$$X_{C\,\text{parallel}} = \frac{1}{2\pi \times 2\,\text{kHz} \times 198\,\text{pF}} = 401.9\,\text{k}\Omega$$

The impedances are:

$$\overline{Z1} = (50\,\Omega\angle 0°)$$

$$\overline{Z2} = (3.617\,\text{M}\Omega\angle -90°)$$

$$\overline{Z3} = (9\,\text{M}\Omega\angle 0°)$$

$$\overline{Z4} = (0\,\Omega\angle 90°)$$

$$\overline{Z5} = (401.9\,\text{k}\Omega\angle -90°)$$

$$\overline{Z6} = (1\,\text{M}\Omega\angle 0°)$$

$$\overline{Z_{56}} = \left[(401.9\,\text{k}\Omega\angle -90°)^{-1} + (1\,\text{M}\Omega\angle 0°)^{-1}\right]^{-1} = (372.9\,\text{k}\Omega\angle -68.1°)$$

$$\overline{Z_{23}} = \left[(3.617\,\text{M}\Omega\angle -90°)^{-1} + (9\,\text{M}\Omega\angle 0°)^{-1}\right]^{-1} = (3.356\,\text{M}\Omega\angle -68.1°)$$

$$\overline{V_{\text{oscilloscope}}} = \frac{(1\,\text{V}_{\text{rms}}\angle 0°) \times (372.9\,\text{k}\Omega\angle -68.1°)}{(50\,\Omega\angle 0°) + (3.356\,\text{M}\Omega\angle -68.1°) + (372.9\,\text{k}\Omega\angle -68.1°)}$$

$$\overline{V_{\text{oscilloscope}}} = (100.00\,\text{mV}_{\text{rms}}\angle 0°)$$

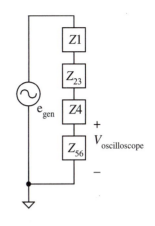

Figure 14-17
Simplified block diagram for Example 14-4

The compensated oscilloscope probe has reduced the voltage by $\frac{1}{10}$.

c. The simulation is given in Figure 14-18, and confirms the manual calculations.

Practice: Repeat Example 14-4 at 40 MHz.

Answer: $\overline{V_{\text{oscilloscope}}} = \left(100.00\,\text{mV}_{\text{rms}} \angle 0°\right)$ The compensated oscilloscope probe's frequency response is flat.

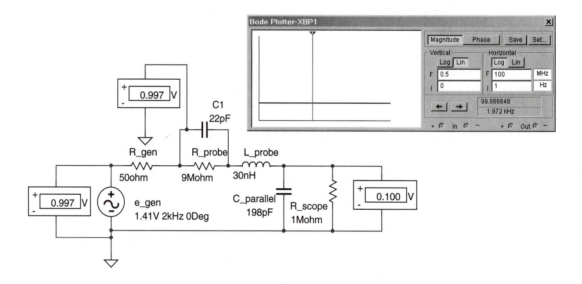

Figure 14-18 Simulation results for Example 14-4

Summary

Many large, single-source networks can be analyzed by combining impedances to reduce the circuit's complexity. Combine and reduce, combine and reduce, combine and reduce until the circuit becomes a simple series or parallel configuration. Next, apply Ohm's law, Kirchhoff's laws, and the voltage divider law to calculate the required parameters for that layer. Then, work your way back out through the layers of increasing complexity. At each new layer, apply the circuit laws as needed.

The big issue is rigorous documentation. Begin by redrawing the circuit, representing each element as a box, labeled Z1, Z2, and so on. At each reduction step, redraw the diagram. Label the combined imped-ances with a clear unambiguous name, e.g. the combination of imped-ances Z6 and Z7 is called Z_{67}. Write the appropriate reduction equation beside each block diagram. Draw a new diagram for each reduction and each equation. When you begin to calculate currents and voltages, the current through Z_{67} is labeled I_{Z67} and the voltage across Z_{67} is V_{Z67}. Continue to draw more complicated block diagrams and equations.

Only *after* you have developed a clear, well documented plan in terms of Zs, should you begin to *calculate* the results. When working with paper and pencil, just go back and add the numbers beside the block diagrams and equations you have already produced.

Several examples were used to practice this impedance combination technique. At high frequencies, cables have series resistance and induc-tance, and shunt capacitance. With the generator's output impedance, wire inductance, and the load, a multi-element low-pass filter is formed.

Oscilloscope $\times 10$ attenuator probes add a 9 MΩ series resistor at their tip to isolate the circuit being probed from the cable's reactances. This forms a severe low-pass filter. A capacitor is *added* across the 9 MΩ resistor to also form a high-pass filter. Properly adjusted, these two filters balance, producing a flat frequency response. Apply a square wave to the probe, and adjust the probe's capacitor until the oscilloscope shows no rounding or overshoot.

Problems

High-frequency Cabling

14-1 The following specifications are given for a coaxial cable:

$R = 26$ Ω/1000 ft, $\quad Z_o = 75$ Ω, $\quad C = 16.9$ pF/ft

    a. Calculate the components in the model for a 100-foot cable.

    b. Calculate the output voltage for

$e_{gen} = 1$ V_{rms}, 50 MHz, $\quad R_{gen} = 75$ Ω, $\quad$ and $R_{load} = 75$ Ω.

14-2 The following specifications are given for a coaxial cable:

$R = 7.6$ Ω/1000 ft, $\quad Z_o = 50$ Ω, $\quad C = 25.5$ pF/ft

    a. Calculate the components in the model for a 25-foot cable.

    b. Calculate the output voltage for

$e_{gen} = 1$ V_{rms}, 30 MHz, $\quad R_{gen} = 50$ Ω, $\quad$ and $R_{load} = 50$ Ω.

The x10 Oscilloscope Probe

14-3   Calculate the voltage at **point a** and at the oscilloscope input for the circuit in Figure 14-19, when C2 is adjusted to 12 pF.

14-4   Repeat Problem 14-3 with C2 properly adjusted.

14-5   Repeat Problem 14-3 at 4 kHz with C2 adjusted to 50 pF.

14-6   Repeat Problem 14-3 at 4 kHz with C2 properly adjusted.

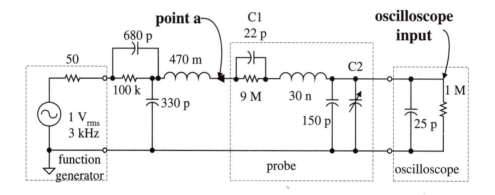

Figure 14-19 Schematic for Problems 14-3 through 14-6

Impedance Combination Lab Exercise

1.   Construct the circuit in Figure 14-20.

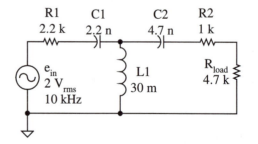

Figure 14-20 Schematic for the Impedance Combination Lab Exercise

2. Measure the actual value of each of the components. Also be sure to measure the internal resistance of the inductor. Record these values in a table similar to Table 14-1.

Table 14-1 Data for the circuit in Figure 14-20

		R1	C1	L1	C2	R2	R_{load}
value	Ω, μF, mH						
theory	V_{rms}						
	degrees						
measured	V_{rms}						
	degrees						
error	% of theory						
	degrees						

3. Calculate the voltage across each of the components using the actual measured values. Be sure to include the inductor's internal resistance in your calculations. Record these values. See Table 14-1.

4. Measure and record the magnitude and phase angle of the voltage across each of the components. You will have to rearrange the circuit for each measurement. To measure phase of the voltage across a component with the oscilloscope, that component must be connected to the oscilloscope's (and therefore the circuit's) common. Be *very* careful to configure the circuit correctly, as indicated in Figures 14-21(a-f).

5. Complete the two error rows. Calculate the magnitude error as a % of the theoretical value. Express the phase error as the difference in degrees between the measured phase and the theoretical phase. Magnitude errors should be less than 5%. Phase errors should be smaller than ±4°.

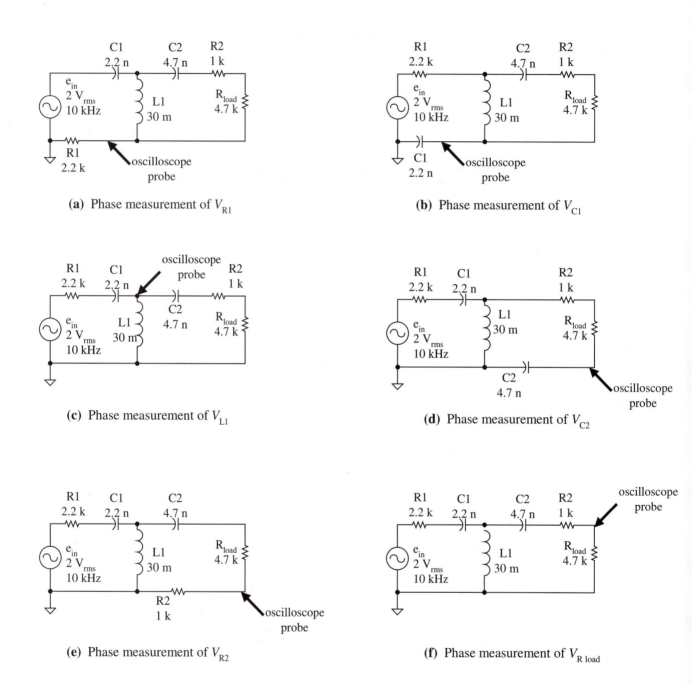

Figure 14-21 Phase measurement schematics

15

Mesh and Nodal Analysis

Introduction

Impedance combination requires *many* steps to determine the effect of a single source. If the circuit has multiple sources, these steps must be repeated for each source, and the results added. At best, this is *very* tedious, and equally error prone.

Mesh analysis uses Kirchhoff's voltage law, and allows you to write a set of simultaneous equations that fully describe the circuit *by inspection*. You can then use your scientific calculations or MATLAB to solve this set of equations. That's it! There are just two steps, one of which you can complete just by looking at the circuit, and the other by entering numbers into your calculator. Mesh analysis is performed on circuits that contain voltage sources. Three-phase electrical power systems commonly require mesh analysis.

For circuits that contain current sources nodal analysis is used. The techniques are parallel to those for mesh analysis, but use Kirchhoff's current law.

Objectives

Upon completion of this chapter, you will be able to do the following:

* Apply Kirchhoff's voltage law to write the simultaneous equations for a circuit driven by voltage sources.

* Write those same equations by inspection.

* Solve the set of simultaneous equations.

* Apply the results to calculate circuit voltages and currents.

* Apply Kirchhoff's current law to write simultaneous equations for a circuit.

* Solve the set of simultaneous equations.

* Apply the results to calculate circuit voltages and currents.

15.1 Mesh Analysis

Mesh analysis is based on Kirchhoff's voltage law. It requires that all sources be *voltage* sources. So, the first step is to convert any current sources to voltage sources. This conversion is shown in Figure 15-1.

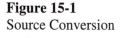

$$\overline{Z} = \frac{1}{\overline{Y}} \qquad\qquad \overline{E} = \overline{I} \times \overline{Z}$$

With the sources all converted to voltage, the circuit can be arranged into multiple loops. The configuration is similar to Figure 15-2.

Figure 15-1
Source Conversion

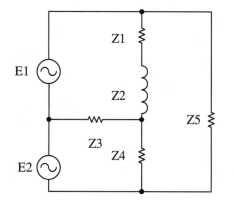

Figure 15-2 Basic circuit ready for mesh analysis

Kirchhoff's Voltage Law Approach

Kirchhhoff's voltage law indicates that the sum of the voltage rises and drops around a closed loop must be zero.

$$\sum \overline{E} = \sum \left(\overline{I} \times \overline{Z} \right)$$

To be sure that you properly account for the sources, the currents, and the voltage drops that they create, it is important to consistently assign loop currents, draw them on the schematic, and indicate the resulting voltage drops. It is traditional to assign a different current to each loop, assumed to be moving in a clockwise direction. If the actual current moves in the opposite direction, the phase you calculate will have an additional 180° phase shift. Look at Figure 15-3.

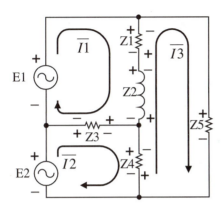

Figure 15-3 Circuit labeled with current and voltage drops

There are several things to notice. All currents are defined in the same direction. Each loop has its own current. Even though that seems to put two currents flowing through some elements, this is just a way of accounting for the voltage drop caused by both loops in which the component is contained. All voltage drop polarities are indicated. Summing the voltage drops and rises clockwise around the loops gives

$$\overline{E1} - \overline{I1}\,\overline{Z1} - \overline{I1}\,\overline{Z2} - \overline{I1}\,\overline{Z3} + \overline{I3}\,\overline{Z1} + \overline{I3}\,\overline{Z2} + \overline{I2}\,\overline{Z3} = 0$$

$$\overline{E2} - \overline{I2}\,\overline{Z3} - \overline{I2}\,\overline{Z4} + \overline{I1}\,\overline{Z3} + \overline{I3}\,\overline{Z4} = 0$$

$$0 - \overline{I3}\,\overline{Z4} - \overline{I3}\,\overline{Z2} - \overline{I3}\,\overline{Z1} - \overline{I3}\,\overline{Z5} + \overline{I1}\,\overline{Z1} + \overline{I1}\,\overline{Z2} + \overline{I2}\,\overline{Z4} = 0$$

These simultaneous equations fully describe the circuit's operation.

However, they can be rearranged to be easier to handle. Like terms are grouped, voltages induced by currents are placed on the left side, and voltages produced by sources are placed on the right.

$$(\overline{Z1} + \overline{Z2} + \overline{Z3})\overline{I1} - (\overline{Z3})\overline{I2} \qquad -(\overline{Z1} + \overline{Z2})\overline{I3} \qquad = \overline{E1}$$

$$-(\overline{Z3})\overline{I1} \qquad +(\overline{Z3} + \overline{Z4})\overline{I2} - (\overline{Z4})\overline{I3} \qquad = \overline{E2}$$

$$-(\overline{Z1} + \overline{Z2})\overline{I1} \qquad -(\overline{Z4})\overline{I2} \qquad +(\overline{Z1} + \overline{Z2} + \overline{Z4} + \overline{Z5})\overline{I3} = 0$$

These equations can also be written as a matrix equation.

$$
\begin{bmatrix}
\left(\overline{Z1}+\overline{Z2}+\overline{Z3}\right) & -\left(\overline{Z3}\right) & -\left(\overline{Z1}+\overline{Z2}\right) \\
-\left(\overline{Z3}\right) & \left(\overline{Z3}+\overline{Z4}\right) & -\left(\overline{Z4}\right) \\
-\left(\overline{Z1}+\overline{Z2}\right) & -\left(\overline{Z4}\right) & \left(\overline{Z1}+\overline{Z2}+\overline{Z4}+\overline{Z5}\right)
\end{bmatrix}
\times
\begin{bmatrix}
\overline{I1} \\ \overline{I2} \\ \overline{I3}
\end{bmatrix}
=
\begin{bmatrix}
\overline{E1} \\ \overline{E2} \\ 0
\end{bmatrix}
$$

By Inspection

You can obtain this same result *by inspection*, without writing any Kirchhoff voltage law loop equations, or performing any algebra manipulation. The schematic is shown again in Figure 15-4.

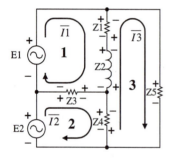

		$\overline{I1}$	$\overline{I2}$	$\overline{I3}$					
Loop 1	$\begin{bmatrix}$	$\left(\overline{Z1}+\overline{Z2}+\overline{Z3}\right)$	$-\left(\overline{Z3}\right)$	$-\left(\overline{Z1}+\overline{Z2}\right)$	$\end{bmatrix}$		$\begin{bmatrix}\overline{I1}\\ \overline{I2}\\ \overline{I3}\end{bmatrix}$	$=$	$\begin{bmatrix}\overline{E1}\\ \overline{E2}\\ 0\end{bmatrix}$

Loop 1, Loop 2, Loop 3 matrix:

$$
\begin{array}{c}
\text{Loop 1} \\ \text{Loop 2} \\ \text{Loop 3}
\end{array}
\begin{bmatrix}
\left(\overline{Z1}+\overline{Z2}+\overline{Z3}\right) & -\left(\overline{Z3}\right) & -\left(\overline{Z1}+\overline{Z2}\right) \\
-\left(\overline{Z3}\right) & \left(\overline{Z3}+\overline{Z4}\right) & -\left(\overline{Z4}\right) \\
-\left(\overline{Z1}+\overline{Z2}\right) & -\left(\overline{Z4}\right) & \left(\overline{Z1}+\overline{Z2}+\overline{Z4}+\overline{Z5}\right)
\end{bmatrix}
\times
\begin{bmatrix}
\overline{I1} \\ \overline{I2} \\ \overline{I3}
\end{bmatrix}
=
\begin{bmatrix}
\overline{E1} \\ \overline{E2} \\ 0
\end{bmatrix}
$$

Figure 15-4 The matrix equation can be written by inspection

- The rows and columns of the matrix equation have been labeled.

- The first element of the first row, Loop 1-$I1$, is the sum of all of the components in Loop 1.

- The second cell in the first row, Loop 1-$I2$, is the sum of those components in Loop 1 that are shared with Loop 2.

- Since $I2$ creates the opposite voltage drop that $I1$ creates, the second element in the first row is negative.

- The third cell in the first row, Loop 1-$I3$, is the sum of those components in Loop 1 that are shared with Loop 3. It too is negative.

The same procedures are used for the second row, Loop 2. The first cell, Loop 2-$I1$, is the sum of the components in Loop 2 that $I1$ also flows through. It is negative. The second element in the row, Loop 2-$I2$, is all of the components in Loop 2. It is positive. Finally, the last element is also negative, and is the sum of the components that are shared between Loop 2 and Loop 3.

The third row, Loop 3, is written the same way. Notice that this time, the last element is positive and the other two are negative.

When written correctly, the coefficient matrix (the first matrix) is symmetric in two ways. First, the elements along the major diagonal, Loop 1-$I1$, Loop 2-$I2$, and Loop 3-$I3$, are positive. All of the other elements are negative (or zero). Secondly, the elements reflect around this major diagonal.

$$\text{Loop 1-}I2 = \text{Loop 2-}I1$$
$$\text{Loop 1-}I3 = \text{Loop 3-}I1$$
$$\text{Loop 3-}I2 = \text{Loop 2-}I3$$

The solution matrix, the last matrix that contains the voltage sources, may also be written by inspection. The first element, for Loop 1, is the sum of all of the voltage sources in Loop 1. A source is *positive* if it *aids* the assigned direction of current flow. It is *negative* if that voltage source *opposes* the loop's current. The second row is for the voltage sources in Loop 2, and the third row is for the sources in Loop 3.

Example 15-1

Write the simultaneous equations for the circuit in Figure 15-5, in matrix form, by inspection.

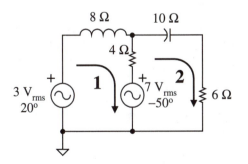

Figure 15-5 Schematic for Example 15-1

Solution

The reactances of the inductor and the capacitor have been provided, the two loops are defined, and the polarities of the voltage sources are noted. The simultaneous equations are

$$\begin{bmatrix} (4\Omega+j8\Omega) & -(4\Omega+j0) \\ -(4\Omega+j0) & (10\Omega-j10\Omega) \end{bmatrix} \times \begin{bmatrix} \overline{I1} \\ \overline{I2} \end{bmatrix} = \begin{bmatrix} (3\,V_{rms}\angle 20°)-(7\,V_{rms}\angle-50°) \\ (7\,V_{rms}\angle-50°) \end{bmatrix}$$

The 3 V_{rms} source *aids* $I1$. So, it is positive. The 7 V_{rms} opposes the flow of $I1$, and is given a negative sign.

Practice: Write the equations if the 7 V_{rms} source and the 6 Ω resistor are exchanged.

Answer:
$$\begin{bmatrix} (10\Omega+j8\Omega) & -(10\Omega+j0) \\ -(10\Omega+j0) & (10\Omega-j10\Omega) \end{bmatrix} \times \begin{bmatrix} \overline{I1} \\ \overline{I2} \end{bmatrix} = \begin{bmatrix} (3\,V_{rms}\angle 20°) \\ -(7\,V_{rms}\angle-50°) \end{bmatrix}$$

Example 15-2

Write the simultaneous equations for the circuit in Figure 15-6, in matrix form, by inspection.

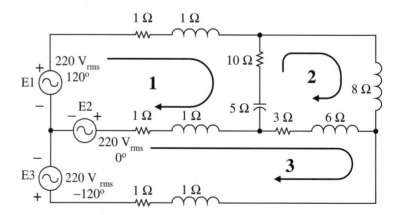

Figure 15-6 Schematic for Example 15-2

Solution

$$\begin{bmatrix} (12\Omega - j3\Omega) & -(10\Omega - j5\Omega) & -(1\Omega + j1\Omega) \\ -(10\Omega - j5\Omega) & (13\Omega + j9\Omega) & -(3\Omega + j6\Omega) \\ -(1\Omega + j1\Omega) & -(3\Omega + j6\Omega) & (5\Omega + j8\Omega) \end{bmatrix} \times \begin{bmatrix} \overline{I1} \\ \overline{I2} \\ \overline{I3} \end{bmatrix} = \begin{bmatrix} (220\,V_{rms}\angle120°) - (220\,V_{rms}\angle0°) \\ 0 \\ (220\,V_{rms}\angle0°) - (220\,V_{rms}\angle-120°) \end{bmatrix}$$

Practice: Write the simultaneous equations if the 5 Ω capacitor is changed to a 7 Ω inductor, the 3 Ω resistor is changed to 6 Ω, and the source (220 $V_{rms}\angle0°$) is changed to 0 V_{rms}.

Answer:

$$\begin{bmatrix} (12\Omega + j5\Omega) & -(10\Omega + j7\Omega) & -(1\Omega + j1\Omega) \\ -(10\Omega + j7\Omega) & (16\Omega + j21\Omega) & -(6\Omega + j6\Omega) \\ -(1\Omega + j1\Omega) & -(6\Omega + j6\Omega) & (8\Omega + j8\Omega) \end{bmatrix} \times \begin{bmatrix} \overline{I1} \\ \overline{I2} \\ \overline{I3} \end{bmatrix} = \begin{bmatrix} (220\,V_{rms}\angle120°) \\ 0 \\ -(220\,V_{rms}\angle-120°) \end{bmatrix}$$

Solving the Simultaneous Equations

Most scientific calculators can solve the equations you have just written. Once you have the solution, these quantities must be related to the *actual* currents you would measure in the circuit. Look at Figure 15-7.

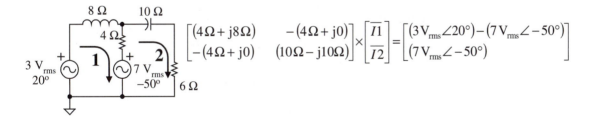

$$\begin{bmatrix} (4\Omega + j8\Omega) & -(4\Omega + j0) \\ -(4\Omega + j0) & (10\Omega - j10\Omega) \end{bmatrix} \times \begin{bmatrix} \overline{I1} \\ \overline{I2} \end{bmatrix} = \begin{bmatrix} (3\,V_{rms}\angle20°) - (7\,V_{rms}\angle-50°) \\ (7\,V_{rms}\angle-50°) \end{bmatrix}$$

Figure 15-7 Schematic and simultaneous equations for Example 15-1

When both equations are properly entered, and the calculator has solved the simultaneous equations, it displays

```
x1=(789.11E-3∠21.27E0)
x2=(604.89E-3∠15.47E0)
```

Look at the schematic again. The current through the 8 Ω inductor is

$$\overline{I_{L=8\Omega}} = \overline{I1} = (789.11\,\text{mA}\angle21.3°)$$

$$\overline{V_{L=8\Omega}} = (789.11\,\text{mA}\angle21.3°)\times(8\Omega\angle90°) = (6.313\,\text{V}_{\text{rms}}\angle111.3°)$$

The current through the 10 Ω capacitor and through the 6 Ω resistor is

$$\overline{I_{C=10\Omega}} = \overline{I_{R=6\Omega}} = \overline{I2} = (604.89\,\text{mA}\angle15.5)$$

$$\overline{V_{C=10\Omega}} = (604.89\,\text{mA}\angle15.5°)\times(10\Omega\angle-90°) = (6.049\,\text{V}_{\text{rms}}\angle-74.5°)$$

$$\overline{V_{R=6\Omega}} = (604.89\,\text{mA}\angle15.5°)\times(6\Omega\angle0°) = (3.629\,\text{V}_{\text{rms}}\angle15.5°)$$

Assuming that the actual current through the 4 Ω resistor in the center of the schematic flows *down*, from top to bottom,

$$\overline{I_{R=4\Omega}} = \overline{I1} - \overline{I2} = (197.66\,\text{mA}\angle39.3°)$$

$$\overline{V_{R=4\Omega}} = (197.66\,\text{mA}\angle39.3°)\times(4\Omega\angle0°) = (0.791\,\text{V}_{\text{rms}}\angle39.3°)$$

Example 15-3

    a. Solve the simultaneous equations for Example 15-2.

    b. Calculate the actual current through each 1 Ω resistor.

    c. Find the voltage drop across the 6 Ω inductor.

Solution

    a.
$$\overline{I1} = (17.42\,\text{A}_{\text{rms}}\angle55.3°)$$

$$\overline{I2} = (38.66\,\text{A}_{\text{rms}}\angle12.7°)$$

$$\overline{I3} = (64.31\,\text{A}_{\text{rms}}\angle-7.7°)$$

    b. Assign current flow through the 1 Ω resistors left to right.

$$\overline{I_{\text{top}1\Omega}} = \overline{I1} = (17.42\,\text{A}_{\text{rms}}\angle55.3°)$$

$$\overline{I_{\text{middle}1\Omega}} = \overline{I3} - \overline{I1} = (58.5\,\text{A}_{\text{rms}}\angle-23.1°)$$

$$\overline{I_{\text{bottom}1\Omega}} = -\overline{I3} = (64.31\,\text{A}_{\text{rms}}\angle172.3°)$$

c.   Assign voltage polarity of + to − across the 6 Ω inductor.

$$\overline{V_{6\Omega}} = \left(\overline{I3} - \overline{I2}\right) \times \left(6\,\Omega\angle90°\right) = \left(186.85\,V_{rms}\angle56.7°\right)$$

Practice: Solve the simultaneous equations from the Practice part of Example 15-2.

Answer: $\overline{I1} = \left(67.26\,A_{rms}\angle58.3°\right)$ $\overline{I2} = \left(51.32\,A_{rms}\angle33.8°\right)$

$\overline{I3} = \left(64.61\,A_{rms}\angle31.4°\right)$

MATLAB can also solve these simultaneous equations. The m file statements and comments to solve Example 15-3a are shown below.

```
Z=[(12-j*3),-(10-j*5),-(1+j*1);-(10-5),(13+j*9),
-(3+j*6);-(1+j*1),-(3+j*6),(5+j*8)];
```
   This is the impedance, coefficient matrix. Elements are entered in a single line, separated by a comma. The rows are separated by semicolons.

```
E1=(220*cos(120/57.3)+j*220*sin(120/57.3));
E2=220;
E3=(220*cos(-120/57.3)+j*220*sin(-120/57.3));
```
   The three lines above convert the three voltages into rectangular form.

```
E=[(E1-E2);0;(E2-E3)];
```
   This is the source or solution matrix.

```
I=Z\E;
```
   This statement solves the simultaneous equation, placing the answers in a matrix called I. This is *not* a division. It is the syntax used to request the solution of simultaneous equations with a coefficient, Z, and a source, E.

```
I1=[abs(I(1)) angle(I(1))*57.3]
I2=[abs(I(2)) angle(I(2))*57.3]
I3=[abs(I(3)) angle(I(3))*57.3]
```
   These three statements extract the individual currents from the matrix I, and convert each into polar form.

```
I1 =
   17.421    55.263
I2 =
   38.665    12.696
I3 =
   64.309   -7.7229
```

$I1 = \left(17.42\,A_{rms}\angle55.3°\right)$

$I2 = \left(38.67\,A_{rms}\angle12.7°\right)$

$I3 = \left(64.31\,A_{rms}\angle-7.7°\right)$

15.2 Nodal Analysis

Nodal analysis uses Kirchhoff's current law to sum the *currents* into and out of the nodes in a circuit. Each node (other than common) produces its own equation. You then solve these simultaneous equations to determine the *voltages* at each node with respect to common.

DC Circuit

The circuit shown in Figure 15-8 is an eight-bit, R-2R digital-to-analog converter. All of the voltage sources may either be turned *on* to the same reference voltage (logic 1) or shorted to common (logic 0).

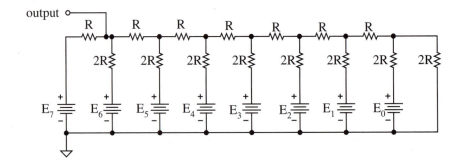

Figure 15-8 Eight-bit, voltage output, R-2R digital-to-analog converter

The output from an *n*-bit converter is

$$V_{output} = \frac{code}{2^n} V_{reference}$$

To analyze the performance of this network, a three-bit version with the code set to 101_{binary} is shown in Figure 15-9. The reference voltages are 10 V_{dc}. The value for R has been set to 500 Ω, making 2R = 1 kΩ. Two nodes have been identified, V_1 and V_2. The currents into and out of each node have been arbitrarily defined. It has been decided to call currents into a node negative. Currents out of a node are positive. The opposite definition would work just as well when applied to *all* nodes.

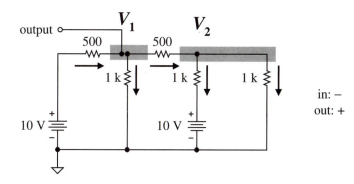

Figure 15-9 Three-bit converter schematic for analysis

$$V_{output} = \frac{101_2}{2^3} \times 10\,V_{dc} = \frac{5}{8} \times 10\,V_{dc} = 6.25\,V_{dc}$$

At node 1:
$$-\frac{10\,V - V_1}{500\,\Omega} + \frac{V_1}{1\,k\Omega} + \frac{V_1 - V_2}{500\,\Omega} = 0$$

Divide through by the 500 Ω and 1 kΩ, converting to conductances.

$$-20\,mA + V_1(2\,mS) + V_1(1\,mS) + V_1(2\,mS) - V_2(2\,mS) = 0$$

$$V_1(5\,mS) - V_2(2\,mS) = 20\,mA \qquad \Longleftarrow$$

At node 2:

$$-\frac{V_1 - V_2}{500\,\Omega} + \frac{V_2 - 10\,V}{1\,k\Omega} + \frac{V_2}{1\,k\Omega} = 0$$

Divide through by the 500 Ω and 1 kΩ, converting to conductances.

$$-V_1(2\,mS) + V_2(2\,mS) + V_2(1\,mS) - 10\,mA + V_2(1\,mS) = 0$$

$$-V_1(2\,mS) + V_2(4\,mS) = 10\,mA \qquad \Longleftarrow$$

Here are two simultaneous equations with the voltages as the unknowns.

$$\begin{array}{l} V_1(5\,mS) - V_2(2\,mS) = 20\,mA \\ -V_1(2\,mS) + V_2(4\,mS) = 10\,mA \end{array} \qquad \text{Or in matrix form} \qquad \begin{bmatrix} 5\,mS & -2\,mS \\ -2\,mS & 4\,mS \end{bmatrix} \times \begin{bmatrix} V_1 \\ V_2 \end{bmatrix} = \begin{bmatrix} 20\,mA \\ 10\,mA \end{bmatrix}$$

This equation exhibits all of the characteristics and symmetry of those you created using mesh analysis. Solving the equations gives

$$V_1 = 6.25 \text{ V}_{dc} = V_{output}$$

$$V_2 = 5.63 \text{ V}_{dc}$$

AC Circuit

The schematic in Figure 15-10 illustrates an interference problem caused by the capacitive coupling between adjacent traces on a printed circuit board (**pcb**). A signal, E_{gen}, is sent through a coaxial cable onto the pcb. It is properly terminated by R_{in} and enters the amplifier as V_1. Current flowing on a trace nearby is delivered to the load, producing V_2.

Without the coupling through C_{couple}, and assuming the following values, calculate the value of V_1 and V_2.

$E_{gen} = (3 \text{ V}_{rms}\angle 0°)$, $R_{gen} = R_{in} = 50 \text{ }\Omega$, $X_{L \text{ cable}} = 40 \text{ }\Omega$, $X_{C \text{ cable}} = 33 \text{ }\Omega$
$I_{trace} = (25 \text{ mA}_{rms}\angle 30°)$, $R_{load} = 75 \text{ }\Omega$

$$\overline{Z_{C//R}} = \left[(0 - j33\,\Omega)^{-1} + (50\,\Omega + j0\,\Omega)^{-1} \right]^{-1} = (27.5\,\Omega\angle -56.6°)$$

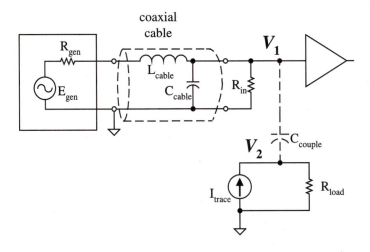

Figure 15-10 AC interference problem schematic

$$\overline{V_1} = \frac{(3\,\text{V}_{\text{rms}}\angle 0°)\times(27.5\Omega\angle-56.6°)}{(50\Omega + j40\Omega)+(27.5\Omega\angle-56.6°)} = (1.23\,\text{V}_{\text{rms}}\angle-71°)$$

$$\overline{V_2} = (25\,\text{mA}_{\text{rms}}\angle 30°)\times(75\Omega\angle 0°) = (1.88\,\text{V}_{\text{rms}}\angle 30°)$$

Because of the close spacing of the traces, part of the signal V_2 can couple through the parasitic capacitance, C_{coupling}, to V_1, and vice versa. This interaction may seriously alter both V_1 and V_2. The coupling may even produce feedback and the system will oscillate.

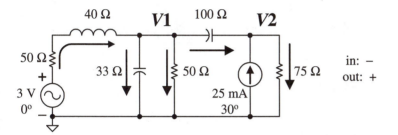

Figure 15-11 Nodal analysis schematic of the AC interference problem

The circuit in Figure 16-12 is a schematic that allows you to apply nodal analysis to calculate these two voltages. At node 1:

$$-\frac{(3\,\text{V}_{\text{rms}}\angle 0°)-\overline{V_1}}{(50\Omega + j40\Omega)} + \frac{\overline{V_1}}{(33\Omega\angle-90°)} + \frac{\overline{V_1}}{(50\Omega\angle 0°)} + \frac{\overline{V_1}-\overline{V_2}}{(100\Omega)\angle-90°} = 0$$

$$-(46.85\,\text{mA}_{\text{rms}}\angle-38.7°)+\overline{V_1}(15.62\,\text{mS}\angle-38.7°)+\overline{V_1}(30\,\text{mS}\angle 90°)$$

$$+\overline{V_1}(20\,\text{mS}\angle 0°)+\overline{V_1}(10\,\text{mS}\angle 90°)-\overline{V_2}(10\,\text{mS}\angle 90°)=0$$

$$\overline{V_1}(44.16\text{mS}\angle 43.2°)-\overline{V_2}(10\,\text{mS}\angle 90°)=(46.85\,\text{mA}\angle-38.7°) \qquad \Longleftarrow$$

At node 2:

$$-\frac{\overline{V_1}-\overline{V_2}}{(100\Omega\angle-90°)} + \frac{\overline{V_2}}{(75\Omega\angle 0°)} - (25\,\text{mA}_{\text{rms}}\angle 30°) = 0$$

$$-\overline{V_1}(10\,\text{mS}\angle 90°)+\overline{V_2}(10\,\text{mS}\angle 90°)+\overline{V_2}(13.3\,\text{mS}\angle 0°)-(25\,\text{mA}_{\text{rms}}\angle 30°)=0 \qquad \Longleftarrow$$

$$-\overline{V_1}\left(10\,\text{mS}\angle 90°\right)+\overline{V_2}\left(16.67\,\text{mS}\angle 36.9°\right)=\left(25\,\text{mA}_{\text{rms}}\angle 30°\right)$$

Written as a matrix, these two simultaneous equations become

$$\begin{bmatrix} \left(44.16\,\text{mS}\angle 43.2°\right) & -\left(10\,\text{mS}\angle 90°\right) \\ -\left(10\,\text{mS}\angle 90°\right) & \left(16.67\,\text{mS}\angle 36.9°\right) \end{bmatrix}\times\begin{bmatrix} \overline{V_1} \\ \overline{V_2} \end{bmatrix}=\begin{bmatrix} \left(46.85\,\text{mA}_{\text{rms}}\angle -38.7°\right) \\ \left(25\,\text{mA}_{\text{rms}}\angle 30°\right) \end{bmatrix}$$

The solution to these equations is

Ignoring interference: Including interference:

$$\overline{V_1}=\left(1.23\,\text{V}_{\text{rms}}\angle -71°\right) \qquad \overline{V_1}=\left(899.2\,\text{mV}_{\text{rms}}\angle -56.3°\right)$$

$$\overline{V_2}=\left(1.88\,\text{V}_{\text{rms}}\angle 30°\right) \qquad \overline{V_2}=\left(2.04\,\text{V}_{\text{rms}}\angle -5.9°\right)$$

Summary

Mesh analysis allows you to solve a circuit with multiple loops and several voltage sources in a few steps.

- Convert all current sources to voltage sources.
- Assign loop currents clockwise around each loop.
- Apply Kirchhoff's voltage law to write the simultaneous equations.
- Check the symmetry of the coefficient matrix.
- Solve the simultaneous equations with your calculator or MATLAB.
- Calculate the actual circuit currents and voltage drops.

With a little practice, the coefficient matrix can be written *by inspection*.

Nodal analysis also results in a set of simultaneous equations.

- Identify the *voltage* nodes.
- Label the currents entering and leaving the nodes.
- Apply Kirchhoff's current law at each node to write a separate equation for that node.
- Algebraically simplify each equation. Converting the resistances to conductances and the impedances to admittances makes this easier.
- Combine the simplified voltage equation from each node into a set of simultaneous equations.
- Solve these equations with your scientific calculator or MATLAB.
- Calculate the actual circuit currents and voltage drops.

Problems

Mesh Analysis

15-1   Calculate the voltage across the resistor in Figure 15-12.

15-2   Use mesh analysis to calculate the voltage across the center of the bridge circuit in Figure 15-13.

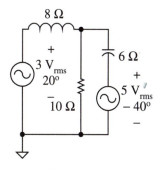

Figure 15-12
Schematic for
Problem 15-1

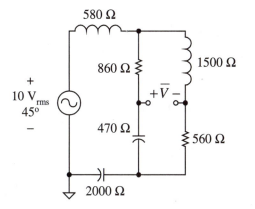

Figure 15-13 Schematic for Problems 15-2 and 15-7

15-3   Calculate all of the voltages in Figure 15-14.

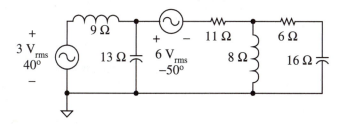

Figure 15-14 Schematic for Problem 15-3

15-4   Calculate all of the voltages in Figure 15-15.

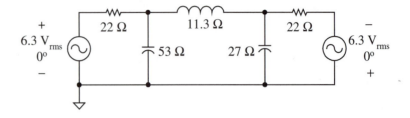

Figure 15-15 Schematic for Problem 15-4 and 15-5

Nodal Analysis

15-5   Use nodal analysis to calculate the voltages across the two capacitors in the circuit in Figure 15-15.

15-6   Use nodal analysis to calculate the voltages across the two resistors in the circuit in Figure 15-16.

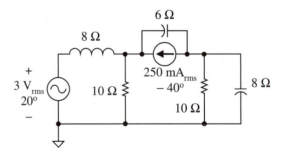

Figure 15-16 Schematic for Problem 15-6

15-7   Use nodal analysis to calculate the voltage across the center of the bridge circuit in Figure 15-13.

15-8   Use nodal analysis to calculate the voltage at each node in the circuit in Figure 15-17.

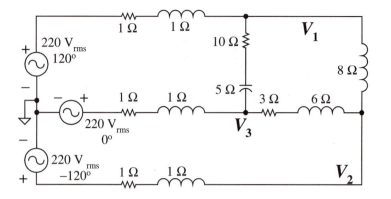

Figure 15-17 Schematic for Problem 15-8

Mesh and Nodal Analysis Lab Exercise

A. Circuit Set-up

   1. Build the circuit in Figure 15-19. Be sure to install the capacitors correctly.

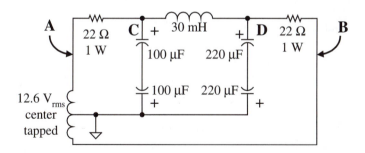

Figure 15-18 Schematic for circuit fabrication

   2. Energize the transformer.

   3. Measure and record the magnitude of the voltage between common and each end of the transformer with a digital multimeter.

B. Calculations
1. Redraw the circuit as indicated in Figure 15-19. On the schematic, indicate the *measured* component values, their impedances, and the voltage source values.

2. Using mesh analysis, calculate the voltage (magnitude and phase) at point **A** and point **B** with respect to common.

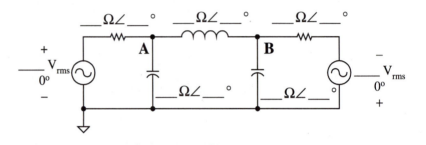

Figure 15-19 Schematic for mesh and nodal analysis

3. Repeat the analysis using nodal analysis.

C. Measurements
1. Energize the circuit.

2. Using a digital multimeter, measure the magnitude of the voltage at point **A** with respect to common, and the voltage at point **B** with respect to common.

3. Connect channel 1 of the oscilloscope between the top of the transformer and circuit common. This is the phase reference, 0°.

4. Using channel 2 of the oscilloscope, measure the phase of the voltage at point **A** (with respect to common) and at point **B** (with respect to common).

5. In a table, compare the magnitudes and phases of these two voltage measurements to the calculations in Section B. Add cells for the % error of the magnitudes and error in degrees.

Power Systems

The production and distribution of electrical energy is one of the most widespread of industries and the common link among all industrial nations. Urban or rural, north or south, east or west, power lines marching cross country announce the arrival of light for the night, heat for the cold, cooling for the heat, refrigeration, contact with the rest of mankind, security, medical assistance, employment, and *power*. They form the threads that bind modern mankind together.

In Chapter 16, you will see that power is the rate of doing work. It tells you about the conversion of energy from one form (electrical) into other forms (heat, light, sound, motion). As with all of the other calculations in ac electronics, the power dissipated by a resistance is simple to calculate and represent. However, during one part of a sinusoid, an inductor *stores* energy in its electromagnetic field. During the other part of the cycle, that energy is *released*, and returns to the generator. Capacitors handle power in a complementary fashion, releasing energy to the circuit during half of a sinusoidal cycle, and storing it in an electrostatic field during the other half.

The result of these leading and lagging currents, this storing and releasing of power, is a complex relationship between the real average power dissipated by the resistance in a circuit, the reactive power of the inductance and capacitance, and the apparent power resulting from their combination. These quantities are best represented with a (phasor) triangle. In fact, properly combining inductance and capacitance actually allows you to *reduce* the real power and the current delivered from the generator. This is called power factor correction.

Three signals (and earth) are produced in commercial power generation, not the single signal you have used so far. Look at the power distribution towers marching across the countryside. There are three large wires, and a small earth. This technique, called three-phase power, makes it much simpler to produce large electrical motors (the prime movers of manufacturing), and to smoothly convert ac to dc without large filtering components. These three generators, and the loads that use their signals, may be connected in a wye (Y) configuration or in a delta arrangement. The impedances of all of the legs may be the same, balanced. Or, they may be different, unbalanced. In Chapter 17, you will apply the circuit analysis techniques of the previous chapters to determine the currents, voltages, and the powers in all of the branches of both wye and delta connected three-phase systems.

16

Single-phase AC Power

Introduction

In dc electronics you learned that $P = I \times E$. Everything was constant, so there was no difference between instantaneous power and average power. Resistance *dissipates* power, converting it to heat, light, sound, or motion. It was simple and straightforward.

However, in ac circuits the voltage and the current vary continuously. The current through a resistance is in phase with the voltage across it. For inductance, the current lags by 90°. For capacitance the current leads the voltage by 90°. So, taking the product $i \times v$ involves multiplying two sine waves at different angles, and results in a sinusoid at twice the frequency and shifted in dc level.

Usually, the *average* power is required, involving the integration of the product of two sinusoids. The result is **real power** dissipated by the resistance, **reactive power** for the inductance and the capacitance, and **apparent power** when they are all combined in a power triangle.

How these all relate has a significant impact on the total current that the generator must produce and must send along the power lines to the loads. In fact, *adding* capacitance to balance load inductance may drastically *reduce* this current requirement, lowering the cost to both the producer and the consumer. This is called **power factor correction**.

Objectives

Upon completion of this chapter, you will be able to do the following:

- Define and calculate the instantaneous and average power for resistance, inductance, and capacitance.

- Define and calculate reactive power and apparent power.

- Draw the power triangle for a given circuit.

- For that circuit, calculate all currents, voltages, powers, and the power factor.

- Calculate the component(s) needed to achieve a given power factor.

16.1 Definitions

Mechanical Work and Power

Work is the effort expended, or the energy used (or gained). If a 150 lb person climbs to the top of a five story building (50 ft), then the work is

$$W = F \times y$$

$$W = 150\,\text{lb} \times 50\,\text{ft} = 7500\,\text{ft} \cdot \text{lb}$$

In the MKS system of measure, that 150 lb person has a mass of 68 kg. Each kg of mass exerts 9.8 newtons of force at the Earth's surface. The 50 ft building is 15.2 m tall. This gives

$$W = 68\,\text{kg} \times \frac{9.8\,\text{N}}{\text{kg}} \times 15.2\,\text{m} = 10.13\,\text{kNm}$$

A newton·meter is a **joule**

$$W = 10.13\,\text{kJ}$$

Walking leisurely, it may take five minutes to climb the five stories. At a full run, that same person may be able to climb the 50 ft in 45 s. In either case, the same work is done. But it is harder to do it in 45 seconds than at a walk. This is the power expended. **Power** is the *rate* of doing work.

$$p = \frac{W}{t}$$

During the slow walk up the stairs,

$$p = \frac{150\,\text{lb} \times 50\,\text{ft}}{5\,\text{min} \times 60\,\dfrac{\text{s}}{\text{min}}} = 25\,\frac{\text{ft} \cdot \text{lb}}{\text{s}}$$

During the run, however, the power being generated is

$$p = \frac{150\,\text{lb} \times 50\,\text{ft}}{45\,\text{s}} = 167\,\frac{\text{ft} \cdot \text{lb}}{\text{s}}$$

In the MKS system, this is

$$p = \frac{10.13\,\text{kJ}}{45\,\text{s}} = 225\,\frac{\text{J}}{\text{s}}$$

A J/s is also called a **watt**. $P = 225$ W.

Even though the person at the top of the stairs has gained (or expended) the same amount of energy whether the stairs were climbed at a slow walk or a dead run, the power dissipated during the climb is much greater for the run. Power gives a measure of the rate at which energy must be delivered to a load, and the heat that might be generated.

Electrical Power

Voltage is defined as the amount of energy given to each coulomb of charge as it is pumped from the negative to the positive terminal of the source by the generator or by the battery. In equation form,

$$v = \frac{W}{Q} \quad \text{J/C or volts.}$$

This indicates how much work each coulomb of charge can do.

It is the rate of moving that charge that makes the difference. How quickly you can deliver charge is defined as current.

$$i = \frac{Q}{t}$$

Power (the rate of doing work) is

$$p = \frac{W}{t}$$

Now, look at the three equations: for v, i, and p.

$$p = \frac{W}{t} = \frac{W}{Q} \times \frac{Q}{t}$$

$$p = v \times i$$

Instantaneous power definition

The voltage is used as the reference, assigning it a phase angle of 0°.

$$v = V_p \sin(\omega t)$$

The current may be leading or lagging this voltage.

$$i = I_p \sin(\omega t + \theta)$$

θ is the phase of the current thorough a component with respect to the voltage across that component

Instantaneous power is the product of these two sinusoids.

$$p = V_p \sin(\omega t) \times I_p \sin(\omega t + \theta)$$

$$p = V_p I_p \sin(\omega t) \sin(\omega t + \theta)$$

You must now simplify the product of two different sine functions

$$\sin(\alpha) \times \sin(\beta)$$

After applying a trig identity, and doing a little algebraic simplification,

$$p = \frac{V_p I_p}{2} \cos\theta [1 - \cos(2\omega t)] + \frac{V_p I_p}{2} \sin\theta \sin(2\omega t)$$

The rms values are generally used when working with power.

$$\frac{V_p I_p}{2} = \frac{V_p}{\sqrt{2}} \frac{I_p}{\sqrt{2}} = V_{rms} I_{rms}$$

$$p = V_{rms} I_{rms} \cos\theta [1 - \cos(2\omega t)] + V_{rms} I_{rms} \sin\theta \sin(2\omega t)$$

The $\cos\theta$ and $\sin\theta$ terms are *fixed* quantities. They do not vary with time. So the power equation has both a dc and a sinusoidal part. The sinusoidal part is at 2ω, twice the frequency of the applied voltage.

General sinusoidal power equation

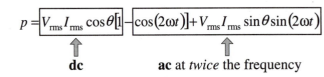

$$p = \boxed{V_{rms} I_{rms} \cos\theta [1} - \boxed{\cos(2\omega t)] + V_{rms} I_{rms} \sin\theta \sin(2\omega t)}$$

          dc   **ac** at *twice* the frequency

Power Dissipated by a Resistor

The current through a resistor is *in phase* with the voltage across the resistor. That is, $\theta = 0°$. The general sinusoidal power equation becomes

$$p = V_{rms} I_{rms} [1 - \cos(2\omega t)]$$

The plot of this equation is shown in Figure 16-1.

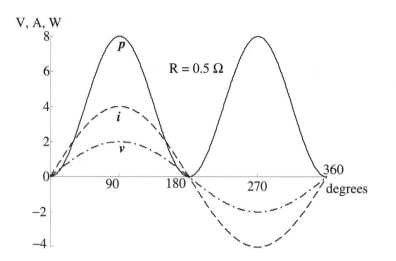

Figure 16-1 Voltage, current, and power for a resistor

```
R=0.5;
Vp=2;

degree=0:1:360;
radians=degree/57.3;
v=Vp*sin(radians);
i=v/R;
p=v.*i;

plot(degree,v,degree,i
,degree,p)
```

The power is at twice the frequency of the applied voltage. It also is all positive. During the *entire cycle* power is delivered to the resistance from the generator and converted into heat, light, sound, or motion.

When you take the average of the equation, the $\cos\theta = 0$. So,

$$P_{\text{ave R}} = V_{\text{rms}} I_{\text{rms}}$$

Average resistor power

Graphically, the average value of the power curve lies halfway between its peak and zero, at 4 W.

$$V_{\text{rms}} = \frac{2\,V_p}{\sqrt{2}} = 1.414\,V_{\text{rms}} \qquad I_{\text{rms}} = \frac{4\,A_{\text{rms}}}{\sqrt{2}} = 2.828\,A_{\text{rms}}$$

$$P_{\text{ave R}} = 1.414\,V_{\text{rms}} \times 2.828\,A_{\text{rms}} = 3.999\,W$$

Power Dissipated by an Inductor

The current through an inductor lags the voltage by 90°.

$$\overline{I_L} = \frac{(V\angle 0°)}{(X_L\angle 90°)} = \left(\frac{V}{X_L}\angle -90°\right) \qquad \theta = -90°$$

Making this substitution into the general power equation gives

$$p = V_{rms}I_{rms}\cos\theta[1-\cos(2\omega t)] + V_{rms}I_{rms}\sin\theta\sin(2\omega t)$$

$$p = -V_{rms}I_{rms}\sin(2\omega t)$$

cos(–90°) = 0   sin(–90°) = –1

This is a sine wave, at twice the frequency of the applied voltage, and with *no* dc offset. Figure 16-2 is the same voltage applied to an inductive reactance of 0.5 Ω. The current is delayed −90°. The average power is at twice the frequency. However, there is no offset.

Positive power occurs as energy is delivered from the generator and is stored in the inductor's electromagnetic field. Negative power results as energy returns from the inductor and the magnetic field collapses.

MATLAB

```
XL=0.5;
Vp=2;

degree=0:1:360;
radians=degree/57.3;
v=Vp*sin(radians);
i=(Vp/XL)*sin(radians-
90/57.3);
p=v.*i;

plot(degree,v,degree,i,
degree,p)
```

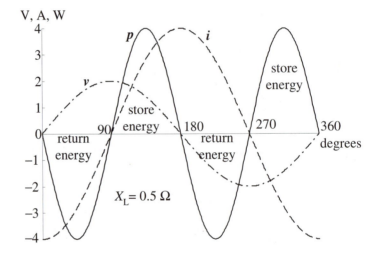

Figure 16-2 Voltage, current, and power for an inductor

On the average, an inductor does not *dissipate* power. It either stores energy in its field, or returns the energy to the generator.

Average inductive power

$$P_{ave\,L} = 0\,W$$

You can see this from the power trace in Figure 16-2. It is *evenly* above and below the *x* axis. Also, the inductor's power equation is a simple sine. The average value of a sine without a dc offset is zero.

Inductance presents several problems. Even though no useful work is being done (power being *dissipated*), the generator must still provide the current required by the inductance, and must ship it down the power lines, wasting energy there in the resistance of the lines. Worse, half of the time, the inductance *returns* energy to the generator.

Power Dissipated by a Capacitor

The current through a capacitor leads the voltage by 90°.

$$\overline{I_C} = \frac{(V\angle 0°)}{(X_C\angle -90°)} = \left(\frac{V}{X_C}\angle 90°\right) \qquad \theta = 90°$$

Making this substitution into the general power equation gives

$$p = V_{rms}I_{rms}\cos\theta[1-\cos(2\omega t)] + V_{rms}I_{rms}\sin\theta\sin(2\omega t)$$

$$p = V_{rms}I_{rms}\sin(2\omega t)$$

cos(90°) = 0   sin(90°) = 1

The plots for the voltage, current, and power are shown in Figure 16-3.

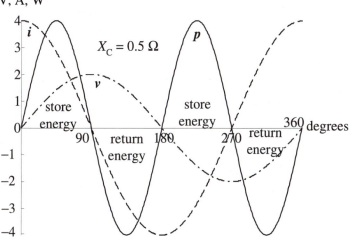

Figure 16-3 Voltage, current, and power for a capacitor

MATLAB

```
XL=0.5;
Vp=2;

degree=0:1:360;
radians=degree/57.3;
v=Vp*sin(radians);
i=(Vp/XL)*sin(radians
+90/57.3);
p=v.*i;

plot(degree,v,degree,
i,degree,p)
```

As with the inductor, the capacitor's instantaneous power is a simple sinusoid, evenly above and below the *x* axis. It's average power, too, is zero. Half of the time the capacitor stores the energy sent to it in an electrostatic field. The other half of the time, it returns this stored energy to the generator.

$$P_{\text{ave C}} = 0\,\text{W}$$

16.2 Real, Reactive, and Apparent Power

For resistance, power is really dissipated, converted into heat, light, sound, or motion. The average of instantaneous power is **real power**.

$$P_R = V_{\text{rms R}} I_{\text{rms R}}$$

Real power is expressed in watts. Ohm's law gives two other forms.

$$P = I_{\text{rms R}}^2 R \qquad\qquad P = \frac{V_{\text{rms R}}^2}{R}$$

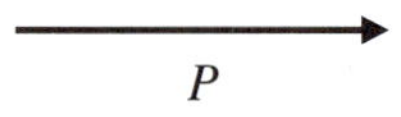

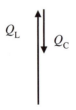

This can be graphically represented as a line of length P, pointing directly to the right, like the phasor of a resistor's impedance, Figure 16-4.

Figure 16-4
Real power

Inductance and capacitance do *not* dissipate power. They either store or release it. However, at any instant in time, energy is indeed flowing from (or to) the generator. **Reactive power, Q,** allows you to keep track of this phenomenon. For the inductance

$$Q_L = I_{\text{rms L}} V_{\text{rms L}}$$

Units: VAR$_L$

$$Q_L = I_{\text{rms L}}^2 X_L \qquad\qquad Q_L = \frac{V_{\text{rms L}}^2}{X_L}$$

Capacitive reactive power is negative.

Units: VAR$_C$

$$Q_C = -I_{\text{rms C}} V_{\text{rms C}}$$

$$Q_C = -I_{\text{rms C}}^2 X_C \qquad\qquad Q_C = -\frac{V_{\text{rms C}}^2}{X_C}$$

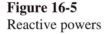

To more clearly distinguish reactive power from real power, Q_L and Q_C are given the units of **volt-amp-reactive, VAR$_L$ and VAR$_C$.**

Figure 16-5
Reactive powers

Reactive power may also be shown as lines. Their lengths are Q_L and Q_C. Since they do no real work, they are drawn vertically, rather than horizontally as real power is drawn. Also, to account for their polarities, Q_L points *up*, and Q_C points *down*. This is just like the phasor for the impedance of the inductor and the capacitor. In Figure 16-5 the reactive power of the inductor is larger than the reactive power of the capacitor.

The real and reactive powers of a circuit may be combined to determine the power that the generator *appears* to have to deliver. This is **apparent power**, because although some of it is dissipated in load resistance, some of the apparent power sent out from the generator during one part of a cycle is returned during the other part of the cycle.

Apparent power may be expressed as a phasor, since it is the combination of quantities that lie at right angles to each other

$$\overline{S} = [P + j(Q_L - Q_C)]$$

Apparent power
Units: VA

Figure 16-6 is called the **power triangle**. It illustrates the relationship between real power, reactive power, and apparent power.

Example 16-1

The circuit in Figure 16-7 consists of a desk lamp, a personal fan, and a small capacitor. Calculate each of the following:

a. The phasor current through each component, and the current delivered by the source.

b. P, Q, and S for each component and the source.

c. P_{total}, Q_{total}, and S_{total} for the entire circuit.

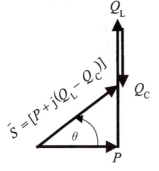

Figure 16-6
The power triangle

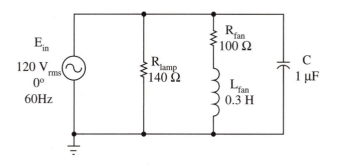

Figure 16-7 Schematic for Example 16-1

Solution

a.

$$\overline{I_{lamp}} = \frac{(120\,V_{rms}\angle 0°)}{(140\,\Omega\angle 0°)} = (0.857\,A_{rms}\angle 0°)$$

$$X_L = 2\pi \times 60\,Hz \times 0.3\,H = 113\,\Omega$$

$$\overline{I_{fan}} = \frac{(120\,V_{rms}\angle 0°)}{(100\,\Omega + j113\,\Omega)} = (0.795\,A_{rms}\angle -48.5°)$$

$$X_C = \frac{1}{2\pi \times 60\,Hz \times 1\mu F} = 2.65\,k\Omega$$

$$\overline{I_{cap}} = \frac{(120\,V_{rms}\angle 0°)}{(2.65\,k\Omega\angle -90°)} = (0.045\,A_{rms}\angle 90°)$$

$$\overline{I_{source}} = (0.857\,A_{rms}\angle 0°) + (0.795\,A_{rms}\angle -48.6°) \\ + (0.045\,A_{rms}\angle 90°) = (1.489\,A_{rms}\angle -21.7°)$$

b. For the lamp:

$$P_{lamp} = (0.857\,A_{rms})^2 \times 140\,\Omega = 102.8\,W$$

$$Q_{lamp} = 0\,VAR$$

$$\overline{S_{lamp}} = (102.8\,W + j0)$$

For the fan:

$$P_{fan} = (0.795\,A_{rms})^2 \times 100\,\Omega = 63.2\,W$$

$$Q_{fan} = (0.795\,A_{rms})^2 \times 113\,\Omega = 71.4\,VAR_L$$

$$\overline{S_{fan}} = (63.2\,W + j71.4\,VAR_L)$$

For the capacitor:

$$P_{cap} = 0\,W$$

$$Q_{cap} = -(0.045\,A_{rms})^2 \times 2.65\,k\Omega = -5.4\,VAR_C$$

$$\overline{S_{cap}} = (0\,W - j5.4\,VAR_L)$$

c.

$$P_{total} = P_{lamp} + P_{fan} + P_{cap}$$

$$P_{total} = 102.8\,\text{W} + 63.2\,\text{W} = 166\,\text{W}$$

$$Q_{total} = 0\,\text{VAR}_R + 71.4\,\text{VAR}_L - 5.4\,\text{VAR}_C = 66\,\text{VAR}_L$$

$$\overline{S_{total}} = (102.8\,\text{W} + \text{j}0) + (63.2\,\text{W} + \text{j}71.4\,\text{VAR}_L)$$
$$+ (0\,\text{W} - \text{j}5.4\,\text{VAR}_C) = (166\,\text{W} + \text{j}66\,\text{VAR}_L)$$
$$\overline{S_{total}} = (178.6\,\text{VA}\angle21.7°)$$

Practice: Calculate S_{total} if a second lamp of equal resistance is added, the fan is $R_{fan} = 40\,\Omega$, $L_{fan} = 0.4\,\text{H}$, and the capacitance is 10 μF.

Answer: $\overline{S_{total}} = (234.6\,\text{VA}\angle8.2°)$

You just found the total apparent power by adding the real power and the reactive power for all of the components. It is also true that:

$$\overline{S} = \overline{V} \times \overline{I}^*$$

Apparent power

where $\overline{I}^*$ is the **complex conjugate** of the current. To form the complex conjugate, reverse the polarity of the angle. In Example 16-1,

$$\overline{E_{source}} = (120\,\text{V}_{rms}\angle0°) \quad \overline{I_{source}} = (1.489\,\text{A}_{rms}\angle-21.7°)$$

$$\overline{S} = (120\,\text{V}_{rms}\angle0°) \times (1.489\,\text{A}_{rms}\angle21.7°) = (178.6\,\text{VA}\angle21.7°)$$

Example 16-1 is a **load-end** analysis. The customer knows what loads are being applied to the generator and can use the calculations to determine the total real, reactive, and apparent powers. However, from the **generator-end**, the supplier has no idea what loads are being connected to the line. Instead, only the generated voltage, current, and the angle between them can be measured. With that information, the total powers can also be determined. Look at Figure 16-8.

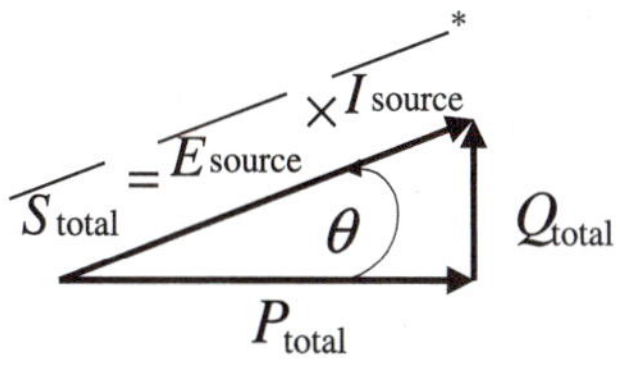

Since the source voltage is the reference, its angle is zero. The angle of the triangle, θ, is the complex conjugate of the current's angle. Knowing the hypotenuse and one angle of a right triangle allows you to calculate the other two sides.

$$Q = |\overline{S}| \times \sin\theta$$

$$P = |\overline{S}| \times \cos\theta$$

Figure 16-8
Generator-end power triangle

Example 16-2

The source produces 120 V$_{rms}$ at 1.49 A$_{rms}$. The current *lags* the voltage by 21.7°. Calculate P, Q, and S.

Solution

$$\bar{I} = \left(1.49\,\text{A}_{rms}\angle - 21.7°\right)$$

$$\bar{S} = \left(120\,\text{V}_{rms}\angle 0°\right) \times \left(1.49\,\text{A}_{rms}\angle 21.7°\right) = \left(179\,\text{VA}\angle 21.7°\right)$$

$$\theta = 21.7°$$

$$Q = 179\,\text{VA} \times \sin\left(21.7°\right) = 66\,\text{VAR}_L$$

$$P = 179\,\text{VA} \times \cos\left(21.7°\right) = 166\,\text{W}$$

Practice: $E_{source} = 120$ V$_{rms}$, $I_{source} = 1.932$ A$_{rms}$. The current lags the voltage by 8.6°. Calculate the apparent, reactive, and real powers.

Answers: $\bar{S} = \left(231.8\,\text{VA}\angle 8.6°\right)$, $Q = 34.7$ VAR$_L$, $P = 229.2$ W

16.3 Power Factor Correction

Real power, P, is the power converted into heat, light, sound, or motion. It does work, and makes the disk of the watt hour meter spin. Reactive power, Q, is stored and returned every cycle. It does *no* work, and does *not* cause the watt hour meter to spin. However, it does require more current to be generated, and to be sent back and forth along the power lines. This increased current *does* dissipate real power as it flows through the resistance of these lines.

To the customer, the reactive power may be of no importance. In the simplest case, they are charged for how much the watt hour meter spins. However, to the utility company, reactive power means that they must generate more current, and absorb the losses when this current is sent through the power lines. Reactive power is a loss to the utility company.

Power Factor Definition

The **power factor, *pf*,** allows these losses to be quantified. It is the unitless ratio of the power dissipated to the apparent power generated.

$$pf = \frac{P}{|S|}$$

Look at Figure 16-9. Large reactive power, Q, means that the apparent power, S, must be considerably larger than the real power, P, This requires much larger current. For lower reactive power, Q, the apparent power, S, is much smaller, requiring much less current to be generated and transported. The best case is when there is *no* reactive power. In Figure 16-9 (c), the apparent power and the real power are equal.

$$pf = \frac{P}{|S|}$$

$$P = |S|\cos\theta$$

$$pf = \cos\theta$$

Remember, θ is the angle between the voltage and the current.

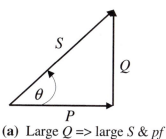

(a) Large Q => large S & pf

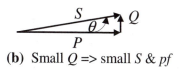

(b) Small Q => small S & pf

(c) No Q => $S = P$ & $pf = 1$

Example 16-3

For the circuit in Figure 16-10, calculate:

$$\overline{I_\mathrm{T}}, \quad P_\mathrm{load}, \quad Q_\mathrm{load}, \quad \overline{S_\mathrm{load}}, \quad pf_\mathrm{load}, \quad \overline{S_\mathrm{generator}}, \quad P_\mathrm{line}$$

Figure 16-9
Effects of Q on S and pf

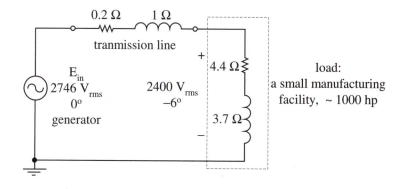
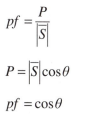

Figure 16-10 Schematic for Example 16-3

Solution

$$\bar{I} = \frac{(2400\,\text{V}_{\text{rms}}\angle -6°)}{(4.4\,\Omega + j3.7\,\Omega)} = (417\,\text{A}_{\text{rms}}\angle -46°)$$

$$P_{\text{load}} = (417\,\text{A}_{\text{rms}})^2 \times 4.4\,\Omega = 765\,\text{kW}$$

$$Q_{\text{load}} = (417\,\text{A}_{\text{rms}})^2 \times 3.7\,\Omega = 643\,\text{kVAR}_{\text{L}}$$

$$\overline{S_{\text{load}}} = (765\,\text{kW} + j643\,\text{kVAR}_{\text{L}}) = (999\,\text{kVA}\angle 40.0°)$$

$$pf = \frac{765\,\text{kW}}{999\,\text{kVA}} = 0.766$$

$$\overline{S_{\text{generator}}} = (2746\,\text{V}_{\text{rms}}\angle 0°) \times (417\,\text{A}_{\text{rms}}\angle 46°)$$
$$= (1.15\,\text{MVA}\angle 46°)$$

$$P_{\text{line}} = (417\,\text{A}_{\text{rms}})^2 \times 0.2\,\Omega = 34.8\,\text{kW}$$

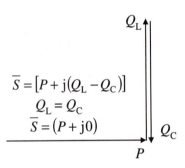

$$\bar{S} = [P + j(Q_{\text{L}} - Q_{\text{C}})]$$
$$Q_{\text{L}} = Q_{\text{C}}$$
$$\bar{S} = (P + j0)$$

Figure 16-11
Result of complete power
factor correction

Full Power Factor Correction

This low of a power factor creates several problems. There is consider-able power loss in the transmission line resistance. The generator must produce larger voltage and current than it would have to without the load inductance.

These problems can be corrected by adding a capacitor in parallel with the load. The resulting power triangle is shown in Figure 16-11. The capacitor is selected so that its reactive power matches and cancels the reactive power of the inductance. The power triangle collapses and

$$|\bar{S}| = P \qquad\qquad pf = 1$$

Example 16-4

For the circuit in Figure 16-12, calculate:

$$\overline{Z_{\text{load}}}\,,\ \overline{I_{\text{load}}}\,,\ \overline{I_{\text{L}}}\,,\ P_{\text{load}},\ Q_{\text{L}},\ Q_{\text{C}},\ \overline{S_{\text{load}}}\,,\ pf,\ \overline{E_{\text{generator}}}\,,\ \overline{S_{\text{generator}}}\,,\ P_{\text{line}}$$

Solution

$$\overline{Z_{\text{load}}} = \left[(4.4\,\Omega + j3.7\,\Omega)^{-1} + (0 - j8.9\,\Omega)^{-1}\right]^{-1}$$

$$\overline{Z_{\text{load}}} = (7.5\,\Omega\angle 0°)$$

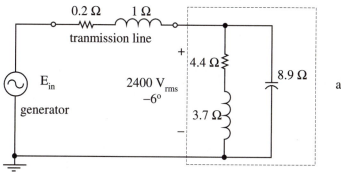

Figure 16-12 Full power factor correction schematic

Adding the capacitor has made the load look purely resistive.

$$\overline{I_{\mathrm{T}}} = \frac{(2400\,\mathrm{V_{rms}}\angle -6°)}{(7.5\,\Omega\angle 0°)} = (319.5\,\mathrm{A_{rms}}\angle -6°)$$

This is almost 100 $\mathrm{A_{rms}}$ *less* than before C was added.

$$\overline{I_{\mathrm{L}}} = \frac{(2400\,\mathrm{V_{rms}}\angle -6°)}{(4.4\,\Omega + \mathrm{j}3.7\,\Omega)} = (417\,\mathrm{A_{rms}}\angle -46°)$$

This is the same as in the previous example, even though the total current is almost 100 $\mathrm{A_{rms}}$ less.

$$P_{\mathrm{load}} = (417\,\mathrm{A_{rms}})^2 \times 4.4\,\Omega = 765\,\mathrm{kW}$$

Correcting the power factor has no effect on the real power.

$$Q_{\mathrm{load\,L}} = (417\,\mathrm{A_{rms}})^2 \times 3.7\,\Omega = +643\,\mathrm{kVAR_L}$$

$$Q_{\mathrm{load\,C}} = \frac{(2400\,\mathrm{V_{rms}})^2}{8.9\,\Omega} = -647\,\mathrm{VAR_C}$$

$$\overline{S_{\mathrm{load}}} = [765\,\mathrm{kW} + \mathrm{j}(643\,\mathrm{kVAR_L} - 647\,\mathrm{kVAR_C})]$$

$$\overline{S_{\mathrm{load}}} = (765\,\mathrm{kVA}\angle -0.3°)$$

$$pf = \frac{765\,\mathrm{kW}}{765\,\mathrm{kVA}} = 1.00$$

$$\overline{E}_{generator} = (319.5\,A_{rms} \angle -6°) \times [(0.2\,\Omega + j1\Omega) + (7.5\,\Omega \angle 0°)] = (2481\,V_{rms} \angle 1.4°)$$

$$P_{line} = (319.5\,A_{rms})^2 \times 0.2\,\Omega = 20.4\,kW$$

Power factor correction has cut the losses in the transmission by 40%.

Partial Power Factor Correction

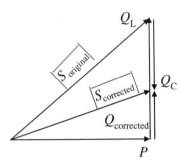

To encourage the customer to provide power factor correction, many utility companies monitor the customers' power factors and add a heavy penalty to the watt hour charges if the power factor falls too low.

Power factor capacitors are large and expensive. A customer often chooses to buy just enough capacitance to avoid the penalty.

Example 16-5

Calculate the Q_C needed to correct the power factor of Example 16-3 to $pf = 0.9$. Before the correction $P = 765$ kW, $pf = 0.765$.

Solution

$$pf = \frac{P}{|\overline{S}|}$$

Figure 16-13
Power triangle for partial *pf* correction

$$\left|\overline{S}_{original}\right| = \frac{P}{pf_{original}} = \frac{765\,kW}{0.765} = 1.0\,MVA$$

$$\left|\overline{S}\right| = \sqrt{P^2 + Q^2}$$

$$Q = \sqrt{\left|\overline{S}\right|^2 - P^2}$$

$$Q_{original} = \sqrt{(1\,MVA)^2 - (765\,kW)^2} = 644\,kVAR_L$$

The corrected power factor is 0.9.

$$\left|\overline{S}_{corrected}\right| = \frac{P}{pf_{corrected}} = \frac{765\,kW}{0.9} = 850\,kVA$$

$$Q_{corrected} = \sqrt{(850\,kVA)^2 - (765\,kW)^2} = 371\,kVAR_L$$

$$Q_C = 371\,kVAR - 644\,kVAR = -273\,kVAR_C$$

Summary

Power is the rate of doing work. Instantaneous power is the product of the voltage at the instant times the current at that instant.

Resistance dissipates power, converting it into heat, light, sound, or motion. Its instantaneous power waveform is a sinusoid that is entirely positive. The average value of this waveform is the real power.

Inductors do *not* dissipate power. They store energy in an electromagnetic field during half of the cycle, and return it to the generator during the other half of the cycle. Capacitors store energy in an electrostatic field, or return it. Their power waveforms are 180° out of phase with each other. This effect results in reactive power, Q_L and Q_C. Apparent power is the phasor combination of the real and the reactive powers. These quantities can be related in a triangle.

The power factor is an indication of the amount of energy that is dissipated by the load compared to the voltage and current that the generator must produce and ship. The closer these two quantities are, the fewer the losses and the lower the cost of generation.

To fully correct the power factor to 1, the customer must add enough Q_C to exactly equal the Q_L produced by the manufacturing machines. Levels of penalties may be charged for low power factor. Partially adjusting the power factor just enough to avoid this penalty is common.

Problems

Definitions

16-1 Using your calculator or a spreadsheet program, plot the instantaneous voltage, current, and power of a 4 Ω resistor connected across $(8 \ V_{rms} \angle 0°)$, 1 kHz.

16-2 Repeat Problem 16-1 for a 4 Ω inductive reactance, and for a 4 Ω capacitive reactance.

Real, Reactive, and Apparent Power

16-3 For the circuit in Figure 16-14,

    a. Calculate for each component the voltage, current, real power, reactive power, and apparent power.

    b. Calculate the current from the supply, the total real power, reactive power, and apparent power, and draw the triangle.

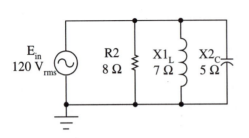

Figure 16-14 Schematic for Problems 16-3, 16-7, and 16-9

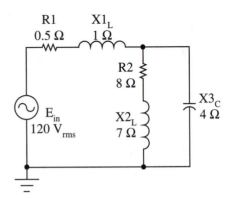

Figure 16-15 Schematic for Problems 16-4, 16-8, and 16-10

16-4 Repeat Problem 16-3 for the circuit in Figure 16-15.

16-5 Calculate $\overline{S}$, P, and Q for a circuit that is driven by 220 V$_{rms}$, at (12 A$_{rms}\angle-30°$).

16-6 Calculate $\overline{S}$, P, and Q for a circuit that is driven by 2400 V$_{rms}$, at (320 A$_{rms}\angle-42°$).

Power Factor Correction

16-7 Calculate the power factor and the VARs needed to fully correct the power factor for the circuit in Figure 16-14.

16-8 Calculate the power factor and the VARs needed to fully correct the power factor for the circuit in Figure 16-15.

16-9 Calculate the power factor and the VARs needed to correct the power factor to 0.92 for the circuit in Figure 16-14.

16-10 Calculate the power factor and the VARs needed to correct the power factor to 0.88 for the circuit in Figure 17-16.

17

Three-phase Systems

Introduction

In homes, classrooms, and businesses, you see ac power delivered from a single-phase outlet. These outlets contain three terminals: hot, neutral, and earth. However, most power is generated and most manufacturing facilities use power from **three-phase systems**. A three-phase system contains three sine waves, equal in amplitude and 120° apart in phase.

Three-phase systems turn the large synchronous machines that provide manufacturing muscle. Three times the amount of energy can be delivered over the same number of wires used to distribute single-phase power. This is a significant savings in the power transmission and distribution. Finally, three-phase signals can be easily converted into relatively smooth dc *without* any filter capacitors. The alternator in your car is really a three-phase generator.

The voltage, current, and power delivered to the three-phase loads depend on both the impedances of the loads, and how these loads are connected. There are two common configurations, the **wye (Y)** and the **delta (Δ)**. All of the impedances may match. This is called **balanced**. Or they may be different, **unbalanced**.

Objectives

Upon completion of this chapter, you will be able to calculate all of the voltages and currents in each of the following three-phase systems:

- Balanced wye with a neutral.

- Unbalanced wye with a neutral.

- Balanced delta.

- Unbalanced delta.

17.1 Three-phase Generator

When a magnetic field cuts a coil of wire, a voltage is induced across the coil. The magnitude of the voltage depends on the strength and polarity of the magnetic field, the speed of the relative motion of the field as it crosses the wire, and the number of turns of wire in the coil.

$$E = N \frac{d\phi}{dt}$$

where E is the generated voltage,
        N is the number of turns of wire in the coil
        ϕ is the magnetic flux

When a magnetic field is *rotated* in relationship to a stationary, single coil of wire, a sine wave voltage is generated. Look at Figure 17-1.

When the north pole is to the right, at 0°, none of the magnetic lines of flux cut the coil, so no voltage is induced into the coil. $E_{AN} = 0$ V. As the magnet rotates counterclockwise, some, and then more and more of the flux cuts the coil. More voltage is induced. E_{AN} rises. At 90°, the maximum flux is cutting the coil, inducing a peak positive voltage. As the magnet rotates from 90° to 180°, less flux cuts the coil, inducing less and less voltage. At 180°, none of the flux cuts the coil, so $E_{AN} = 0$ V.

Between 180° and 270°, the magnet's south pole is approaching the coil. This induces the opposite polarity voltage, that grows larger as the flux cuts the coil more perpendicularly. At 270° E_{AN} reachs a maximum negative. Finally, as the magnet returns to its starting position, fewer and fewer of the flux lines cut the coil, inducing less and less voltage. At 360°, the magnet returns to its starting position, and $E_{AN} = 0$ V again.

One rotation of the shaft creates one cycle of the sine wave, setting its period. So the angular velocity of the generator's shaft sets the frequency. This velocity and the strength of the field and the number of turns of wire in the coils set the signal's amplitude.

Attaching a second coil of wire to the stator (the frame) of the generator allows a second voltage to be produced. The frequency of the two signals will be the same. If the coils are identical, the two voltages also have the same magnitude. However, since the coils are placed mechanically 120° apart, the resulting sine waves are 120° out of phase. Look at Figure 17-2.

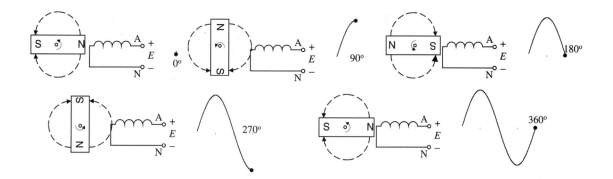

Figure 17-1 Single-phase ac generator

When the rotating magnetic field is at 0°, as shown, no voltage is induced into the A-N coil. However, the B-N coil, −120° behind, is being approached by the south pole. So, E_{BN} is negative. In just another 30° of rotation, the south pole aligns with the B-N coil. This induces a maximum negative voltage of E_{BN}. At that instant, E_{AN} is just beginning to be cut by a little of the north pole. E_{AN} is slightly positive when E_{BN} is at its negative peak. The final result of placing coil B-N at −120° behind coil A-N is to make E_{BN} electrically lag −120° behind E_{AN}.

$$\overline{E_{BN}} = \left[\left|\overline{E_{AN}}\right| \angle (\theta_{AN} - 120°)\right]$$

There is a hole on both the stator and in the symmetry of the voltages. Adding a third coil −120° behind B-N, (+120° ahead of A-N) fills this hole. This is the C-N winding. Look at the result in Figure 17-3.

The phasor diagram, at the bottom of Figure 17-3, shows three voltages, equal in magnitude, but displaced by 120° from each other. Calling the magnitude of the voltage the **phase voltage**, E_ϕ, these three voltages are

$$\overline{E_{AN}} = \left(E_\phi \angle 0°\right)$$

$$\overline{E_{BN}} = \left(E_\phi \angle -120°\right)$$

$$\overline{E_{CN}} = \left(E_\phi \angle 120°\right)$$

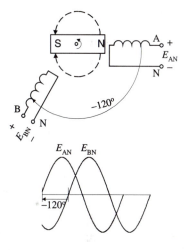

Figure 17-2
Two-phase generation:
B lags A by −120°

**Normal phase sequence
A-B-C-A-B-C- ...**

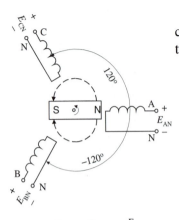

These three sources are often arranged in a **wye (Y)** configuration by connecting their **neutrals (N)** together. This neutral connection is tied to earth near the generator. Look at Figure 17-4.

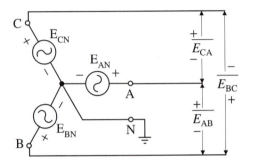

Figure 17-4 Wye connected generator

The signals across each of these generators are called the **phase** voltages. Loads may be connected between each phase and neutral, *or* the loads may be connected *between* the phases, from one line to another, $\overline{E_{AB}}$, $\overline{E_{BC}}$ and $\overline{E_{CA}}$. These are called **line** voltages.

The line voltages may be calculated from the phase voltages by applying Kirchhoff's voltage law. Starting at neutral, sum the voltages around the lower loop. Remember, the voltage at the node designated by the first subscript is positive with respect to the second subscript.

$$\overline{E_{AN}} - \overline{E_{AB}} - \overline{E_{BN}} = 0$$

$$\overline{E_{AB}} = \overline{E_{AN}} - \overline{E_{BN}}$$

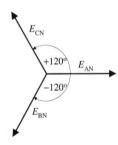

Similarly

$$\overline{E_{BC}} = \overline{E_{BN}} - \overline{E_{CN}}$$

$$\overline{E_{CA}} = \overline{E_{CN}} - \overline{E_{AN}}$$

In general, the magnitude of the line voltage is

Figure 17-3
Three-phase generator

Line voltage magnitude

$$E_{\text{line}} = \sqrt{3} \times E_{\phi}$$

Example 17-1

a. Calculate the phasor version of the three phase voltages and the three line voltages for $E_\phi = 120\ V_{rms}$, sequence A-B-C.

b. Draw the phasor diagram of all six voltages.

Solution

a. $\overline{E_{AN}} = \left(120\,V_{rms}\angle 0°\right)$ $\overline{E_{BN}} = \left(120\,V_{rms}\angle -120°\right)$

$\overline{E_{CN}} = \left(120\,V_{rms}\angle 120°\right)$

$\overline{E_{AB}} = \left(120\,V_{rms}\angle 0°\right) - \left(120\,V_{rms}\angle -120°\right) = \left(208\,V_{rms}\angle 30°\right)$

$\overline{E_{BC}} = \left(120\,V_{rms}\angle -120°\right) - \left(120\,V_{rms}\angle 120°\right) = \left(208\,V_{rms}\angle -90°\right)$

$\overline{E_{CA}} = \left(120\,V_{rms}\angle 120°\right) - \left(120\,V_{rms}\angle 0°\right) = \left(208\,V_{rms}\angle 150°\right)$

b.   The phasors are shown in Figure 17-5. 1 inch = 200 V_{rms}

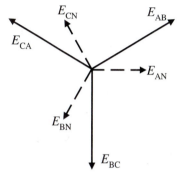

Practice: Calculate phase and line voltages for $E_\phi = 220\ V_{rms}$, $\theta = 45°$.

Answers: (220 $V_{rms}\angle 45°$), (220 $V_{rms}\angle -75°$), (220 $V_{rms}\angle 165°$), (381 $V_{rms}\angle 75°$), (381 $V_{rms}\angle -45°$), (381 $V_{rms}\angle -165°$)

Figure 17-5
Phase and line phasors for Example 17-1

17.2 Wye Connected Loads

Three-phase voltages are connected to three-phase loads. The wye connected load *with* a neutral is shown in Figure 17-6.

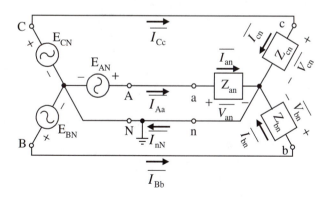

Figure 17-6 Wye connected load with a neutral

The subscript convention is important. The source subscripts are capital letters and the load subscripts are lowercase. For voltages, the second subscript is the reference node for the first. That is, $\overline{V_{an}}$ is the voltage at node **a** with respect to node **n.** Currents are assigned to flow from the first subscript to the second. Current $\overline{I_{an}}$ flows through the load, from node **a** to node **n.**

Each phase voltage is applied *directly* across its phase load.

$$\overline{V}_{an} = \overline{E}_{AN}$$

$$\overline{V}_{bn} = \overline{E}_{BN}$$

$$\overline{V}_{cn} = \overline{E}_{CN}$$

The load phase currents may be calculated with Ohm's law.

$$\overline{I}_{an} = \frac{\overline{V}_{an}}{\overline{Z}_{an}}$$

$$\overline{I}_{bn} = \frac{\overline{V}_{bn}}{\overline{Z}_{bn}}$$

$$\overline{I}_{cn} = \frac{\overline{V}_{cn}}{\overline{Z}_{cn}}$$

Each phase creates a single, simple circuit. All of the currents within that circuit are equal. The same current flows from E_{AN}, through the line Aa, through the load Z_{an}, and returns along the neutral, nN. So, each **line current** is the same as its **phase current**.

$$\overline{I}_{Aa} = \overline{I}_{an}$$

$$\overline{I}_{Bb} = \overline{I}_{bn}$$

$$\overline{I}_{Cc} = \overline{I}_{cn}$$

These phase currents combine and return to the generator along the neutral. From Kirchhoff's current law,

$$\overline{I}_{nN} = \overline{I}_{an} + \overline{I}_{bn} + \overline{I}_{cn}$$

Example 17-2

a. Calculate the line voltages and all of the quantities shown in Figure 17-6 for a 100 hp three-phase motor.

$$E_\phi = 277 \text{ V}_{\text{rms}}, \quad \overline{Z_{an}} = \overline{Z_{bn}} = \overline{Z_{cn}} = (1.9\,\Omega + j1.2\,\Omega)$$

b. Verify your results by simulation.

Solution

a. $\overline{E_{AN}} = (277 \text{ V}_{\text{rms}} \angle 0°)$ $\qquad \overline{E_{BN}} = (277 \text{ V}_{\text{rms}} \angle -120°)$

$\overline{E_{CN}} = (277 \text{ V}_{\text{rms}} \angle 120°)$

$\overline{E_{AB}} = (277 \text{ V}_{\text{rms}} \angle 0°) - (277 \text{ V}_{\text{rms}} \angle -120°) = (480 \text{ V}_{\text{rms}} \angle 30°)$

$\overline{E_{BC}} = (277 \text{ V}_{\text{rms}} \angle -120°) - (277 \text{ V}_{\text{rms}} \angle 120°) = (480 \text{ V}_{\text{rms}} \angle -90°)$

$\overline{E_{CA}} = (277 \text{ V}_{\text{rms}} \angle 120°) - (277 \text{ V}_{\text{rms}} \angle 0°) = (480 \text{ V}_{\text{rms}} \angle 150°)$

$\overline{V_{an}} = \overline{E_{AN}} = (277 \text{ V}_{\text{rms}} \angle 0°)$

$\overline{V_{bn}} = \overline{E_{BN}} = (277 \text{ V}_{\text{rms}} \angle -120°)$

$\overline{V_{cn}} = \overline{E_{CN}} = (277 \text{ V}_{\text{rms}} \angle 120°)$

$\overline{I_{an}} = \dfrac{(277 \text{ V}_{\text{rms}} \angle 0°)}{(1.9\,\Omega + j1.2\,\Omega)} = (123 \text{ A}_{\text{rms}} \angle -32°)$

$\overline{I_{bn}} = \dfrac{(277 \text{ V}_{\text{rms}} \angle -120°)}{(1.9\,\Omega + j1.2\,\Omega)} = (123 \text{ A}_{\text{rms}} \angle -152°)$

$\overline{I_{cn}} = \dfrac{(277 \text{ V}_{\text{rms}} \angle 120°)}{(1.9\,\Omega + j1.2\,\Omega)} = (123 \text{ A}_{\text{rms}} \angle 88°)$

$\overline{I_{Aa}} = \overline{I_{an}} = (123 \text{ A}_{\text{rms}} \angle -32°)$

$\overline{I_{Bb}} = \overline{I_{bn}} = (123 \text{ A}_{\text{rms}} \angle -152°)$

$\overline{I_{Cc}} = \overline{I_{cn}} = (123 \text{ A}_{\text{rms}} \angle 88°)$

$$\overline{I}_{nN} = \overline{I}_{an} + \overline{I}_{bn} + \overline{I}_{cn}$$

$$= \left(123\,\mathrm{A}_{rms}\angle - 32°\right) + \left(123\,\mathrm{A}_{rms}\angle - 152°\right) + \left(123\,\mathrm{A}_{rms}\angle 88°\right)$$

$$\overline{I}_{nN} = 0\,\mathrm{A}_{rms}$$

b. The simulation schematic and edited output file are shown in Figure 17-7.

how big can you dream?™

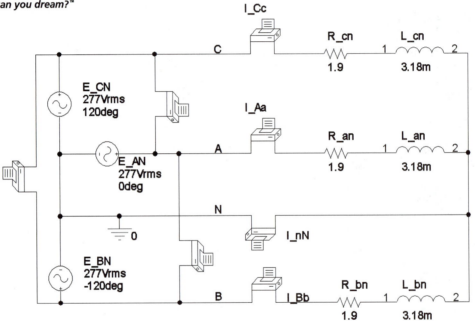

Figure 17-7 (a) Simulation schematic for Example 17-2

```
****    AC ANALYSIS              TEMPERATURE =  27.000 DEG C
*****************************************************************
FREQ        IM(V_I_Aa)  IP(V_I_Aa)           VM(A,B)    VP(A,B)
6.000E+01   1.233E+02   -3.225E+01           4.798E+02  3.000E+01

            IM(V_I_Bb)  IP(V_I_Bb)           VM(B,C)    VP(B,C)
            1.233E+02   -1.523E+02           4.798E+02  -9.000E+01

            IM(V_I_Cc)  IP(V_I_Cc)           VM(C,A)    VP(C,A)
            1.233E+02   8.775E+01            4.798E+02  1.500E+02

            IM(V_I_nN)  IP(V_I_nN)
            7.105E-14   -3.687E+01
```

Figure 17-7 (b) Simulation results for Example 17-2

Only in theory can a load truly be balanced. Manufacturing tolerances assure that three-phase motors present slightly different impedances to each phase. Single-phase power is distributed to homes and businesses by sending E_{AN} to about one-third of the residents, E_{BN} to about one-third, and E_{CN} to the last third. So, an **unbalanced** load, where each phase impedance is different, is normal.

Sending each phase *and* the neutral to the load assures that each phase of the load receives the same voltage, regardless of the phase impedance. Example 17-3 illustrates an unbalanced wye connected load.

Example 17-3

Calculate the line voltages and all of the quantities shown in Figure 17-6 for a residential neighborhood. $E_\phi = 220$ V_{rms},

$$\overline{Z_{an}} = (6.6\Omega + j3.1\Omega) \quad \text{heavy air conditioning load}$$

$$\overline{Z_{bn}} = (4.4\Omega + j0.5\Omega) \quad \text{resistive heating and electric dryers}$$

$$\overline{Z_{cn}} = (15\Omega + j0\Omega) \quad \text{lighting, TV, computers}$$

Solution

$$\overline{E_{AN}} = (220\,V_{rms}\angle 0°) \qquad \overline{E_{BN}} = (220\,V_{rms}\angle -120°)$$

$$\overline{E_{CN}} = (220\,V_{rms}\angle 120°)$$

$$\overline{E_{AB}} = (220\,V_{rms}\angle 0°) - (220\,V_{rms}\angle -120°) = (381\,V_{rms}\angle 30°)$$

$$\overline{E_{BC}} = \left(220\,V_{rms}\angle-120°\right) - \left(220\,V_{rms}\angle120°\right) = \left(381\,V_{rms}\angle-90°\right)$$

$$\overline{E_{CA}} = \left(220\,V_{rms}\angle120°\right) - \left(220\,V_{rms}\angle0°\right) = \left(381\,V_{rms}\angle150°\right)$$

$$\overline{V_{an}} = \overline{E_{AN}} = \left(220\,V_{rms}\angle0°\right) \qquad \overline{V_{bn}} = \overline{E_{BN}} = \left(220\,V_{rms}\angle-120°\right)$$

$$\overline{V_{cn}} = \overline{E_{CN}} = \left(220\,V_{rms}\angle120°\right)$$

$$\overline{I_{an}} = \frac{\left(220\,V_{rms}\angle0°\right)}{\left(6.6\Omega + j3.1\Omega\right)} = \left(30.2\,A_{rms}\angle-25°\right)$$

$$\overline{I_{bn}} = \frac{\left(220\,V_{rms}\angle-120°\right)}{\left(4.4\Omega + j0.5\Omega\right)} = \left(49.7\,A_{rms}\angle-126°\right)$$

$$\overline{I_{cn}} = \frac{\left(220\,V_{rms}\angle120°\right)}{\left(15\Omega + j0\Omega\right)} = \left(14.7\,A_{rms}\angle120°\right)$$

$$\overline{I_{Aa}} = \overline{I_{an}} = \left(30.2\,A_{rms}\angle-25°\right) \qquad \overline{I_{Bb}} = \overline{I_{bn}} = \left(49.7\,A_{rms}\angle-126°\right)$$

$$\overline{I_{Cc}} = \overline{I_{cn}} = \left(14.7\,A_{rms}\angle120°\right)$$

$$\overline{I_{nN}} = \overline{I_{an}} + \overline{I_{bn}} + \overline{I_{cn}}$$

$$= \left(30.2\,A_{rms}\angle-25°\right) + \left(49.7\,A_{rms}\angle-126°\right) + \left(14.7\,A_{rms}\angle120°\right)$$

$$\overline{I_{nN}} = \left(41.3\,A_{rms}\angle-103°\right)$$

17.3 Delta Connected Loads

Wye connected loads are tied between each line and neutral. Loads may also be connected from line to line. This is called a **delta** connection. A delta connection applies a larger magnitude voltage to each leg of the load, and removes the need for a neutral connection. See Figure 17-8.

$$\overline{V_{ab}} = \overline{E_{AB}} = \overline{E_{AN}} - \overline{E_{BN}} \qquad\qquad \overline{V_{bc}} = \overline{E_{BC}} = \overline{E_{BN}} - \overline{E_{CN}}$$

$$\overline{V_{ca}} = \overline{E_{CA}} = \overline{E_{CN}} - \overline{E_{AN}}$$

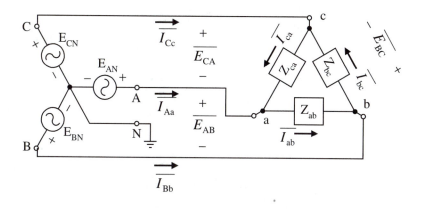

Figure 17-8 Delta connected load

$$\overline{I_{ab}} = \frac{\overline{V_{ab}}}{\overline{Z_{ab}}} \qquad \overline{I_{bc}} = \frac{\overline{V_{bc}}}{\overline{Z_{bc}}} \qquad \overline{I_{ca}} = \frac{\overline{V_{ca}}}{\overline{Z_{ca}}}$$

Line currents are calculated by applying Kirchhoff's current law at the node where the line connects to the load. The current that enters a node must equal the current leaving the node. The subscript convention tells you which way current is assumed to be flowing.

$$\overline{I_{Aa}} + \overline{I_{ca}} = \overline{I_{ab}} \qquad\qquad \overline{I_{Aa}} = \overline{I_{ab}} - \overline{I_{ca}}$$

$$\overline{I_{Bb}} + \overline{I_{ab}} = \overline{I_{bc}} \qquad\qquad \overline{I_{Bb}} = \overline{I_{bc}} - \overline{I_{ab}}$$

$$\overline{I_{Cc}} + \overline{I_{bc}} = \overline{I_{ca}} \qquad\qquad \overline{I_{Cc}} = \overline{I_{ca}} - \overline{I_{bc}}$$

Example 17-4

Calculate all of the quantities shown in Figure 17-8 for a 100 hp three-phase motor. $E_\phi = 277\ V_{rms}$, $\overline{Z} = (1.9\,\Omega + j1.2\,\Omega)$

Solution

$$\overline{V_{ab}} = \overline{E_{AB}} = \overline{E_{AN}} - \overline{E_{BN}}$$

$$\overline{V_{ab}} = \left(277\ V_{rms}\angle 0°\right) - \left(277\ V_{rms}\angle -120°\right) = \left(480\ V_{rms}\angle 30°\right)$$

$$\overline{V_{bc}} = \overline{E_{BC}} = \overline{E_{BN}} - \overline{E_{CN}}$$

$$\overline{V_{bc}} = (277\,\text{V}_{rms}\angle -120°) - (277\,\text{V}_{rms}\angle 120°) = (480\,\text{V}_{rms}\angle -90°)$$

$$\overline{V_{ca}} = (277\,\text{V}_{rms}\angle 120°) - (277\,\text{V}_{rms}\angle 0°) = (480\,\text{V}_{rms}\angle 150°)$$

$$\overline{I_{ab}} = \frac{\overline{V_{ab}}}{Z_{ab}} = \frac{(480\,\text{V}_{rms}\angle 30°)}{(1.9\,\Omega + j1.2\,\Omega)} = (214\,\text{A}_{rms}\angle -2.3°)$$

$$\overline{I_{bc}} = \frac{\overline{V_{bc}}}{Z_{bc}} = \frac{(480\,\text{V}_{rms}\angle -90°)}{(1.9\,\Omega + j1.2\,\Omega)} = (214\,\text{A}_{rms}\angle -122.3°)$$

$$\overline{I_{ca}} = \frac{\overline{V_{ca}}}{Z_{ca}} = \frac{(480\,\text{V}_{rms}\angle 150°)}{(1.9\,\Omega + j1.2\,\Omega)} = (214\,\text{A}_{rms}\angle 117.7°)$$

$$\overline{I_{Aa}} = (214\,\text{A}_{rms}\angle -2.3°) - (214\,\text{A}_{rms}\angle 117.7°) = (370\,\text{A}_{rms}\angle -32°)$$

$$\overline{I_{Bb}} = (214\,\text{A}_{rms}\angle -122.3°) - (214\,\text{A}_{rms}\angle -2.3°) = (370\,\text{A}_{rms}\angle -152°)$$

$$\overline{I_{Cc}} = (214\,\text{A}_{rms}\angle 117.7°) - (214\,\text{A}_{rms}\angle -122.3°) = (370\,\text{A}_{rms}\angle 88°)$$

how big can you dream?™

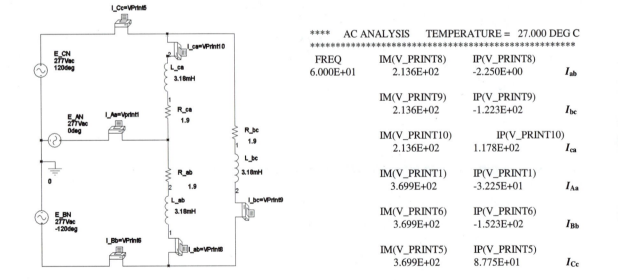

```
****    AC ANALYSIS    TEMPERATURE =  27.000 DEG C
*********************************************************
  FREQ          IM(V_PRINT8)      IP(V_PRINT8)
6.000E+01        2.136E+02        -2.250E+00          I_ab

               IM(V_PRINT9)      IP(V_PRINT9)
                2.136E+02        -1.223E+02          I_bc

               IM(V_PRINT10)     IP(V_PRINT10)
                2.136E+02         1.178E+02          I_ca

               IM(V_PRINT1)      IP(V_PRINT1)
                3.699E+02        -3.225E+01          I_Aa

               IM(V_PRINT6)      IP(V_PRINT6)
                3.699E+02        -1.523E+02          I_Bb

               IM(V_PRINT5)      IP(V_PRINT5)
                3.699E+02         8.775E+01          I_Cc
```

Figure 17-9 Simulation results for Example 17-4

Practice: Calculate the six currents from Figure 17-8, if

$$E_\phi = 120 \text{ V}_{\text{rms}}, \quad \overline{Z_{an}} = \overline{Z_{bn}} = \overline{Z_{cn}} = (22\Omega + j18\Omega)$$

Answer: $\overline{I_{ab}} = (7.32 \text{ A}_{\text{rms}} \angle -9°)$, $\quad \overline{I_{bc}} = (7.32 \text{ A}_{\text{rms}} \angle -129°)$,

$\overline{I_{ca}} = (7.32 \text{ A}_{\text{rms}} \angle -81°)$, $\quad \overline{I_{Aa}} = (12.7 \text{ A}_{\text{rms}} \angle -39°)$,

$\overline{I_{Bb}} = (12.7 \text{ A}_{\text{rms}} \angle -159°)$, $\quad \overline{I_{Cc}} = (12.7 \text{ A}_{\text{rms}} \angle 81°)$

From the results in Example 17-4, you can see that for a *balanced* delta load, the magnitudes of load currents are equal.

$$I_\phi = \left| \overline{I_{ab}} \right| = \left| \overline{I_{bc}} \right| = \left| \overline{I_{ca}} \right|$$

Balanced delta load phase currents

The load currents' angles are 120° apart.

Also, for a *balanced* delta load, the magnitudes of the *line* currents are equal and are related to the load currents.

$$I_{\text{line}} = \sqrt{3} \times I_\phi = \left| \overline{I_{Aa}} \right| = \left| \overline{I_{Bb}} \right| = \left| \overline{I_{Cc}} \right|$$

Balanced delta load line currents

Of course, this symmetry does *not* apply when the loads are unbalanced.

Example 17-5

Calculate the line voltages and all of the quantities shown in Figure 17-8 for $\quad E_\phi = 277 \text{ V}_{\text{rms}}, \quad \overline{Z_{ab}} = (1.9\Omega + j1.2\Omega)$,

$\overline{Z_{bc}} = (0\Omega + j0.8\Omega), \quad \overline{Z_{ca}} = (1.5\Omega + j0\Omega)$

Solution

$$\overline{V_{ab}} = \overline{E_{AB}} = \overline{E_{AN}} - \overline{E_{BN}}$$

$$\overline{V_{ab}} = (277 \text{ V}_{\text{rms}} \angle 0°) - (277 \text{ V}_{\text{rms}} \angle -120°) = (480 \text{ V}_{\text{rms}} \angle 30°)$$

$$\overline{V_{bc}} = \overline{E_{BC}} = \overline{E_{BN}} - \overline{E_{CN}}$$

$$\overline{V_{bc}} = (277 \text{ V}_{\text{rms}} \angle -120°) - (277 \text{ V}_{\text{rms}} \angle 120°) = (480 \text{ V}_{\text{rms}} \angle -90°)$$

$$\overline{V_{ca}} = \overline{E_{CA}} = \overline{E_{CN}} - \overline{E_{AN}}$$

$$\overline{V_{ca}} = (277 \text{ V}_{\text{rms}} \angle 120°) - (277 \text{ V}_{\text{rms}} \angle 0°) = (480 \text{ V}_{\text{rms}} \angle 150°)$$

$$\overline{I_{ab}} = \frac{\overline{V_{ab}}}{\overline{Z_{ab}}} = \frac{(480 \text{ V}_{\text{rms}} \angle 30°)}{(1.9\Omega + j1.2\Omega)} = (214 \text{ A}_{\text{rms}} \angle -2.3°)$$

$$\overline{I_{bc}} = \frac{\overline{V_{bc}}}{\overline{Z_{bc}}} = \frac{(480\,V_{rms}\angle-90°)}{(0\Omega+j0.8\Omega)} = (600\,A_{rms}\angle180°)$$

$$\overline{I_{ca}} = \frac{\overline{V_{ca}}}{\overline{Z_{ca}}} = \frac{(480\,V_{rms}\angle150°)}{(1.5\Omega+j0\Omega)} = (320\,A_{rms}\angle150°)$$

$$\overline{I_{Aa}} = (214\,A_{rms}\angle-2.3°)-(320\,A_{rms}\angle150°) = (519\,A_{rms}\angle-19°)$$

$$\overline{I_{Bb}} = (600\,A_{rms}\angle180°)-(214\,A_{rms}\angle-2.3°) = (814\,A_{rms}\angle179°)$$

$$\overline{I_{Cc}} = (320\,A_{rms}\angle150°)-(600\,A_{rms}\angle180°) = (360\,A_{rms}\angle26°)$$

Practice: Calculate the six currents from Figure 17-8, if $E_\phi = 120\,V_{rms}$, $\overline{Z_{ab}} = (22\Omega+j18\Omega)$, $\overline{Z_{bc}} = (0\Omega+j25\Omega)$, $\overline{Z_{ca}} = (17\Omega+j5\Omega)$

Answer: $\overline{I_{ab}} = (7.32\,A_{rms}\angle-9°)$, $\quad \overline{I_{bc}} = (8.3\,A_{rms}\angle180°)$,

$\overline{I_{ca}} = (11.7\,A_{rms}\angle134°)$, $\quad \overline{I_{Aa}} = (18.1\,A_{rms}\angle-32°)$,

$\overline{I_{Bb}} = (15.5\,A_{rms}\angle176°)$, $\quad \overline{I_{Cc}} = (8.4\,A_{rms}\angle89°)$

Summary

Single-phase ac power may be generated by spinning a magnet past a coil of wire. For each revolution of the magnet a single cycle of the sine wave is generated in the coil. Placing two additional coils on the frame, 120° apart, allows the generation of three-phase ac. Each sine wave lags the one behind it by −120°, and leads the one after it by 120°.

The three generators are called **phase voltages**. Normally their neutrals are connected, and tied to earth, producing E_{AN}, E_{BN}, and E_{CN}. This is the **wye** (Y) configuration. The voltage *between* the phases is called the **line** voltages, E_{AB}, E_{BC}, and E_{CA}. By Kirchhoff's voltage law,

$$\overline{E_{AB}} = \overline{E_{AN}} - \overline{E_{BN}} \qquad \overline{E_{BC}} = \overline{E_{BN}} - \overline{E_{CN}} \qquad \overline{E_{CA}} = \overline{E_{CN}} - \overline{E_{AN}}$$

The line voltage magnitudes are larger than the phase magnitudes

$$E_{line} = \sqrt{3} \times E_\phi$$

Line voltage phases are 120° apart, and 30° ahead of the phase voltage.

Three phase loads may be connected between each phase and neutral, in a wye. The generator's phase voltages are across each phase of the load. Ohm's law may be used to calculate each phase current. The line current is the same as the phase current.

When the loads are all equal, that is, balanced, the magnitudes of these currents are all equal, and their phases are separated by 120°. They add to form the return neutral current. When the loads are balanced, the neutral current is zero, but it may be very significant if the loads are not balanced. Theoretically, for balanced loads, the neutral may be removed without affecting the circuit's performance. However, without a neutral, any mismatch among the load's phases causes the load phase voltages to vary widely, with potentially disastrous results.

The load phases may be connected *between* the generator lines. This is a **delta** connection. Each leg of the load receives a higher voltage, and the neutral is not needed. Ohm's law may be used to calculate each of the load's currents. Kirchhoff's current law is required to calculate the *line* currents.

$$\overline{I_{Aa}} = \overline{I_{ab}} - \overline{I_{ca}} \qquad \overline{I_{Bb}} = \overline{I_{bc}} - \overline{I_{ab}} \qquad \overline{I_{Cc}} = \overline{I_{ca}} - \overline{I_{bc}}$$

For balanced delta loads, the magnitude of the line current is

Problems

Three-phase Generator

17-1 **a.** Calculate the phasor version of the three phase voltages for $E_\phi = 220\ V_{rms}$. The A-N phase is at −60° to a reference.

b. Draw the phasor diagram of the voltages. 1 in. = 100 V_{rms}

17-2 **a.** Calculate the phasor version of the three phase voltages for $E_\phi = 2.7\ kV_{rms}$. The A-N phase is at 45° to a reference.

b. Draw the phasor diagram of the voltages. 1 in. = 1 kV_{rms}

17-3 **a.** Calculate the line voltages for $E_\phi = 220\ V_{rms}$. The A-N phase is at −60° to a reference.

b. Draw the phasor diagram of the line and the phase voltages. 1 in. = 100 V_{rms}

17-4 **a.** Calculate the line voltages for $E_\phi = 2.7\ kV_{rms}$. The A-N phase is at 45° to a reference.

b. Draw the phasor diagram of the line and the phase voltages. 1 in. = 1 kV_{rms}

Wye Loads

17-5 Calculate the line voltages and all of the quantities shown in Figure 17-6 for $E_\phi = 220$ V$_{rms}$, $\overline{Z_{an}} = \overline{Z_{bn}} = \overline{Z_{cn}} = (17\,\Omega + j9\,\Omega)$

17-6 Calculate the line voltages and all of the quantities shown in Figure 17-6 for $E_\phi = 2.7$ kV$_{rms}$, $\overline{Z_{an}} = \overline{Z_{bn}} = \overline{Z_{cn}} = (53\,\Omega + j42\,\Omega)$

17-7 Calculate the line voltages and all of the quantities shown in Figure 17-6 for $E_\phi = 220$ V$_{rms}$, $\overline{Z_{an}} = (17\,\Omega + j9\,\Omega)$, $\overline{Z_{bn}} = (2\,\Omega + j12\,\Omega)$, $\overline{Z_{cn}} = (18\,\Omega + j0\,\Omega)$

17-8 Calculate the line voltages and all of the quantities shown in Figure 17-6 for $E_\phi = 2.7$ kV$_{rms}$, $\overline{Z_{an}} = (53\,\Omega + j42\,\Omega)$, $\overline{Z_{bn}} = (53\,\Omega + j0\,\Omega)$, $\overline{Z_{cn}} = (0\,\Omega + j42\,\Omega)$

Delta Loads

17-9 Calculate the line voltages and all of the quantities shown in Figure 17-8 for $E_\phi = 220$ V$_{rms}$, $\overline{Z_{ab}} = \overline{Z_{bc}} = \overline{Z_{ca}} = (17\,\Omega + j9\,\Omega)$

17-10 Calculate the line voltages and all of the quantities shown in Figure 17-8 for $E_\phi = 2.7$ kV$_{rms}$, $\overline{Z_{ab}} = \overline{Z_{bc}} = \overline{Z_{ca}} = (53\,\Omega + j42\,\Omega)$

17-11 Calculate the line voltages and all of the quantities shown in Figure 17-8 for $E_\phi = 220$ V$_{rms}$, $\overline{Z_{ab}} = (17\,\Omega + j9\,\Omega)$, $\overline{Z_{bc}} = (2\,\Omega + j12\,\Omega)$, $\overline{Z_{ca}} = (18\,\Omega + j0\,\Omega)$

17-12 Calculate the line voltages and all of the quantities shown in Figure 17-8 for $E_\phi = 2.7$ kV$_{rms}$, $\overline{Z_{ab}} = (53\,\Omega + j42\,\Omega)$, $\overline{Z_{bc}} = (53\,\Omega + j0\,\Omega)$, $\overline{Z_{ca}} = (0\,\Omega + j42\,\Omega)$

Three-phase Lab Exercise

A. Three-phase Generator

The three-phase generator used in lab is shown in Figure 17-10

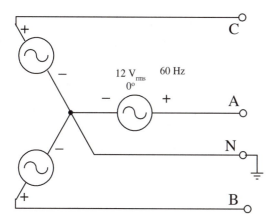

Figure 17-10 Three-phase generator

Each phase has been reduced by a step-down transformer for your safety. As with most electrical power generators, the neutral has been connected to earth at the generator. Your oscilloscope's common and chassis ground are connected to earth through its power cord. So, your oscilloscope is measuring with respect to neutral, even without connecting its common to neutral. You can *only* use the oscilloscope to make measurements *with respect to neutral*.

1. Measure and record the *phase* voltages. Measure the magnitude with your voltmeter. Place channel 1 of the oscilloscope on E_{AN}. It will be used as the phase reference, $0°$. Tabulate each and compare it with the theoretical magnitude and angle.

2. Measure and record the magnitude of the *line* voltages. You can only measure phase with respect to neutral. So, do *not* try to measure the angle. Tabulate each and compare it with the theoretical magnitude.

B. Balanced Wye Connection
1. Add the loads shown in Figure 17-11.

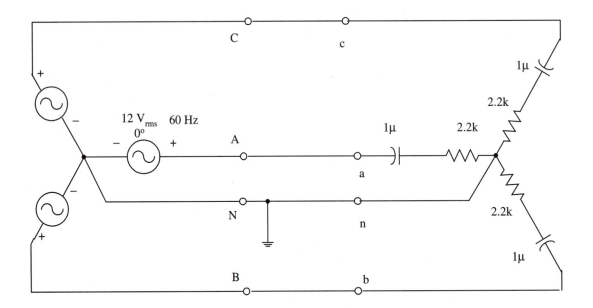

Figure 17-11 Balanced wye load

2. Measure the voltage across each 2.2 kΩ resistor. From each measured voltage, calculate the current through that branch.

3. Tabulate each and compare it with the theoretical magnitude and angle of each phase current.

4. Remove the connection between the load's neutral and the generator's neutral.

5. Measure and record the voltage at the load's neutral with respect to the generator's neutral using the digital multimeter.

6. With the neutral still removed, repeat the measurements and comparisons from steps B2 and 3 above.

C. Unbalanced Wye Connection

1. Reconnect the neutral from the generator to the load.

2. Alter the loads:

    Z_{an} is a 1 µF capacitor in series with a 47 Ω resistor,
    Z_{bn} is a 2.2 kΩ resistor only,
    Z_{cn} is a 1 µF capacitor in series with a 2.2 kΩ resistor.

3. Measure the voltage across each resistor. From each measured voltage, calculate the current through that branch.

4. Measure the neutral current with an ac ammeter.

5. Tabulate each and compare it with the theoretical magnitude and angle of each phase current.

6. Remove the connection between the load's neutral and the generator's neutral.

7. Measure and record the voltage at the load's neutral with respect to the generator's neutral using the digital multimeter.

8. With the neutral still removed, repeat the measurements and comparisons from steps C3 and 5 above (but *not* the neutral current). You will have to complete a mesh analysis to determine the theoretical load phase currents.

D. Balanced Delta Connection

1. Rearrange the load as indicated in Figure 17-12.

2. Measure the magnitude of the six load and line currents.

3. Tabulate each and compare it with the theoretical magnitude of each current.

E. Unbalanced Delta Connection

1. Alter the loads:

    Z_{ab} is a 1 µF capacitor in series with a 47 Ω resistor,
    Z_{bc} is a 1 µF capacitor in series with a 2.2 kΩ resistor,
    Z_{ca} is only a 2.2 kΩ resistor.

2. Measure the magnitude of the six load and line currents.

3. Tabulate each and compare it with the theoretical magnitude of each current.

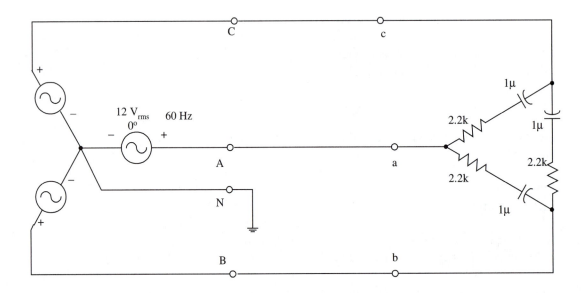

Figure 17-12 Balanced delta load

Appendix A

Answers to Odd-Numbered Problems

Chapter 1

1-1   **a.** 0.866   **b.** −0.707   **c.** 0.707   **d.** −0.5

1-3   **a.** 30°   **b.** −135°

1-5   **a.** 2.36 rad   **b.** 2.79 rad   **c.** −0.7 rad   **d.** 115°   **e.** 131°   **f.** −74°

1-7 $12\ \text{A}_\text{p}*\sin(45°) = 8.485\ \text{A}$

1-9 $207.9\ \text{V}_\text{rms}$

1-11   16.67 ms, 377 rad/sec

1-13   **a.** $28.28\ \text{V}_\text{rms}$ **b.** 2 kHz   **c.** 500 μs   **d.** 12.57 krad/sec   **e.** 23.51 V

1-15   20.11 μs

1-17   **a.** 83.33 μs   **b.** Lag

1-19   omitted

1-21   **a.** omitted   **b.** 167 μs

1-23   omitted

1-25   omitted

Chapter 2

2-1   **a.** $(219.91\ \text{V}_\text{rms}\ \angle 0°)$ **b.** $(38.89\ \text{V}_\text{rms}\ \angle{-}36°)$ **c.** $(247\ \text{mV}_\text{rms}\ \angle 90°)$ **d.** $(5.09\ \text{A}_\text{rms}\ \angle{-}150°)$ **e.** $(12.02\ \mu\text{A}_\text{rms}\ \angle{-}120°)$

2-3   **a.** $11.74\ \text{V}_\text{p}*\sin(4.4\ \text{k}*t + 32°)$ **b.** $169.71\ \text{V}_\text{p}*\sin(377*t - 45°)$ **c.** $2.12\ \text{A}_\text{p}*\sin(3.39\ \text{M}*t + 135°)$ **d.** $45.25\ \mu\text{A}_\text{p}*\sin(664\ \text{M}*t - 100°)$

2-5   omitted

2-7   omitted

2-9   **a.** $(7.972\ \text{V}_\text{rms} + \text{j}\ 4.981\ \text{V}_\text{rms})$ **b.** $(77.78\ \text{V}_\text{rms} - \text{j}\ 77.78\ \text{V}_\text{rms})$ **c.** $(-1.556\ \text{A}_\text{rms} + \text{j}\ 1.556\ \text{A}_\text{rms})$ **d.** $(-7.814\ \mu\text{A}_\text{rms} - \text{j}\ 44.32\ \mu\text{A}_\text{rms})$

2-11   **a.** $(13.58\ \text{V}_\text{rms}\ \angle{-}11°)$ **b.** $(2.448\ \text{A}_\text{rms}\ \angle 67.44°)$ **c.** $(4.902\ \text{V}_\text{rms}\ \angle{-}159.7°)$ **d.** $(2.437\ \text{A}_\text{rms}\ \angle 6.038°)$

Chapter 3

3-1   omitted

3-3 $-2\pi*V_\text{p}*f*\sin(2\pi*f*t)$

3-5   **a.** $(16\ \Omega\ \angle 0°)$, $(16\ \Omega + \text{j}\ 0\ \Omega)$ **b.** omitted

3-7   **a.** $(16\ \Omega\ \angle 90°)$, $(0\ \Omega + \text{j}\ 16\ \Omega)$ **b.** omitted

3-9   **a.** $(16\ \Omega\ \angle{-}90°)$, $(0\ \Omega - \text{j}\ 16\ \Omega)$ **b.** omitted

3-11   **a.** $(426\ \Omega\ \angle 90°)$, $(0\ \Omega + \text{j}\ 426\ \Omega)$ **b.** omitted

3-13   **a.** $(11.4\ \text{k}\Omega\ \angle{-}89.9°)$, $(22.6\ \Omega - \text{j}\ 11.4\ \text{k}\Omega)$ **b.** omitted   **c.** omitted

3-15   **a.** $(191.7\ \Omega\ \angle 41.4°)$ $(143.8\ \Omega + \text{j}\ 126.8\ \Omega)$ **b.** omitted

Chapter 4

4-1   **a.** $(2\ \text{A}_\text{rms}\ \angle{-}48°)$ **b.** $(16.125\ \text{V}_\text{rms}\ \angle{-}55.1°)$ **c.** $(8.062\ \Omega\ \angle{-}7.1°)$

4-3   omitted

4-5   **a.** omitted   **b.** 5.305 Ω   **c.** $Z_\text{R} = (8\ \Omega\ \angle 0°)$, $Z_\text{C} = (5.305\ \Omega\ \angle{-}90°)$ **d.** $(9.599\ \Omega\ \angle{-}33.55°)$ **e.** omitted   **f.** $I_\text{R} = I_\text{C} = (4.167\ \text{A}_\text{rms}\ \angle 33.55°)$ **g.** $V_\text{R} = (33.34\ \text{V}_\text{rms}\ \angle 33.55°)$

$V_C = (22.11 \text{ V}_{rms} \angle -56.45°)$
h. omitted,   **i.** omitted

4-7 **a.** Figure 4-7
b. $R = 4 \, \Omega, \ L = 1.27 \text{ mH}$
c. $V_{R \ 100 \text{ Hz}} = (29.4 \text{ V}_{rms} \angle -11°)$
$V_{R1: \ 500 \text{ Hz}} = (21.2 \text{ V}_{rms} \angle -45°)$
$V_{R1: \ 1000 \text{ Hz}} = (13.4 \text{ V}_{rms} \angle -63°)$

4-9 $V_a = 2.5 \text{ V}_{dc}, (81 \text{ mV}_{rms} \angle 36°)$
$V_b = V_c = 2.5 \text{ V}_{dc}, (0 \text{ V}_{rms} \angle 0°)$
$V_d = 2.5 \text{ V}_{dc}, (1.215 \text{ V}_{rms} \angle -144°)$
$V_e = 0 \text{ V}_{dc,} (1.187 \text{ V}_{rms} \angle -132°)$

4-11 **a.** omitted,
b. $X_C = 0.91 \, \Omega, X_L = 5.09 \, \Omega$
c. $Z_R = (8 \, \Omega \angle 0°), Z_C = (0.91 \Omega \angle -90°)$
$Z_L = (5.09 \, \Omega \angle 90°)$
d. $Z_{total} = (9.03 \, \Omega \angle 27.6°)$
e. omitted   **f.** $I = (4.43 \text{ A}_{rms} \angle -27.6°)$
g. $V_R = (35.45 \text{ V}_{rms} \angle -27.6°)$
$V_C = (4.03 \text{ V}_{rms} \angle -117.6°)$
$V_L = (22.55 \text{ V}_{rms} \angle 62.4°)$
h. omitted

4-13 **a.** omitted
b. $X_c = 0.343 \, \Omega$
$X_L = 13.57 \, \Omega$
c. $Z_R = (8 \, \Omega \angle 0°), Z_L = (13.6 \, \Omega \angle 90°)$
$Z_C = (0 \, \Omega - j \, 0.343 \, \Omega)$
d. $(8 \, \Omega + j \, 13.23 \, \Omega)$
e. omitted
f. $I = (2.587 \text{ A}_{rms} \angle -58.8°)$
g. $V_R = (20.70 \text{ V}_{rms} \angle -58.8°)$
$V_L = (35.12 \text{ V}_{rms} \angle 31.2°)$
$V_C = (0.887 \text{ V}_{rms} \angle -148.8°)$
h. omitted

4-15 500 Hz
$V_a = V_b = 2.5 \text{ V}_{dc}, \ (0 \text{ V}_{rms} \angle 0°)$
$V_c = 2.5 \text{ V}_{dc}$
$(284.9 \text{ mV}_{rms} \angle -103.8°)$
$I_a = I_b = 0 \text{ A}_{dc}$
$(5.088 \text{ μA}_{rms} \angle 76.16°)$

1500 Hz
$V_a = V_b = 2.5 \text{ V}_{dc}, \ (0 \text{ V}_{rms} \angle 0°)$

$V_c = 2.5 \text{ V}_{dc}$
$(1.18 \text{ V}_{rms} \angle -172°)$
$I_a = I_b = 0 \text{ A}_{dc}$
$(21.07 \text{ μA}_{rms} \angle 8.0°)$

4000 Hz
$V_a = V_b = 2.5 \text{ V}_{dc}, \ (0 \text{ V}_{rms} \angle 0°)$
$V_c = 2.5 \text{ V}_{dc}$
$(368.8 \text{ mV}_{rms} \angle 108°)$
$I_a = I_b = 0 \text{ A}_{dc}$
$(6.586 \text{ μA}_{rms} \angle -72.0°)$

4-17 $(22 \text{ mV}_{rms} \angle -91°)$

Chapter 5
5-1 **a.** $(220 \text{ V}_{rms} \angle 60°)$
b. $I_{Z1} = (0.9 \text{ A}_{rms} \angle 60°)$
$I_{Z2} = (6 \text{ A}_{rms} \angle 150°)$
$I_{Z3} = (4.889 \text{ A}_{rms} \angle 90°)$
$I_{Z4} = (5.5 \text{ A}_{rms} \angle 0°)$
c. $I_1 = (3 \text{ A}_{rms} \angle 120°)$
$I_2 = (6.475 \text{ A}_{rms} \angle 69.6°)$
$I_T = (8.701 \text{ A}_{rms} \angle 85.0°)$
d. omitted
e. $(25.28 \, \Omega \ a \angle -25.0°)$
f. $(0.03955 \text{ S} \angle 25.0°)$
g. $Y_1 = (4.091 \text{ mS} \angle 0°)$
$Y_2 = (27.27 \text{ mS} \angle 90°)$
$Y_3 = (22.22 \text{ mS} \angle 30°)$
$Y_4 = (25 \text{ mS} \angle -60°)$
h. omitted
i. omitted

5-3 **a.** omitted
b. $X_c = 8.842 \, \Omega$
c. $Z_C = (8.842 \, \Omega \angle -90°)$
$Z_R = (5 \, \Omega \angle 0°)$
d. $Y_C = (113.1 \text{ mS} \angle 90°)$
$Y_R = (200 \text{ mS} \angle 0°)$
e. $Y_T = (229.8 \text{ mS} \angle 29.5°)$
f. $Z_T = (4.352 \, \Omega \angle -29.49°)$
g. omitted
h. $I_C = (1.425 \text{ A}_{rms} \angle 90°)$
$I_R = (2.52 \text{ A}_{rms} \angle 0°)$
i. omitted

5-5 **a.** omitted

b. $Z_{\text{Lighting}} = (48\ \Omega\ \angle\ 0°)$
$Z_{\text{Refrig}} = (60\ \Omega\ \angle\ 53.1°)$
$Z_{\text{Fan}} = (92.31\ \Omega\ \angle\ 60°)$
$Y_{\text{Lighting}} = (20.83\ \text{mS}\ \angle\ 0°)$
$Y_{\text{Refrig}} = (16.67\ \text{mS}\ \angle\ -53.1°)$
$Y_{\text{Fan}} = (10.83\ \text{mS}\ \angle\ -60°)$
c. $Y_T = (42.78\ \text{mS}\ \angle\ -32.1°)$
d. omitted
e. $I_{\text{Lighting}} = (2.5\ \text{A}_{\text{rms}}\ \angle\ 0°)$
$I_{\text{Refrig}} = (2\ \text{A}_{\text{rms}}\ \angle\ -53.1°)$
$I_{\text{Fan}} = (1.3\ \text{A}_{\text{rms}}\ \angle\ -60°)$
f. $I_T = (5.133\ \text{A}_{\text{rms}}\ \angle\ -32.1°)$
g. omitted

5-7 **a.** $V_{\text{IN}} = (10\ \text{mV}_{\text{rms}}\ \angle\ 0°)$
b. $I_{\text{Ri}} = (10\ \mu\text{A}_{\text{rms}}\ \angle\ 0°)$
c. $I_{\text{Rf // Cf}} = (10\ \mu\text{A}_{\text{rms}}\ \angle\ 0°)$
d. $X_C = 48.23\ \text{k}\Omega$
e. $Y_C = (20.73\ \mu\text{S}\ \angle\ 90°)$
f. $Z_{\text{Rf // Cf}} = (33.66\ \text{k}\Omega\ \angle\ -44.3°)$
g. $V_{\text{Rf // Cf}} = (336.6\ \text{mV}_{\text{rms}}\ \angle\ -44.3°)$
h. $V_{\text{out}} = (343.8\ \text{mV}_{\text{rms}}\ \angle\ -43.1°)$

5-9 $I_{\text{total}} = (198.5\ \text{A}_{\text{rms}}\ \angle\ -24.9°)$

5-11 **a.** $V_T = (1.139\ \text{V}_{\text{rms}}\ \angle\ 14.3°)$
b. $I_R = (242.3\ \mu\text{A}_{\text{rms}}\ \angle\ 14.3°)$
$I_C = (330.6\ \mu\text{A}_{\text{rms}}\ \angle\ 104.3°)$
$I_L = (392.5\ \mu\text{A}_{\text{rms}}\ \angle\ -75.7°)$
c. omitted

Chapter 6
6-1 **a.**

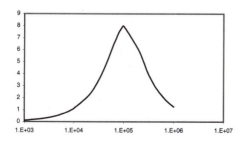

b. Band-pass
c. omitted

6-3

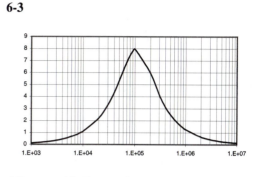

6-5 **a.** −40 dB **b.** −3.012 dB

6-7 $P_{\text{load}} = 1.902\ \text{mW}$

6-9 $P_{\text{in}} = 351.7\ \text{W}$

6-11 **a.** 16.99 dBm **b.** 13.98 dBW
c. −6.90 dBu **d.** 17.27 dBV

6-13 **a.**

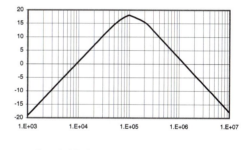

b. 18.02 dB

6-15 **a.** −60 dB/decade, **b.** 12 dB/octave
c. −120 dB/decade **d.** 30 dB/octave

6-17 $f_{-3\text{dB}} \sim 50\ \text{kHz and} \sim 210\ \text{kHz}$

6-19 $A_{@\,f-3\text{dB}} = 16.3$

Chapter 7
7-1 $A_o = 1$, Filter Order = First Order,
Roll off rate = −20 dB/dec or −6dB/oct
$f_{-3\text{dB}} = 2.842\ \text{kHz}$, $f_o = 2.842\ \text{kHz}$
$A_{f-3\text{dB}} = 0.707$, $\theta_{@\,fo} = -45°$

7-3 $R = 10\ \text{k}\Omega$, $C = 57\ \text{nF}$

7-5 **a.** $V_{\text{out}\,f=0} = E_{\text{in}}$, $V_{\text{out}\,f=\infty} = E_{\text{in}}/5$

b. mag. $= \dfrac{\sqrt{R^2 + X_C^2}}{\sqrt{25R^2 + X_C^2}}$

phase $= \tan^{-1}\left(\dfrac{X_C}{5R}\right) - \tan^{-1}\left(\dfrac{X_C}{R}\right)$

c. $f_{-3dB} = \dfrac{1}{2\pi\sqrt{23}\,RC}$

7-7 $A_o = 1$,   Filter Order = First Order
Roll off rate = -20 dB/dec & -6dB/oct
$f_{-3dB} = f_o = 106.1$ MHz
$A_{f-3dB} = 0.707$, $\theta_{fo} = 45°$

7-9 $L = 14.7$ μH, $R = 50\ \Omega$

7-11 $A_o = -22$,   Filter Order = First Order
Roll off rate = -20 dB/dec & -6dB/oct
$f_{-3dB} = 4.823$ kHz, $f_o = 4.823$ kHz
$A_{f-3dB} = -15.56$, $\theta_{fo} = 135°$

7-13 $R_i = 1$ kΩ, $R_f = 33$ kΩ, $C_f = 3.2$ nF

7-15 **a.** $V_{out\,f=0} = -\infty\ \ V_{out\,f=\infty} = 0$

  b. mag. $= \dfrac{X_C}{R}$,   phase $= 90°$

7-17 $n = 6$, $R = 50\ \Omega$, L1 = 41 nH,
C1 = 48 pF, L2 = 0.19 μH, C2 = 99 pF,
L3 = 0.28 μH, C3 = 98 μH,

Chapter 8
8-1 $A_o = 1$,   Filter Order = First Order
Roll off = 20 dB/dec or 6dB/oct
$f_{-3dB} = f_o = 2.842$ kHz
$A_{f-3dB} = 0.707$, $\theta_{fo} = 45°$

8-3 $C = 57$ nF, $R = 10$ kΩ

8-5 **a.** $V_{out\,f=0} = 0\ \ V_{out\,f=\infty} = \dfrac{4E_{in}}{5}$

  b. mag. $= \dfrac{4R}{\sqrt{25R^2 + X_C^2}}$

  phase $= \tan^{-1}\left(\dfrac{X_C}{5R}\right)$

c & d. $f_{-3dB} = f_o = \dfrac{1}{2\pi\,5RC}$

8-7 $A_o = 1$,   Filter Order = First Order
Roll off rate = 20 dB/dec or 6 dB/oct
$f_{-3dB} = f_o = 106.1$ MHz
$A_{f-3dB} = 0.707$, $\theta_{fo} = 45°$

8-9 $L = 14.7$ μH, $R = 50\ \Omega$

8-11 $A_o = -22$,   Filter Order = First Order
Roll off rate = 20 dB/dec or 6dB/oct
$f_{-3dB} = f_o = 106$ kHz
$A_{f-3dB} = -15.56$, $\theta_{fo} = 45°$

8-13 $R_i = 1$ kΩ, $C_i = 106$ nF, $R_f = 33$ kΩ

8-15 **a.** $V_{out\,f=0} = 0\ \ V_{out\,f=\infty} = \infty$

  b. mag. $= \dfrac{R}{X_C}$,   phase $= -90°$

8-17 $n = 6$, $R = 50\ \Omega$, C1 = 6.2 nF,
L1 = 5.2 μH, C2 = 1.3 nF, L2 = 2.6 μH,
C3 = 0.9 nF, L3 = 2.6 μH

Chapter 9
9-1 d.

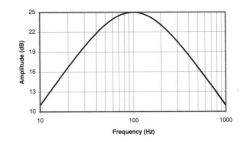

  e. Omitted
  f. $f_{center} = 100$ Hz, $A_o = 17.8$
    $f_{low} = 40$ Hz, $f_{high} = 245$ Hz
    $\Delta f = 205$ Hz, $Q = 0.488$

9-3 $A_o = -4.7$, $f_{hi} = 3.386$ kHz
$f_{low} = 159$ Hz, $Q = 0.227$

9-5 $R_i = 10$ kΩ, $R_f = 330$ kΩ,
$C_i = 31.8$ nF, $C_f = 40.2$ pF

9-7 Omitted

9-9

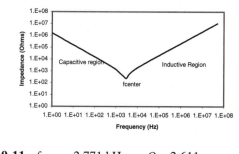

9-11 f_{center} = 2.771 kHz, Q = 2.611

9-13 L = 23.9 µH, C = 10.6 pF, R = 75 Ω

9-15 f_{center} = 2.771 kHz, Q = 2.13

9-17

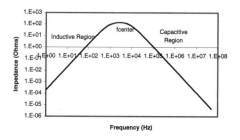

9-19 f_{center} = 2.771 kHz, Q = 0.383

9-21 R = 75 Ω, L = 59.7 nH, C = 4.24 nF

9-23 f_{center} = 2.771 kHz, Q = 0.37

Chapter 10

10-1 R_f = 220 kΩ, C_f = 2.41 nF

10-3 A_o = −14.24, f_{-3dB} = 1.539 kHz, A_{-3dB} = 10.07

10-5 R_i = 2.136 kΩ, C_i = 248 nF

10-7 A_o = −14.2, f_{-3dB} = 21.92 kHz A_{-3dB} = 10.0

10-9 omitted

10-11 A_o = −14.24, f_{-3dB} = 67.2 kHz $f_{10\%}$ = 33.6 kHz, $f_{5\%}$ = 22.4 kHz, $f_{1\%}$ = 9.61 kHz

10-13 GBW = 5.1 MHz

10-15 omitted

10-17 GBW = 5 MHz

10-19 omitted

10-21 t_{rise} = 3.252 µs

10-23 v = 49.2 mV

10-25 SR = 0.16 V/µs

10-27 f_{max} = 7.83 kHz

10-29 $V_{p\,max}$ = 3.91 V_p

10-31 SR = 1.885 V/µs

Chapter 11

11-1 omitted

11-3 3.18 V_{dc} + 5 V_p*sin(377t) + 2.12 V_p*sin(754 t–90°)

11-5 2 V_{dc} + 2.60 V_p sin(9425t +17.7°) + 0.733 V_p sin(4712t − 71.9°)

Chapter 12

12-1 8.602*sin(6283t + 54.5°)

12-3 3.500 V_{rms}

12-5 **a.** 6 V_{dc} + 7.639 V_p sin(942 kt) + 2.546 V_p sin(3 x 942 kt) + 1.528 V_p sin(5 x 942 kt) + 1.091 V_p sin(7 x 942 kt) + 0.849 V_p sin(9 x 942 kt)

b.

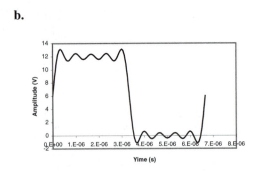

b.

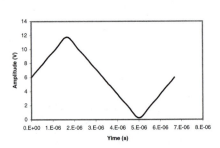

12-7 a. 3 V$_{dc}$ +
5.402 V$_p$ sin(942 kt +90°) +
3.820 V$_p$ sin(2 × 942 kt +90°) +
1.801 V$_p$ sin(3 × 942 kt + 90°) +
1.080 V$_p$ sin(5 × 942 kt – 90°) +
1.273 V$_p$ sin(6 × 942 kt –90°) +
0.772 V$_p$ sin(7 × 942 kt – 90°) +
0.600 V$_p$ sin(9 × 942 kt + 90°)

b.

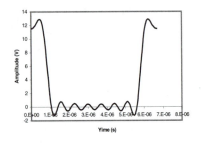

12-9 a. 6 V$_{dc}$
4.863 V$_p$ sin (942 kt)
0.540 V$_p$ sin(3 × 942 kt + 180°)
0.195 V$_p$ sin(5 × 942 kt)
0.099 V$_p$ sin (7 × 942 kt + 180°)
0.060 V$_p$ sin(9 × 942 kt)

12-11 a. 6 V$_{dc}$ +
3.820 V$_p$ sin(942 kt) +
1.910 V$_p$ sin(2 × 942 kt + 180°) +
1.273 V$_p$ sin(3 × 942 kt) +
0.955 V$_p$ sin(4 × 942 kt + 180°) +
0.764 V$_p$ sin(5 × 942 kt) +
0.637 V$_p$ sin(6 × 942 kt + 180°) +
0.546 V$_p$ sin(7 × 942 kt) +
0.478 V$_p$ sin(8 × 942 kt + 180°) +
0.424 V$_p$ sin(9 × 942 kt)

b.

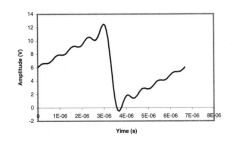

12-13 a. 3.820 V$_{dc}$ +
6.000 V$_p$ sin(942 kt) +
2.546 V$_p$ sin(2 × 942 kt – 90°) +
0.509 V$_p$ sin(4 × 942 kt – 90°) +
0.218 V$_p$ sin(6 × 942 kt – 90°) +
0.121 V$_p$ sin(8 × 942 kt – 90°)

b.

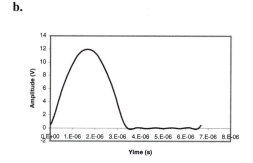

12-15 a. 7.639 V_{dc} +
  5.093 V_p sin(2 × 942 kt − 90°) +
  1.019 V_p sin(4 × 942 kt − 90°) +
  0.437 V_p sin(6 × 942 kt − 90°) +
  0.242 V_p sin(8 × 942 kt − 90°)
b.

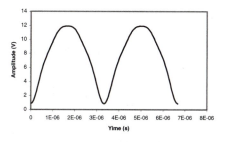

Chapter 13
13-1 6 V_{dc} +
  4.368 V_p sin (12.57 kt −26.1°) +
  0.303 V_p sin(3 × 12.57 kt + 124 2°) +
  0.073 V_p sin(5 × 12.57 kt −67.8°) +
  0.028 V_p sin(7 × 12.57 kt +106.3°) +
  0.013 V_p sin(9 × 12.57 kt − 77.3°)

13-3 5.093 V_{dc} +
  2.371 V_p sin (377t −72.8°) +
  0.516 V_p sin(2 × 377t − 171.3°) +
  0.052 V_p sin(4 × 377t −175.6°) +
  0.015 V_p sin(6 × 377t − 177.1°) +
  0.006 V_p sin(8 × 377t − 177.8°)

13-5 0 V_{dc} +
  9.928 V_p sin (12.57 kt + 90°) +
  0.368 V_p sin(3 × 12.57 kt − 90°) +
  0.079 V_p sin(5 × 12.57 kt + 90°) +
  0.029 V_p sin(7 × 12.57 kt − 90°) +
  0.014 V_p sin(9 × 12.57 kt + 90°)

13-7 0 V_{dc} +
  18.412 V_p sin (377t + 90°) +
  2.046 V_p sin(3 × 377t +90°) +
  0.738 V_p sin(5 × 377t + 90°) +
  0.378 V_p sin(7 × 377t + 90°) +
  0.228 V_p sin(9 × 377t + 90°)

13-9 0 V_{dc} +
  4.794 V_p sin (12.57 kt + 9.6°) +
  0.540 V_p sin(3 × 12.57 kt − 176.8°) +
  0.192 V_p sin(5 × 12.57 kt + 1.9°) +
  0.096 V_p sin(7 × 12.57 kt − 178.6°) +
  0.060 V_p sin(9 × 12.57 kt + 1.1°)

13-11 0 V_{dc} +
  0.787 V_p sin (12.57 kt − 90.5°) +
  0.262 V_p sin(3 × 12.57 kt + 88.4°) +
  0.152 V_p sin(5 × 12.57 kt − 92.6°) +
  0.113 V_p sin(7 × 12.57 kt + 86.3°) +
  0.087 V_p sin(9 × 12.57 kt − 94.6°)

13-13 4 V_{dc} +
  7.596 V_p sin (3.14 kt + 2.7°) +
  2.544 V_p sin(3 × 3.14 kt + 0.9°) +
  1.528 V_p sin(5 × 3.14 kt + 0.5°) +
  1.091 V_p sin(7 × 3.14 kt + 0.4°) +
  0.848 V_p sin(9 × 3.14 kt + 0.3°)

Chapter 14

14-1 **a.** R = 2.6 Ω, C = 1.69 nF, L = 9.506 μH
   b. V_{out} = (0.630 mV$_{rms}$ ∠ −177.1°)

14-3 V_a = (745.7 mV$_{rms}$ ∠ −10.5°)
   V_{in} = (78.2 mV$_{rms}$ ∠ −9.8°)

14-5 V_a = (717.4 mV$_{rms}$ ∠ −9.1°)
   V_{in} = (64.2 mV$_{rms}$ ∠ −10.3°)

Chapter 15

15-1 $(6.673 \text{ V}_{rms} \angle 0.7°)$

15-3 $V_{9\,\Omega} = (6.140 \text{ V}_{rms} \angle 149.5°)$
$V_{13\,\Omega} = (7.676 \text{ V}_{rms} \angle -8.9°)$
$V_{11\,\Omega} = (2.805 \text{ V}_{rms} \angle 1°)$
$V_{8\,\Omega} = (3.486 \text{ V}_{rms} \angle 74.6°)$
$V_{6\,\Omega} = (1.224 \text{ V}_{rms} \angle 144°)$
$V_{16\,\Omega} = (3.264 \text{ V}_{rms} \angle 54°)$

15-5 $V_{53\,\Omega} = (5.030 \text{ V}_{rms} \angle -30.9°)$
$V_{27\,\Omega} = (5.625 \text{ V}_{rms} \angle -32.9°)$

15-7 $V = (787.3 \text{ mV}_{rms} \angle 65.6°)$

Chapter 16

16-1 omitted

16-3 a. $V_R = V_C = V_L = (120 \text{ V}_{rms} \angle 0°)$
$I_R = (15 \text{ A}_{rms} \angle 0°)$
$I_C = (24 \text{ A}_{rms} \angle 90°)$
$I_L = (17.1 \text{ A}_{rms} \angle -90°)$
$P_R = 1800 \text{ W} \quad Q_r = 0$
$P_C = 0. \quad Q_c = 2880 \text{ VAR}_c$
$P_L = 0, \quad Q_l = .2047 \text{ VAR}_l$
b. $I_{total} = (16.5 \text{ A}_{rms} \angle 24.7°)$
$P_{total} = 1800 \text{ W}$
$Q_{total} = -833 \text{ VAR}$
$S_{app} = (1983 \text{ VA} \angle -24.8°)$

16-5 $S = (2640 \text{ VA} \angle 30°)$
$Q = 1320 \text{ VAR}$
$P = 2286.31 \text{ W}$

16-7 $PF = 0.908 \quad Q_{cor} = 833 \text{ VAR}_L$

16-9 $PF = 0.52, \quad Q_{cor} = 1904 \text{ VAR}$

Chapter 17

17-1 a. $E_{AN} = (220 \text{ V}_{rms=} \angle -60°)$
$E_{BN} = (220 \text{ V}_{rms} \angle -180°)$
$E_{CN} = (220 \text{ V}_{rms} \angle 60°)$
b. omitted

17-3 a. $E_{AB} = (381 \text{ V}_{rms} \angle -30°)$
$E_{BC} = (381 \text{ V}_{rms} \angle -150°)$
$E_{CA} = (381 \text{ V}_{rms} \angle 90°)$
b. omitted

17-5 $E_{AB} = (381 \text{ V}_{rms} \angle 30°)$
$E_{BC} = (381 \text{ V}_{rms} \angle -90°)$
$E_{CA} = (381 \text{ V}_{rms} \angle 150°)$
$I_{an} = (11.44 \text{ A}_{rms} \angle -27.9°)$
$I_{bn} = (11.44 \text{ A}_{rms} \angle -147.9°)$
$I_{cn} = (11.44 \text{ A}_{rms} \angle 92.1°)$
$I_{Aa} = I_{an} = I_{Bb} = I_{bn} = I_{Cc} = I_{cn}$
$I_{nN} = 0$

17-7 $E_{AB} = (381 \text{ V}_{rms} \angle 30°)$
$E_{BC} = (381 \text{ V}_{rms} \angle -90°)$
$E_{CA} = (381 \text{ V}_{rms} \angle 150°)$
$I_{an} = (11.44 \text{ A}_{rms} \angle -27.9°)$
$I_{bn} = (18.08 \text{ A}_{rms} \angle 159.5°)$
$I_{cn} = (6.67 \text{ A}_{rms} \angle 120°)$
$I_{Aa} = I_{an}, \quad I_{Bb} = I_{bn}, \quad I_{Cc} = I_{cn}$
$I_{nN} = (12.2 \text{ A}_{rms} \angle 146.4°)$

17-9 $E_{AB} = (381 \text{ V}_{rms} \angle 30°)$
$E_{BC} = (381 \text{ V}_{rms} \angle -90°)$
$E_{CA} = (381 \text{ V}_{rms} \angle 150°)$
$I_{ab} = (19.8 \text{ A}_{rms} \angle 2.1°)$
$I_{bc} = (19.8 \text{ A}_{rms} \angle -117.9°)$
$I_{ca} = (19.8 \text{ A}_{rms} \angle 122.1°)$
$I_{Aa} = (34.3 \text{ A}_{rms} \angle -27.9°)$
$I_{Bb} = (34.3 \text{ A}_{rms} \angle -147.9°)$
$I_{Cc} = (34.3 \text{ A}_{rms} \angle 92.1°)$

17-11 $E_{AB} = (381 \text{ V}_{rms} \angle 30°)$
$E_{BC} = (381 \text{ V}_{rms} \angle -90°)$
$E_{CA} = (381 \text{ V}_{rms} \angle 150°)$
$I_{ab} = (19.8 \text{ A}_{rms} \angle 2.1°)$
$I_{bc} = (31.3 \text{ A}_{rms} \angle -170.5°)$
$I_{ca} = (21.2 \text{ A}_{rms} \angle 150°)$
$I_{Aa} = (39.4 \text{ A}_{rms} \angle -14.5°)$
$I_{Bb} = (51.0 \text{ A}_{rms} \angle -173.3°)$
$I_{Cc} = (20.1 \text{ A}_{rms} \angle 51.6°)$

Appendix B

Laboratory Parts

	Phase Angle Measurement (Chapter 1, pp. 26 – 27)	Phasor Measurement (Chapter 2, pp. 44 – 45)	Impedance Measurement (Chapter 3, pp. 71 – 73)	Series Tuned Amplifier (Chapter 4, pp. 104 – 106)	Parallel Circuits (Chapter 5, pp. 130 – 133)	Filter Response (Chapter 6, pp. 160 – 162)	Low-pass Filters (Chapter 7, pp. 190 – 191)	High-pass Filters (Chapter 8, pp. 216 – 217)	Band-pass Filters (Chapter 9, pp. 247 – 248)	Op Amp Speed (Chapter 10, pp. 276 – 278)	Harmonics (Chapter 12, pp. 323 – 328)	Transient Analysis (Chapter 13, pp. 357 – 359)	Impedance Combination (Chapter 14, pp. 380 – 382)	Mesh and Nodal (Chapter 15, pp. 399 – 401)	Three Phase (Chapter 17, pp. 437 – 440)
resistors (1/4 W)															
22 (1 W)														2	
47															1
56								1							
100					3		1			1	1				
220				1	1										
330						1									
470			1			1									
1 k							1	1	1	2	1	1	1		
2.2 k		1				1							1		3
3.3 k					1										
4.7 k													1		
5.6 k				1	1										
6.2 k				1	1										
10 k										1	8	1			
22 k	1											1			
33 k							1	1							
56 k											2				
100 k				2	2										
120 k									1						
220 k										1					
potentiometers															
500											1				
1 k											1				

	Phase Angle Measurement (Chapter 1, pp. 26 – 27)	Phasor Measurement (Chapter 2, pp. 44 – 45)	Impedance Measurement (Chapter 3, pp. 71 – 73)	Series Tuned Amplifier (Chapter 4, pp. 104 – 106)	Parallel Circuits (Chapter 5, pp. 130 – 133)	Filter Response (Chapter 6, pp. 160 – 162)	Low-pass Filters (Chapter 7, pp. 190 – 191)	High-pass Filters (Chapter 8, pp. 216 – 217)	Band-pass Filters (Chapter 9, pp. 247 – 248)	Op Amp Speed (Chapter 10, pp. 276 – 278)	Harmonics (Chapter 12, pp. 323 – 328)	Transient Analysis (Chapter 13, pp. 357 – 359)	Impedance Combination (Chapter 14, pp. 380 – 382)	Mesh and Nodal (Chapter 15, pp. 399 – 401)	Three Phase (Chapter 17, pp. 437 – 440)
capacitors															
30 p											1				
270 p									1						
2.2 n													1		
3.3 n							1								
4.7 n													1		
10 n		1			1							1			
22 n	1														
33 n									1						
0.1 μ			1	2	2	3		1	2	2	5				
0.22 μ								1	1						
0.47 μ							1								
1 μ (film)															3
5.6 μ							1								
100 μ														2	
220 μ														2	
inductor															
33 m		1	1	1	1	1	1	1	1				1	1	
transformer	12.6 V CT													1	
diode															
1N914 or 1N3600											1				
IC															
CD4046 - pll											1				
LM324 - op amp			1	1	1	1	1	1	1			1			
MF10 - filter											1				

Index